Delivery of Therapeutics for Biogerontological Interventions

From Concepts to Experimental Design

Delivery of Therapeutics for Biogerontological Interventions

From Concepts to Experimental Design

Wing-Fu Lai
Department of Applied Biology and Chemical Technology,
The Hong Kong Polytechnic University, Hong Kong

Academic Press is an imprint of Elsevier
125 London Wall, London EC2Y 5AS, United Kingdom
525 B Street, Suite 1650, San Diego, CA 92101, United States
50 Hampshire Street, 5th Floor, Cambridge, MA 02139, United States
The Boulevard, Langford Lane, Kidlington, Oxford OX5 1GB, United Kingdom

British Library Cataloguing-in-Publication Data
A catalogue record for this book is available from the British Library

Library of Congress Cataloging-in-Publication Data
A catalog record for this book is available from the Library of Congress

ISBN: 978-0-12-816485-3

For Information on all Academic Press publications
visit our website at https://www.elsevier.com/books-and-journals

Publisher: Stacy Masucci
Acquisition Editor: Stacy Masucci
Editorial Project Manager: Rebeka Henry
Production Project Manager: Kiruthika Govindaraju
Cover Designer: Miles Hitchen
Typeset by MPS Limited, Chennai, India

Contents

Foreword

I am delighted to write the foreword for the book *Delivery of Therapeutics for Biogerontological Interventions*. It has also been a pleasure to go over the contents of this book, which forms a much valued compendium of knowledge for those interested in the design and applications of delivery technologies in antiaging medicine.

Due to the vast practical potential of therapeutics delivery in antiaging medicine, there was a need for a comprehensive book relating advances in delivery technologies to the development and execution of biogerontological interventions. This book meets this need, and sufficiently demonstrates the importance of delivery technologies in tackling aging. The present book is organized into four parts. Part I provides the introductory knowledge of diverse technologies used in therapeutics delivery, Parts II and III cover topics that are related to the development, optimization, and applications of the technologies for the design and execution of biogerontological interventions. The last part of this book discusses challenges to be tackled for transforming the development of biogerontological interventions from concepts to reality. Dr. Lai has done a fine job of bringing together such a diverse range of topics effectively to highlight major issues and advances that are related to the field. It is highly organized, starting from discussing the basic theoretical frameworks for designing an intervention to introducing various technologies that can bring the design into reality. Major topics relating to current delivery technologies, ranging from viral gene delivery and polyfection to direct mitochondrial transfer, have been covered. In addition to presenting the latest delivery technologies for genetic manipulation, there are chapters on topics such as tissue engineering and herbal medicine. The specific details and discussions on every specific topic make this book appealing to all readers in every age group.

Compared to a plethora of existing books relating to antiaging medicine, this book has a much higher practical value. In Parts II and III of this book, each chapter contains sample protocols for experimental design. These protocols are intentionally written for researchers who have just entered the field to plan their experiments to translate the knowledge into practice. At the end of each chapter, there is a section called "Directions for Intervention Development," in which main points discussed in the chapter have been summarized and guidelines for the design of practicable interventions have been offered. Other sections such as "Highlights for Experimental Design" and "Outstanding Questions for Clinical Translation" are also provided to inspire readers to transform the knowledge and concepts into executable experimental designs. This book is therefore both educational and practical. It presents a basic overview of therapeutics delivery and referenced resources to biogerontologists, researchers, and postgraduate students; indeed, anyone who is fascinated with the world of antiaging medicine. It is also a unique tribute to the many professionals who are involved in aging research and geriatric medicine.

In closing, I warmly congratulate Dr. Wing-Fu Lai for bringing this unique book to fruition. Although there are books and review articles available to date covering various areas of antiaging medicine and therapeutics delivery, this book will stand out from the crowd because of its unique approach to integrating the two fields by presenting the opportunities and challenges of translating the advances in therapeutics delivery into practicable interventions to combat aging. I am very satisfied with the material presented as well as the science- and engineering-related examples tailored for the readers. I am sure that this book will be an excellent asset to the literature on aging. It is a very valuable reference and tool for those entering the field from different disciplines, as well as those seeking systemic and advanced knowledge in the use of delivery technologies for designing and developing biogerontological interventions to tackle aging. It will inspire many young and senior people to advocate the translation of delivery technologies into practicable antiaging interventions.

Wing-Tak Wong

Dean and Chair Professor of Chemical Technology, Faculty of Applied Sciences & Textiles, Hong Kong Polytechnic University, Hong Kong

Preface

As early as the Middle Ages, alchemists in the West experimented with chemicals in an attempt to create life. A similar situation existed in ancient China, where Ying Zheng (also known as Qin Shi Huang) sent Taoists overseas to search for an elixir of immortality. From these, we can see that it is a common desire of mankind to combat aging and to prolong longevity. For centuries, human life span has been remarkably extended. Genetic studies on life span determination over the years has also reached a critical mass, warranting possible modulation of the aging process, repositioning the notion of life span extension from simply a fantasy in the past to a viable reality at present. Despite this, human life expectancy has thus far been improved primarily through better living conditions and more effective medical care, the role played by interventive biogerontology is little. The situation is worsened by the virtual absence of executable interventions for life span extension. This is partly caused by the deep-seated notion circulating in scientific societies that life span is a natural trait and any quest to intervene with it is bound to fail. This and the consequences thereof (including the unjustified reluctance among researchers to go anywhere beyond basic aging research, and the hesitancy of funding agencies to invest in intervention development for life span extension) have obstructed technological progress in interventive biogerontology.

Over the last several decades, significant advances have been made in the fundamentals and methodologies of therapeutics delivery. The discoveries and advancements made in the field have been documented in an increasing number of book publications and journal articles. Unfortunately, detailed discussions of the relevance of these technologies to the development of biogerontological interventions have been missing until now. I was fundamentally motivated with the hopes to fill this gap by providing the most up-to-date description of possible methods, with the support from recent advances reported in the literature, of taking advantage of delivery technologies to develop biogerontological interventions, from concept to practice.

This book has several special features that make it ideal to be a guide for designing and developing biogerontological interventions based on current advances in delivery technologies. First, it covers essential, up-to-date information on therapeutics delivery, and exploits hands-on techniques for using carriers in intervention biogerontology. It can be used by researchers in the field to refresh their working knowledge for developing and selecting delivery systems. In addition, with the support of excellent visual elements (including photographs and figures) and references, this book presents comprehensive coverage of diverse topics (including gene silencing, tissue engineering, and herbal medicine) to guide readers through the whole process of intervention development. Importantly, this book is the first of its kind to explore seriously possible translation of bench works to practicable tactics to retard the aging process.

The contents of this book are separated into four parts. The first part provides an introduction to existing delivery technologies, and describes the fundamental theoretical frameworks from which biogerontological interventions can be designed. In the second part, I extensively discuss the designing principles, important variables, and practical strategies for developing, engineering, and optimizing different types of carriers, ranging from viral vectors to hydrogel-based particles, for intervention execution. In the third part, I describe the use of delivery technologies in manipulating the aging process at the molecular, cellular, and tissue levels. The final part of this book is devoted to highlighting some currently unsolved challenges that have been hindering the execution and clinical translation of interventions to combat aging. These challenges should be the major directions for further research in interventive biogerontology. It is anticipated that the information provided by this book can facilitate planning of research studies, and may point to clearer directions for translation of delivery technologies from the laboratory to real-world applications. This book can, therefore, be a reference and a starting point for advanced undergraduate- and graduate-level students training in nanomedicine and nanotechnology. It should also be an appeal to clinicians and medical engineers in geriatric medicine, to researchers in chemistry and materials science, and to the general audience who is working or interested in the field of biogerontology and nanoscience.

Here I would like to thank all of the very many individuals who have given their time and expertise for helpful discussions during the writing of this book. Thanks are extended to Runyu Wu, Minjian Huang, Wai-Sum Lo, Cheungshen Hu, and Guoxing Deng for administrative assistance during manuscript preparation. Finally, a number of figures presented in this book have been adapted from published articles. The publishers and authors, who have granted the permission for reprinting these materials, are gratefully acknowledged.

Wing-Fu Lai

Department of Applied Biology and Chemical Technology, The Hong Kong Polytechnic University, Hong Kong

Abbreviations

2-hy-β-CD	(2-hydroxypropyl)-β-cyclodextrin
2-hy-γ-CD	(2-hydroxypropyl)-γ-cyclodextrin
5-FU	5-fluorouracil
5-MeCyt	5-methylcytosine
β-CD	β-cyclodextrin
γ-CD-NMA	acrylamidomethyl-γ-cyclodextrin
AAV	adeno-associated virus
AEAPTMS	*N*-[3-(trimethoxysilyl)propyl]ethylenediamine
AETMAC	2-acryloxyethyltrimethylammonium chloride
ACh	acetylcholine
ACR	albumin/creatinine ratio
ACS	alkylated CS
AD	Alzheimer's disease
Ad-RGD	RGD-conjugated adamantane
ADP	adenosine diphosphate
ADSC	adipose-derived stem cells
AF680	Alexa Fluor 680
AFM	atomic force microscope
Alg	alginate
AMD	age-related macular degeneration
AMO	anti-miRNA oligonucleotide
AMPK	AMP-activated kinase
APC	antigen-presenting cells
APEF	adult porcine ear skin fibroblasts
APP	amyloid precursor protein
ARE	Au-rich element
ATI	alveolar epithelial type I
ATP	adenosine triphosphate
ATRP	atom transfer radical polymerization
BB	benzyl benzoate
BBB	blood−brain barrier
BCG	*bacillus* Calmette-Guérin
BLA	*Bacillus licheniformis* α-amylase
BMMSC	bone marrow-derived mesenchymal stem cell
BMP2	bone morphogenetic protein-2
BSA	bovine serum albumin
calcein-AM	calcein acetoxymethyl ester
CAMK	calcium/calmodulin-dependent protein kinase
CBDL	common bile duct ligation
CCl_4	tetrachlorocarbon
CD	cyclodextrins
CDI	1,1′-carbonyldimidazole
CDK	cyclin-dependent protein kinase
Ce6	chlorine6
C. elegans	*Caenorhabditis elegans*
CMC	carboxymethylcellulose
CMV	cytomegalovirus
CNS	central nervous system
CNTF	ciliary neurotrophic factor
CNTs	carbon nanotubes
CR	caloric restriction
CS	chitosan
CSC	cancer stem cell
CSH	thiolated chitosan
CT	computed tomography
CW	continuous-wave
DA	dopamine
DBC	dental bud cell
DC	direct current
DCA	deoxycholic acid
DDAB	dimethyldioctadecylammonium bromide
DDBAC	dodecyl dimethyl benzyl ammonium chloride
DEAAm	*N,N*-diethylacrylamide
DHEA	dehydroepiandosterone
DLS	dynamic light scattering
DMA	*N,N*-dimethylaminoethyl methacrylate
DMAP	4-(dimethylamino)-pyridine
DMSO	dimethyl sulfoxide
D. melanogaster	*Drosophila melanogaster*
DMNPE	4,5-dimethoxy-2-nitroacetophenone
DOPE	1,2-dioleoyl-*sn*-glycero-3-phosphoethanolamine
DOTA	1,4,7,10-tetraazacyclododecane-1,4,7,10-tetraacetic acid
DOTMA	*N*-[1-(2,3-dioleyloxy)propyl]-*N,N,N*-trimethylammonium chloride
DOTAP	1,2-dioleoyl-3-trimethylammonium-propane
DOX	doxorubicin
dPDLSCs	stem cells extracted from the periodontal ligament of deciduous teeth
DPPC	1,2-dipalmitoyl-*sn*-glycero-3-phosphocholine
DQA	dequalinium
D_2R	D_2 receptor
DS	degrees of substitution
DTT	dithiothreitol
EBV	Epstein Barr virus
ECM	extracellular matrix
EDC	1-(3-dimethylaminopropyl)-3-ethylcarbodiimide hydrochloride
EDCI	1-ethyl-3-(3-dimethylaminopropyl)-carbodiimide hydrochloride
EGDMA	ethylene glycol dimethacrylate
EGFP	enhanced green fluorescent protein
EPR	enhanced permeability and retention
ER	endoplasmic reticulum
ERT	enzyme replacement therapy
ESA	excited-state absorption

ESC	embryonic stem cell
ETC	electron transport chain
ETFs	ear tip fibroblasts
EthD-1	ethidium homodimer-1
ETU	energy transfer upconversion
FA	folic acid
FACS	fluorescence-activated cell sorting
FBS	fetal bovine serum
FdUrd	5-fluoro-2′-deoxyuridine
FEV1	forced expiratory volume in 1s
FGF	fibroblast growth factor
FG-repeat	phenylalanine-glycine-repeat
FITC	fluorescein isothiocyanate
FRET	Förster resonance energy transfer
Gal-LMWCS	galactosylated low-molecular-weight chitosan
GCP	chitosan-graft-poly(ethylene glycol)
Gd	gadolinium
GelMA	gelatin methacryloyl
GFP	green fluorescent protein
GLUT1	glucose transporter 1
GTPase	guanosine triphosphatase
GWAS	genome-wide association studies
H&E	hematoxylin and eosin
HA	hemagglutinin
HC	hypromellose-*g*-chitosan
HCC	hepatocellular carcinoma
HCV	hepatitis C virus
HDF	human diploid fibroblasts
HEA	2-hydroxyethylacrylate
HEMA	2-hydroxyethyl methacrylate
HEPES	4-(2-hydroxyethyl)-1-piperazineethanesulfonic acid
hFVIII	human factor VIII
HHV-7	human herpesvirus 7
HIV	human immunodeficiency virus
HITS-CLIP	high-throughput sequencing of RNA isolated by crosslinking immunoprecipitation
HKR	HEPES-buffered Krebs Ringer
HRQoL	health-related quality of life
HSPC	hematopoietic stem and progenitor cell
hTERT	human telomerase reverse transcriptase
HTM	human trabecular meshwork
HVR	hypervariable regions
ICH	immunocytochemistry
IFN-β	interferon-β
IFN-γ	interferon-γ
lg β	chelate constant
Ig	immunoglobulin
IGF	insulin-like growth factor
IGFBP	insulin-like growth factor-binding protein
IL	interleukin
i.p.	intraperitoneal
iPS cells	induced pluripotent stem cells
IRS-1	insulin receptor substrate-1
i.v.	intravenous
JC-1	5,5′,6,6′-tetrachloro-1,1′,3,3′-tetraethylbenzimidazolyl-carbocyanine iodide
LB	Lewy bodies
LbL	layer-by-layer
LCST	lower critical solution temperature
LED	liposome-encapsulated doxorubicin
LLLT	low-level light therapy
LNA	locked nucleic acid
lPEI	linear poly(ethylenimine)
LPH	liposome-polycation-hyaluronic acid
LPS	lipopolysaccharide
LRET	luminescence resonance energy transfer
macroRAFT agents	macromolecular reversible addition-fragmentation chain transfer agents
MAPK	p38 mitogen-activated protein kinase
MB	methylene blue
MBA	*N,N*′-methylene bisacrylamide
MBCP	macroporous biphasic calcium phosphate
Mbd2	murine β-defensin 2
MC	multicompartment
MCS	mannosylated chitosan
MDA	malondialdehyde
MEND	multifunctional envelope-type nanodevice
MERRF	myoclonic epilepsy with ragged-red fibers
MES	2-(*N*-morpholino) ethanesulfonic acid
MILES	Metformin in Longevity Study
miRISC	miRNA-induced silencing complex
miRNA	microRNA
mitoK(ATP) channels	mitochondrial ATP-sensitive potassium channels
MLV	murine leukemia virus
MMP2	matrix metalloproteinase-2
MMP	matrix metalloproteinase
MPF	M-phase-promoting factors
MPE	malignant pleural effusions
MPM	malignant pleural mesothelioma
MPP^+	*N*-methyl-4-phenylpyridinium
MRI	magnetic resonance imaging
mRNA	messenger RNA
MSCs	mesenchymal stem cells
mtDNA	mitochondrial DNA
NAPI	*N*-(3-aminopropyl) imidazole
NBU	*o*-nitrobenzyl urethane
NCT	nucleocytoplasmic transport
NHS	*N*-hydroxysuccinimide
NIR	near-infrared
NLS	nuclear localization signal
NMBD	nuclear membrane breakdown
NMN	nicotinamide mononucleotide
NMP	*N*-methyl-2-pyrrolidone
NMR	nuclear magnetic resonance
NPC	nuclear pore complex
NPCP	3-naphthyl-1-phenyl-5-(4-carboxyphenyl)-2-pyrazoline
Nrf2	nuclear factor erythroid 2–related factor 2
NSCLC	nonsmall cell lung cancer
Nups	nucleoporins
NZB	New Zealand Black
OA	oleic acid
OAB	overactive bladder
OA-IL	oleic acid-ionic liquid

OAm oleylamine
ODE 1-octadecence
ODN oligodeoxynucleotide
OXPHOS oxidative phosphorylation
P407 Poloxamer407
PA photon avalanche
PAE poly(β-amino ester)
PAMAM poly(amidoamine)
P. anserine *Podospora anserine*
PAR-CLIP photoactivatable-ribonucleoside-enhanced crosslinking and immunoprecipitation
PBMC peripheral blood mononuclear cell
PBS phosphate buffered saline
PCL poly(E-caprolactone)
PCC premature chromosome condensation
PD Parkinson's disease
pDADMAC polydiallyldimethylammonium chloride
PDEAAm poly(*N*,*N*-diethylacrylamide)
PDK-1 3-phosphoinositide-dependent protein kinase-1
PDL periodontal ligament
PDMA poly(*N*,*N*-dimethylaminoethyl methacrylate)
PDMS polydimethylsiloxane
PDT photodynamic therapy
P(EO-r-PO) poly((ethylene oxide)-ran-(propylene oxide))
PECT poly(ε-caprolactone-*co*-1,4,8-trioxa[4.6]spiro-9-undecanone)-poly(ethylene glycol)-poly (ε-caprolactone-*co*-1,4,8-trioxa[4.6]spiro-9-undecanone)
PEDF pigment epithelium-derived factor
PEF porcine embryonic fibroblasts
PEG poly(ethylene glycol)
PEG-AD poly(ethylene glycol)-adamantane
PEGDA poly(ethylene glycol)-diacrylate
PEI poly(ethylenimine)
PEO poly(ethylene oxide)
PEO-*b*-PDMAAm poly(ethylene oxide)-*b*-poly(N,N-dimethylacrylamide)
PFV prototype foamy virus
PG propylene glycol
FG-repeat phenylalanine-glycine-repeat
Phe phenamil
PHEMA poly(2-hydroxy ethyl methacrylate)
PI3K phosphatidylinositol 3-kinase
PKA protein kinase A
PKC protein kinase C
PL photoluminescence
PLA poly(D,L-lactide)
PLGA poly(lactic-co-glycolic acid)
PLL poly(L-lysine)
PN pronuclei
PNA peptide nucleic acid
PNIPAAm poly(*N*-isopropylacrylamide)
POCG poly(1,8-octanedio-citric acid)-co-poly(ethylene glycol)
POEOMA poly(oligo(ethylene oxide) monomethyl ether methacrylate)
poly(NIPAAm) poly(*N*-isopropylacrylamide)
pPDLSC stem cell extracted from the periodontal ligament of permanent teeth
PRF platelet-rich fibrin
PSI polysuccinimide
PTD protein transduction domain
PVA polyvinyl alcohol
PVI poly(*N*-vinyl imidazole)
PVP polyvinylpyrrolidone
QDs quantum dots
QSAR quantitative structure–activity relationship
rAd5 replication-incompetent adenovirus serotype 5
RanGAP Ran GTPase activating protein
RanGEF Ran guanine nucleotide exchange factor
rF 2,4-difluorotoluyl ribonucleoside
RGC retinal ganglion cell
RISC RNA-induced silencing complex
RNAi RNA interference
ROP ring-opening polymerization
ROS reactive oxygen species
RPE retinal pigment epithelium
RPP arginine-rich cell-penetrating peptide
RT-PCR reverse transcriptase polymerase chain reaction
SA-β-gal senescence-associated β-galactosidase
SAR structure–activity relationship
S. cerevisiae *Saccharomyces cerevisiae*
SCID severe combined immune deficiency
SCNT somatic cell nuclear transfer
SCPL solvent casting and particulate leaching
SDS sodium dodecyl sulfate
SENS Strategies for Engineered Negligible Senescence
S-FIL step and flash imprint lithography
SH3 Src homology 3
shRNA small hairpin RNA
SIPS stress-induced premature senescence
siRNA small inferring RNA
SLN-gel solid-lipid-nanoparticle-enriched hydrogel
SLN solid lipid nanoparticles
SNP single nucleotide polymorphism
SOD superoxide dismutase
SPDP *N*-succinimidyl 3-(2-pyridyldithio)propionate
SRT substrate reduction therapy
SS Szeto-Schiller
SSQ Speech, Spatial, and Qualities of Hearing Scale
TA triamcinolone acetonide
TALEN transcription activator-like effector nuclease
t-BHP *tert*-butyl hydroperoxide
TCID50 50% tissue culture infectious dose
TCM traditional Chinese medicine
TEM transmission electron microscopy
Th2 T helper 2
TH tetracycline hydrochloride
TOR target of rapamycin
TORC 1 target of rapamycin complex 1
TPGS D-α-tocopheryl poly(ethylene glycol) 1000 succinate
TPP triphenylphosphonium
TREM tetramethylrhodamine, ethyl ester

TRH thyrotropin releasing hormone
TRITC tetramethylrhodamine isothiocyanate
Ts-CD mono-6-(*p*-toluenesulfonyl)-6-deoxy-cyclodextrin
TUNEL terminal deoxynucleotidyl transferase-mediated dUTP nick-end labeling
UCL upconversion luminescence
UCNP upconversion nanoparticle
UTR untranslated region
UV ultraviolet
VAP viral attachment protein
VDT 2-vinyl-4,6-diamino-1,3,5-triazine
VEGF vascular endothelial growth factor
VLU venous leg ulcer
VSMC vascular smooth muscle cell
VSV vesicular stomatitis virus
WAR water absorption ratio
WAS Wiskott–Aldrich syndrome
WS Werner syndrome
ZFR zinc-finger recombinase

Part I

From Concepts to Plans

Chapter 1

Theoretical frameworks for intervention development

Introduction

Aging is a biological process that is not only mediated by environmental factors but is also determined genetically. The latter has been supported by several studies [1–4], which have examined the concordance of longevity in monozygous and dizygous twins. Upon analysis of 2872 pairs of nonemigrant like-sex twins, the heritability of longevity has been estimated to be around 0.26 for males and 0.23 for females [5]. Until now, numerous molecular determinants of longevity and aging have been identified. One example is *age-1*. It involves nondauer development and normal senescence [6], and is one of the first gene mutants being shown to extend longevity [7]. Other examples of genes found to influence lifespan include *daf-16, smk-1, hcf-1, AGTR1,* and *sir-2.1* [8–10].

By increasing understanding of the aging process, over the years a diversity of tactics [ranging from caloric restriction (CR) [11] to hormone replacement [12,13]] have been proposed for combating aging [14]. Now, the concept of antiaging has become more practical than that in the past, thanks to the advancement of genetic engineering and molecular technologies. In 2002, the idea of "Strategies for Engineered Negligible Senescence" (SENS) was proposed. SENS has revolutionized the biogerontological field by offering a framework for developing strategies to reverse pathogenic age-associated damage in a fragmented approach [15,16]. Based on this, a number of coping strategies to combat and reverse age-associated molecular and cellular changes (e.g., cellular senescence, nuclear mutations, mitochondrial mutations, lysosomal aggregates, extracellular aggregates and cross-links, and cell loss) have been designed [15,16]. Besides in vivo and clinical studies, research has been carried out on plant longevity (which is partly contributed by stem cell immortality, vascular autonomy, and epicormic branching) [17]. With years of research, not only has the aging mechanism been elucidated, but the fantasy of lifespan prolongation has also been made possible. Apart from SENS, other longevity strategies have been reported in literature [12,13,18–28]. Representative examples of these strategies are shown in Table 1.1.

Despite these advances, at the moment clinically applicable interventions for combating aging are lacking, owing partly to the lack of efforts devoted to intervention development. This has been shown by an earlier article [29], which is the result of a database search on PubMed and Web of Science. During the database search, only seven articles have been found to work directly on the development of nucleic acid therapy for longevity enhancement and/or aging retardation. Other retrieved articles have been devoted to studying only one to two facets of the aging process (such as delayed angiogenesis [30], erectile dysfunction [31], memory impairment [32], thymic involution [33], and vascular dysfunction [34]) rather than tackling aging as a whole. Only two out of these seven articles have touched upon lifespan extension. One of them is Boghossian et al.'s [35] work, which has successfully extended the lifespan of the ob/ob mice from 55.5 weeks to 106.5 weeks by injecting a recombinant adeno-associated virus (AAV) encoding the leptin gene intracerebroventricularly into the mice. The other one is Chung et al.'s [36] study, which has attempted to prolong lifespan simply at the cellular level rather than the organismal level. Such a deficiency of research may partly be attributed to the challenge of manipulating lifespan as a polygenic trait [37]. Furthermore, being able to change the expression of a few genes in in vivo models does not necessarily mean that lifespan prolongation can be achieved in humans, which physiologically are much more complex than fruit flies and yeasts. Along with the fact that the long-term physiological price paid by genetic manipulation still has not been completely determined, the extent of experimental works to intervene with the aging process is limited. This has greatly impeded the development of biogerontological interventions.

Delivery of Therapeutics for Biogerontological Interventions. DOI: https://doi.org/10.1016/B978-0-12-816485-3.00001-5

TABLE 1.1 Examples of approaches to enhancing longevity.

Approach	Underlying principles	Reference(s)
Treatment with herbal medicine	Using herbs (e.g., *Lycium barbarum*) and some traditional formulae (e.g., *Sip-Jeon-Dae-Bo-Tang*) to prevent the occurrence of age-associated diseases	[18–20]
Cell injection	Replacing the lost or dead cells	[21,22]
Stem cell intervention	Rejuvenating aged cells and restoring normal expression of antiaging genes	[23]
Hormone replacement	Restoring the normal hormonal state that has been changed by aging	[12,13,24]
Caloric restriction	Reducing the production of reactive oxygen metabolites during carbohydrate metabolism so as to reduce age-associated damage in the body. Furthermore, with persistent CR, more energy can be directed to body maintenance, thereby enhancing longevity	[25]
Drug treatment	Using chemical agents to restore the normal functioning of cells and tissues, and to combat the aging phenotypes	[25–27]
Genetic manipulation	Intervening with the aging process directly at the genetic level	[28]

A reductionist approach to understand aging

To design a biogerontological intervention, identifying an ideal intervention point is vital. This requires deep understanding of the aging mechanism. The reductionist approach is a common method to comprehend the aging process. This approach breaks the aging network into pieces (called subprocesses) and tries to understand how the pieces work at smaller and smaller levels of organization [38,39]. The viability of understanding aging in this manner has been supported scientifically by the discovery of an increasing number of subprocesses that have contributed to aging. One example of these subprocesses is DNA methylation. The involvement of DNA methylation in aging was first proposed by Vanyushin and coworkers [40], who have observed that the 5-methylcytosine (5-MeCyt) content of DNA from different organs (e.g., spleen, heart, and brain) has changed with age in mice. This observation has been supported by a subsequent study that has reported the correlation between aging and loss of global DNA methylation [41]. Recent studies on monozygotic twins have found that epigenetic changes are related to aging but are independent of the genetic sequence [42]. This has illustrated the complexity of aging by revealing how environmental factors and gene function may interact. Besides DNA methylation, posttranscriptional histone modifications (via ubiquitination, methylation, acetylation, and phosphorylation) are involved in epigenetic signaling. This has been reviewed in other articles [39,43].

Since the turn of the last century, the identification of "longevity" or "antiaging" genes has been facilitated by the advent of high-throughput technologies and genetic techniques [44–46]. Examples of studies on the genetics of aging are listed in Table 1.2 [38,47–55]. As our understanding of various age-associated pathways (e.g., insulin/insulin-like growth factor (IGF)-1 signaling) and genes (including those encoding Sod2, Ras, protein kinase A, Msn2, Msn4, and adenylate cyclase) has increased [56], lifespan prolongation is no longer a hypothetical notion. In fact, the feasibility of manipulating the aging network at the genetic level has already been revealed in literature. For instance, Wang and coworkers have reported that, though IDH4 human fibroblasts have shown signs of senescence upon the suppression of the activity of HuR (a ubiquitously expressed Elav-like RNA-binding protein), the senescent fibroblasts have been rejuvenated when the cellular level of HuR has been escalated [57]. In the in vivo context, a study performed by Hsieh and colleagues has also demonstrated that mutation of Pit1 has prolonged lifespan in Snell dwarf mice [58]. More recently, the relationship between longevity and gene function has been corroborated by Copeland et al. [59], who, by silencing the expression of selected genes encoding components of mitochondrial respiratory complexes I, III, IV, and V, have prolonged lifespan in *Drosophila melanogaster (D. melanogaster)*. All evidence presented previously has not only enriched our understanding of genetics on longevity, but has also revealed the apparent reversibility of the aging process, thereby imbuing genetic manipulation with striking potential in antiaging medicine [37].

TABLE 1.2 Studies on the genetics of aging.

Source	Gene	Model	Results and implications	References
Holzenberger et al. (2003)	*Igf1r*	Mouse	Heterozygous IGF-1R knockout mice have displayed greater resistance to oxidative stress, with their lifespan being approximately 30% longer than that of their wild-type littermates	[38]
Fujii et al. (2011)	*oxy5*	Nematode	An increase in the activity of Oxy5 has promoted the sensitivity of the nematode to oxygen, causing a reduction in lifespan	[47]
Berdichevsky et al. (2010)	*kat-1*	Nematode	The loss-of-function mutation of *kat-1* has shortened lifespan in nematodes, and has elicited abnormalities that are characteristic of premature aging	[48]
Chen et al. (2010)	*ATM*	Human	Longevity has been shown to be affected by a functional single nucleotide polymorphism (SNP) in the promoter of *ATM*.	[49]
Tang et al. (2009)	*dOpa1*	Fruit fly	The production of reactive oxygen species (ROS) has been enhanced after *dOpa1* has undergone a mutation, leading to the shortening of lifespan	[50]
Madia et al. (2008)	*SCH9*	Yeast	Premature genomic instability and recombination errors have been shown to be reduced upon the mutation of *SCH9* (homologous to *AKT* and *S6K*). This has provided insights into the possibility of protecting mammals from premature aging by modulating the IGF-I-Akt-56K pathway.	[51]
Zhao et al. (2008)	*WRN*	Human	Mutation of *WRN* has been associated with premature aging	[52]
Zheng et al. (2007)	*Il-2* and *fas*	Mouse	The lifespan of Scurfy mice has been extended after the knockout of *Il-2* or the mutation of *fas*	[53]
Roux et al. (2006)	*pka1* and *sck2*	Yeast	Pka1 and Sck2 have involved in regulating chronicle aging in *Schizosaccharomyces pombe*	[54]
Fabrizio et al. (2003)	*SOD2, SOD1,* and *RAS2*	Yeast	The yeast has survived for a longer period upon the overexpression of *SOD2* and/or *SOD1*. The mean lifespan has been doubled when the *RAS2* gene has been deleted.	[55]

Reliability theory of aging

Although the reductionist approach to understand aging has been well-supported by scientific evidence, it fails to explain the late-life mortality plateaus [60–62] and the compensation law of mortality [63]. The concept of the reliability theory has, therefore, been proposed as an alternative approach to understanding the aging process [64]. Reliability theory comprises a series of mathematical models and ideas to predict, estimate, and optimize the lifespan distribution of a system or its components. To describe the reliability of a system at time x, one may use the reliability function $S(x)$. The definition of $S(x)$ is provided below, where X is the failure time and $F(x)$ is the standard cumulative distribution function in the probability theory:

$$S(x) = P(X > x) = 1 - P(X \leq x) = 1 - F(x) \qquad (1.1)$$

To describe the relative rate for the decline of $S(x)$, the hazard rate $h(x)$ [also known as failure rate $\lambda(x)$] can be adopted. In demography, this rate is regarded as the mortality force $\mu(x)$.

$$h(x) = -\frac{dS(x)}{S(x)dx} = -\frac{d\left[\log_e S(x)\right]}{dx} \quad (1.2)$$

A system in reality usually has a failure rate that comprises both aging and nonaging terms. A good example is the Gompertz–Makeham law of mortality, which contains the age-dependent Gompertz function ($Re^{\alpha,x}$) as well as the age-independent Makeham parameter (A) [63]. $Re^{\alpha,x}$ designates the age-associated factors of mortality (e.g., age-associated diseases), whereas A represents those deaths led by age-independent causes (e.g., accidents):

$$\mu(x) = A + Re^{\alpha,x} \quad (1.3)$$

As predicted by reliability theory, even if a human body is constructed from entirely nonaging elements where the failure rate does not change with age, the body will still deteriorate with age because it is redundant in irreplaceable elements. Owing to this redundancy, death may not happen at once, even as damage occurs. This allows all sorts of damage possibly to be accumulated. Here the accumulation of aging-independent defects in the human body, as suggested by the concept of reliability theory, could be the cause, rather than the consequence, of the aging process. When damage accumulates, the redundancy in the number of elements in a body decreases. In the end, the body will have no more redundancy. Any new damage imposed to the body will result in death.

Highlights for experimental design

1. The aging process can be comprehended by using either the reductionist approach or the reliability theory. The former dissects a system into different subprocesses, whereas the latter comprises diverse mathematical models and ideas to predict, estimate, and optimize the lifespan distribution of a system or its components.
2. The causal relationship between aging and body damage is ambiguous and is heavily related to the approach adopted to perceive aging.
3. Human mortality involves both age-dependent and age-independent factors. A nonaging body will still deteriorate with age if it is redundant in irreplaceable elements.

Hypothetical frameworks for tackling aging

No matter whether the reductionist approach or the reliability theory is used to understand aging, there are three major approaches available to resist the aging process (Fig. 1.1) [65]. As prevention is generally more cost-effective than treatment [66], the geriatric approach (which aims to address age-associated damage when diseases occur) is less desirable than the gerontological and engineering approaches, which are more proactive in nature. The underlying principle of the gerontological approach is to manipulate the fundamental metabolic pathways to retard or ward off aging. The technical viability of this approach has been increased by the

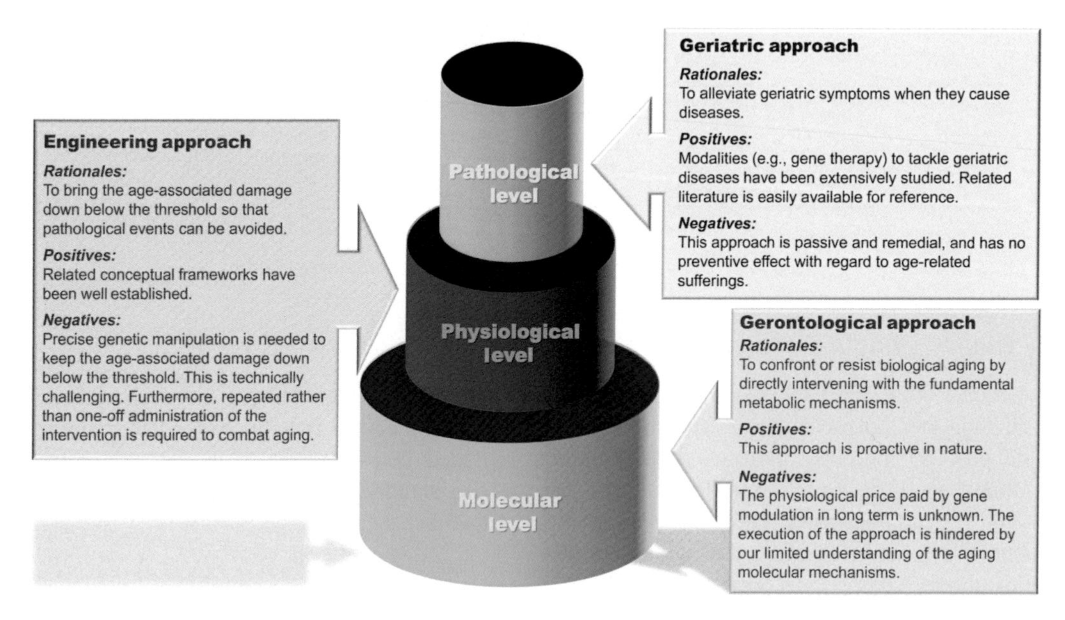

FIGURE 1.1 Three major approaches targeting different levels of the aging process. *Reproduced from W.F. Lai, Nucleic acid delivery: roles in biogerontological interventions, Ageing Res. Rev. 12 (2013) 310–315 with permission from Elsevier B.V., [65].*

cumulative research efforts devoted to the identification of candidate genes for manipulation of the aging network [67,68]. Experimental evaluation of the antiaging effects of a gene or an intervention, however, necessitates years of dedication in mice [69], and is basically nonviable in humans (whose long lifespan and low birth rate make the experimental time too long to be practical). Because of this, diverse simpler eukaryotes [e.g., *D. melanogaster*, *Caenorhabditis elegans (C. elegans)*, and *Saccharomyces cerevisiae (S. cerevisiae)*] have been deployed as alternative models. In these models, the plausible use of transgenic manipulation to extend lifespan has been demonstrated. For instance, by overexpressing the *D-GADD45* gene in the nervous system, Plyusnina and colleagues have successfully prolonged lifespan *in D. melanogaster*, and have maintained the flies' locomotor activity and fecundity [70]. Along with other successful instances documented in literature [71–73], lifespan and aging have been shown to be manipulable at the genetic level. Despite this, age-associated damage is intrinsically caused by metabolic processes in a human body (or actually any body of a homeotherm). To halt the emergence of such damage, the metabolic network has to be redesigned as a whole. Owing to the high complexity of the aging process, there is a long way to go before the gerontological approach can be effectively executed in antiaging medicine.

Contrary to the gerontological approach that attempts directly to manipulate the metabolic network, the engineering approach prevents the occurrence of pathological events by bringing age-associated damage below the threshold (Table 1.3) [16]. For example, with the use of induced pluripotent stem (iPS) cells derived from somatic cells, cell replacement therapy for Parkinson's disease (an age-associated degenerative disorder characterized by bradykinesia, tremor, muscle rigidity, and a progressive loss of dopamine neurons in the substantia nigra pars compacta) has been made plausible [74]. Via hematopoietic cell transplantation, replacement of malfunctional stem cells with normal ones has also been successfully achieved over the last several decades [75]. All these have heightened the plausibility of combating cell loss and tissue atrophy caused by aging.

Highlights for experimental design

1. Tackling aging can be achieved at the molecular, physiological, or pathological levels.
2. Prevention is often more cost-effective than is treatment. Compared to the geriatric approach, the gerontological and engineering approaches are more proactive in nature and hence more desirable.
3. Owing to the difficulty in redesigning and manipulating the metabolic network, the engineering approach is easier to be achieved as compared to the gerontological.

TABLE 1.3 Selected age-associated changes and the feasible methods for reversal.

Change	Possible method for reversal
Accumulation of nuclear mutations	Angiostasis; delivery and expression of agents to mediate suicide gene therapy
Cell senescence	Elimination of senescent cells
Cell loss	Stem cell therapy; growth factor-induced cell replacement
Accumulation of mitochondrial mutations	Allotopic expression of mitochondrial DNA (mtDNA)
Accumulation of lysosomal aggregates	Delivery and expression of bacterial hydrolase genes to degrade lysosomal aggregates
Accumulation of extracellular aggregates and cross-links	Stimulated phagocytosis; delivery of agents for removing the cross-links and aggregates
Immune system decline	IL-7-stimulated thymopoiesis
Hormone secretion decline	Delivery and expression of genes encoding hormones to combat the decline in hormone secretion

As far as aging is concerned, the final outcome is death, which is characterized by the irreversible and complete cessation of all vital functions [76]. According to common knowledge, both aging and death are also permanent and irreversible processes. Before we discuss more deeply whether aging is irreversible, we first look at an imaginary scenario. Here a living body with normal functions is obtained. An intuitively fatal procedure is given to the subject by opening up his chest and carefully taking out the pumping heart. The body is then examined regularly until all vital signs (including the heartbeat, breathing rate, temperature, and blood pressure) are absent. The body stays in an open area for another 12 hours so as to ensure that all vital functions are completely lost. After 12 hours, the heart is transplanted back to the same body, which has lost all vital functions. When the operation is complete, the signs of the body are reexamined regularly for an infinite period to see whether the vital signs will be resumed at any time point. In this scenario, the probability of surgical failure is assumed to be zero. The surgical skills adopted to transplant the heart back to the body are assumed to be perfect so that any difference between the pretransplanted heart and the transplanted heart is absent.

There are two possible outcomes after heart transplantation. The first outcome is that the subject shows vital signs again after the surgical procedure, either immediately or after an unknown period. This outcome may be counter-intuitive to our general knowledge of death; however, if this scenario holds, death can be reversed simply by repairing the damage that directly leads to the state of death. The second outcome is that the subject's state of death persists after heart transplantation. This scenario matches better with our common sense. Here the state of death is defined as the state in which all vital signs cease [76]. We hypothesize that, if a subject's *biological* body, which can be interpreted as simply an aggregate of compounds and is now in a state of death, can be turned back to be *genuinely* identical, in all physical senses, to the state that has been held by the subject before his/her death, the subject shall be able to live again. This hypothesis is consistent with the logic that, when there are two states *A* and *B*, if *A* changes to *B* in a way that every single aspect of *A* is equivalent to that of *B*, A shall be equivalent to *B*. In other words, all properties of *B* shall be possessed by *A*. This concept also applies to aging. If all age-associated changes can be reversed so that the aged body, in every single aspect, becomes the same as the one in the young state, an aged body shall be rejuvenated.

Reasons for the failure of an intervention

Taken the imaginary scenario above into consideration, the apparent irreversibility of aging is solely because of the failure of existing interventions genuinely to turn the state of the altered body back to the original. Such failure may be caused by the fact that some of the biological change caused by, or led to, the damage has either been overlooked or has not yet been completely repaired. It is the unrepaired damage that continues to lock the body into the altered state. Taken the previous imaginary scenario as an example; it is possible that the procedure of heart removal causes damage not only to the heart but also to other parts of the body. For instance, active cell death has been reported in tissues during hormone ablation [77], which may result from the disruption of the neuro-endocrinological system experienced by the subject when he/she is in the state of death. In addition, protein degradation and various other degenerative processes can be caused by hypoxia and other factors [78–82], which are experienced by the subject during the death period. In other words, to ensure that the body of the subject can be changed from the state of death to the state of living, simply repairing the heart is not sufficient. Further damage (at the molecular, cellular, and physiological levels) that has been caused during the death period has to be sufficiently repaired. It is predicted that the longer the period in which the subject has been locked into the state of death, the more the damage appears in the body, and the harder the damage can be reversed.

This concept has been partly supported by some of the reported cases of resuming the vital signs of some patents when first-aid interventions are promptly administered [83,84]. It is, however, worth noting that in all of these successful cases, the vital signs of the patients have ceased only for a very short period (so that further damage caused by entering the state of death is minimal), or the cause of the cessation of the vital signs is comparatively mild (e.g., mild heart failure which can be promptly amended by procedures such as cardiopulmonary resuscitation and hence can shorten the period of death experienced by the subject). In our imaginary scenario, the death of the subject is, however, caused by more severe damage. Not only does the damage take a longer time to repair, but repairing the damage is also more technically demanding. This locks the subject into the altered state for a much longer period, thereby tremendously increasing the amount of damage needed to be repaired for genuinely resuming the body back to the original state. Here we would like to emphasize that, when an altered body is attempted to be converted into the original, some damage (particularly nonvital damage) can be tolerated to be unrepaired. For example, the failure of repairing the damage in body parts (such as limbs) should not affect the success rate of the intervention if all vital changes in a dead body can be resumed.

Highlights for experimental design

1. Rejuvenation is possible, at least theoretically, provided that the body damaged by aging can be genuinely converted back to its original state.
2. Failure of an intervention for rejuvenation is caused by the fact that some age-associated changes have not been repaired.
3. Attaining a full picture of the aging network is helpful when developing an intervention to rejuvenate age-associated phenotypes.

Summary

As written by Williams [85], "it would be not only always better to live, but better to live always, that is, never to die." Because the final outcome of aging is death, warding off aging is a long-held desire of humans. In this chapter, we have discussed the reductionist approach and the reliability theory for comprehending the aging process, and have presented theoretical frameworks for designing antiaging interventions at different levels. Based on the frameworks presented, we envisage that if technological advances in the future can enable sophisticated repairs of biological damage, rejuvenation of a body is viable. When a biological body is resumed back to the desired state, it is, however, important not to overlook nonobvious nonphysical changes caused by aging. This could be explained by using the case of a toy robot, which is physically the same regardless of whether it is full of battery or out of battery. The difference between the "battery-full" state and the "battery-out" state is only manifested by subtle changes in concentrations of chemicals around the anode and the cathode. For this, during the design of a biogerontological intervention, the aged body should be converted back to the undamaged state not only at the macroscopic level but also at the microscopic level. Achieving this goal may be challenging at the moment because the aging network still has not been fully elucidated; however, while humbly recognizing the impossibility of knowing all, scientists hold fast to the assumption that all things are inherently and ultimately knowable.

Directions for intervention development

Based on the accumulated knowledge gathered from aging research, various theoretical frameworks for tackling aging have been proposed. With these as the foundation, the following steps help us get closer to attaining executable interventions in practice:

1. Select an appropriate theoretical framework for designing an intervention.
2. Gain knowledge of the biological pathways and mechanisms that are related to the framework.
3. Modify the framework if theoretical gaps and flaws are identified.
4. When a gap in the framework cannot be satisfactorily rectified, replace the framework with another one that is deemed more practicable.
5. Repeat the steps above until a theoretically sound framework are established.
6. List possible routes to manipulate the biological pathways and mechanisms that are related to the ultimately selected framework.

References

[1] P.E. Slagboom, M. Beekman, W.M. Passtoors, J. Deelen, A.A. Vaarhorst, J.M. Boer, et al., Genomics of human longevity, Philos. Trans. R. Soc. Lond. B. Biol. Sci. 366 (2011) 35–42.

[2] J. vB Hjelmborg, I. Iachine, A. Skytthe, J.W. Vaupel, M. McGue, M. Koskenvuo, et al., Genetic influence on human lifespan and longevity, Hum. Genet. 119 (2006) 312–321.

[3] A. Skytthe, N.L. Pedersen, J. Kaprio, M.A. Stazi, J.V. Hjelmborg, I. Iachine, et al., Longevity studies in GenomEUtwin, Twin. Res. 6 (2003) 448–454.

[4] H. Gudmundsson, D.F. Gudbjartsson, M. Frigge, J.R. Gulcher, K. Stefansson, Inheritance of human longevity in Iceland, Eur. J. Hum. Genet. 8 (2000) 743–749.

[5] A.M. Herskind, M. McGue, N.V. Holm, T.I.A. Sorensen, B. Harvald, J.W. Vaupel, The heritability of human longevity: a population-based study of 2872 Danish twin pairs born 1870-1900, Hum. Genet. 97 (1996) 319–323.

[6] J.Z. Morris, H.A. Tissenbaum, G. Ruvkun, A phosphatidylinositol-3-OH kinase family member regulating longevity and diapause in Caenorhabditis elegans, Nature 382 (1996) 536–539.

[7] D.B. Friedman, T.E. Johnson, A mutation in the age-1 gene in caenorhabditis-elegans lengthens life and reduces hermaphrodite fertility, Genetics 118 (1988) 75–86.

[8] A. Benigni, S. Orisio, M. Noris, P. Iatropoulos, D. Castaldi, K. Kamide, et al., Variations of the angiotensin II type 1 receptor gene are associated with extreme human longevity, Age 35 (2013) 993–1005.

[9] G. Rizki, T.N. Iwata, J. Li, C.G. Riedel, C.L. Picard, M. Jan, et al., The evolutionarily conserved longevity determinants HCF-1 and SIR-2.1/SIRT1 collaborate to regulate DAF-16/FOXO, PLoS Genet. 7 (2011) e1002235.

[10] S. Wolff, H. Ma, D. Burch, G.A. Maciel, T. Hunter, A. Dillin, SMK-1, an essential regulator of DAF-16-mediated longevity, Cell 124 (2006) 1039–1053.

[11] G. Balazsi, Network reconstruction reveals new links between aging and calorie restriction in yeast, HFSP J. 4 (2010) 94–99.

[12] C.A. Allan, B.J. Strauss, H.G. Burger, E.A. Forbes, R.I. McLachlan, Testosterone therapy prevents gain in visceral adipose tissue and loss of skeletal muscle in nonobese aging men, J. Clin. Endocrinol. Metab. 93 (2008) 139–146.

[13] D. Heutling, H. Lehnert, Hormone therapy and anti-aging: is there an indication? Internist. (Berl). 49 (2008) 570. 572–576, 578–579.

[14] W.F. Lai, Z.C. Chan, Beyond sole longevity: a social perspective on healthspan extension, Rejuvenation Res. 14 (2011) 83–88.

[15] A.D. de Grey, The SENS challenge: $20,000 says the foreseeable defeat of aging is not laughable, Rejuvenation Res. 8 (2005) 207–210.

[16] A.D. de Grey, B.N. Ames, J.K. Andersen, A. Bartke, J. Campisi, C.B. Heward, et al., Time to talk SENS: critiquing the immutability of human aging, Ann. N.Y. Acad. Sci. 959 (2002) 452–462. discussion 463-455.

[17] R.M. Borges, Phenotypic plasticity and longevity in plants and animals: cause and effect? J. Biosci. 34 (2009) 605–611.

[18] I.M. Chang, Anti-aging and health-promoting constituents derived from traditional oriental herbal remedies: information retrieval using the TradiMed 2000 DB, Ann. N.Y. Acad. Sci. 928 (2001) 281–286.

[19] R.C. Chang, K.F. So, Use of anti-aging herbal medicine, Lycium barbarum, against aging-associated diseases. What do we know so far? Cell. Mol. Neurobiol. 28 (2008) 643–652.

[20] L.W. Xu, L. Kluwe, T.T. Zhang, S.N. Li, Y.Y. Mou, Z. Sang, et al., Chinese herb mix Tiao-Geng-Tang possesses antiaging and antioxidative effects and upregulates expression of estrogen receptors alpha and beta in ovariectomized rats, BMC Complement. Altern. Med. 11 (2011) 137.

[21] K. Ebisawa, R. Kato, M. Okada, Y. Kamei, A.L. Mazlyzam, Y. Narita, et al., Cell therapy for facial anti-aging, Med. J. Malaysia 63 (Suppl A) (2008) 41.

[22] B.S. Park, K.A. Jang, J.H. Sung, J.S. Park, Y.H. Kwon, K.J. Kim, et al., Adipose-derived stem cells and their secretory factors as a promising therapy for skin aging, Dermatol. Surg. 34 (2008) 1323–1326.

[23] M. Ullah, Z. Sun, Stem cells and anti-aging genes: double-edged sword-do the same job of life extension, Stem Cell Res. Ther. 9 (2018) 3.

[24] E. Diamanti-Kandarakis, M. Dattilo, D. Macut, L. Duntas, E.S. Gonos, D.G. Goulis, et al., Mechanisms In Endocrinology: aging and anti-aging: a Combo-Endocrinology overview, Eur. J. Endocrinol. 176 (2017) R283–R308.
[25] C.S. Catana, A.G. Atanasov, I. Berindan-Neagoe, Natural products with anti-aging potential: affected targets and molecular mechanisms, Biotechnol. Adv. 36 (2018) 1649–1656.
[26] S. Hosseini, M. Abdollahi, G. Azizi, M.J. Fattahi, N. Rastkari, F.T. Zavareh, et al., Anti-aging effects of M2000 (beta-D-mannuronic acid) as a novel immunosuppressive drug on the enzymatic and non-enzymatic oxidative stress parameters in an experimental model, J. Basic Clin. Physiol. Pharmacol. 28 (2017) 249–255.
[27] J. Lei, X. Gu, Z. Ye, J. Shi, X. Zheng, Antiaging effects of simvastatin on vascular endothelial cells, Clin. Appl. Thromb. Hemost. 20 (2014) 212–218.
[28] J. Kim, S.Y. Cho, S.H. Kim, D. Cho, S. Kim, C.W. Park, et al., Effects of Korean ginseng berry on skin antipigmentation and antiaging via FoxO3a activation, J. Ginseng. Res. 41 (2017) 277–283.
[29] W.F. Lai, Nucleic acid therapy for lifespan prolongation: present and future, J. Biosci. 36 (2011) 725–729.
[30] H. Wang, J.A. Keiser, B. Olszewski, W. Rosebury, A. Robertson, I. Kovesdi, et al., Delayed angiogenesis in aging rats and therapeutic effect of adenoviral gene transfer of VEGF, Int. J. Mol. Med. 13 (2004) 581–587.
[31] A. Melman, G. Biggs, K. Davies, W. Zhao, M.T. Tar, G.J. Christ, Gene transfer with a vector expressing Maxi-K from a smooth muscle-specific promoter restores erectile function in the aging rat, Gene Ther. 15 (2008) 364–370.
[32] A. Mouravlev, J. Dunning, D. Young, M.J. During, Somatic gene transfer of cAMP response element-binding protein attenuates memory impairment in aging rats, Proc. Natl. Acad. Sci. U.S.A. 103 (2006) 4705–4710.
[33] J.A. Phillips, T.I. Brondstetter, C.A. English, H.E. Lee, E.L. Virts, M.L. Thoman, IL-7 gene therapy in aging restores early thymopoiesis without reversing involution, J. Immunol. 173 (2004) 4867–4874.
[34] K.A. Brown, Y. Chu, D.D. Lund, D.D. Heistad, F.M. Faraci, Gene transfer of extracellular superoxide dismutase protects against vascular dysfunction with aging, Am. J. Physiol. Heart Circ. Physiol. 290 (2006) H2600–H2605.
[35] S. Boghossian, N. Ueno, M.G. Dube, P. Kalra, S. Kalra, Leptin gene transfer in the hypothalamus enhances longevity in adult monogenic mutant mice in the absence of circulating leptin, Neurobiol. Aging. 28 (2007) 1594–1604.
[36] S.A. Chung, A.Q. Wei, D.E. Connor, G.C. Webb, T. Molloy, M. Pajic, et al., Nucleus pulposus cellular longevity by telomerase gene therapy, Spine 32 (2007) 1188–1196.
[37] W.F. Lai, M.C.M. Lin, Chemical derivatization of chitosan for plasmid DNA delivery: present and future, in: S.K. Kim (Ed.), Chitin, Chitosan and Their Derivatives: Biological Activities and Industrial Applications, CRC Press, Boca Raton, FL, 2010, pp. 69–82.
[38] M. Holzenberger, J. Dupont, B. Ducos, P. Leneuve, A. Geloen, P.C. Even, et al., IGF-1 receptor regulates lifespan and resistance to oxidative stress in mice, Nature 421 (2003) 182–187.
[39] S.F. Gilbert, Ageing and cancer as diseases of epigenesis, J. Biosci. 34 (2009) 601–604.
[40] B.F. Vanyushin, L.E. Nemirovsky, V.V. Klimenko, V.K. Vasiliev, A.N. Belozersky, The 5-methylcytosine in DNA of rats. Tissue and age specificity and the changes induced by hydrocortisone and other agents, Gerontologia 19 (1973) 138–152.
[41] V.L. Wilson, P.A. Jones, DNA methylation decreases in aging but not in immortal cells, Science 220 (1983) 1055–1057.
[42] M.F. Fraga, E. Ballestar, M.F. Paz, S. Ropero, F. Setien, M.L. Ballestar, et al., Epigenetic differences arise during the lifetime of monozygotic twins, Proc. Natl. Acad. Sci. U.S.A. 102 (2005) 10604–10609.
[43] M.F. Fraga, M. Esteller, Epigenetics and aging: the targets and the marks, Trends Genet. 23 (2007) 413–418.
[44] B.K. Kennedy, The genetics of ageing: insight from genome-wide approaches in invertebrate model organisms, J. Intern. Med. 263 (2008) 142–152.
[45] S.S. Lee, Whole genome RNAi screens for increased longevity: important new insights but not the whole story, Exp. Gerontol. 41 (2006) 968–973.
[46] N. Minois, P. Sykacek, B. Godsey, D.P. Kreil, RNA interference in ageing research—a mini-review, Gerontology 56 (2010) 496–506.
[47] M. Fujii, K. Shikatani, K. Ogura, Y. Goshima, D. Ayusawa, Mutation in a mitochondrial ribosomal protein causes increased sensitivity to oxygen with decreased longevity in the nematode Caenorhabditis elegans, Genes. Cells 16 (2011) 69–79.
[48] A. Berdichevsky, S. Nedelcu, K. Boulias, N.A. Bishop, L. Guarente, H.R. Horvitz, 3-Ketoacyl thiolase delays aging of Caenorhabditis elegans and is required for lifespan extension mediated by sir-2.1, Proc. Natl. Acad. Sci. U.S.A. 107 (2010) 18927–18932.
[49] T. Chen, B. Dong, Z. Lu, B. Tian, J. Zhang, J. Zhou, et al., A functional single nucleotide polymorphism in promoter of ATM is associated with longevity, Mech. Ageing. Dev. 131 (2010) 636–640.
[50] S. Tang, P.K. Le, S. Tse, D.C. Wallace, T. Huang, Heterozygous mutation of Opa1 in Drosophila shortens lifespan mediated through increased reactive oxygen species production, PLoS One. 4 (2009) e4492.
[51] F. Madia, C. Gattazzo, M. Wei, P. Fabrizio, W.C. Burhans, M. Weinberger, et al., Longevity mutation in SCH9 prevents recombination errors and premature genomic instability in a Werner/Bloom model system, J. Cell. Biol. 180 (2008) 67–81.
[52] N. Zhao, F. Hao, T. Qu, Y.G. Zuo, B.X. Wang, A novel mutation of the WRN gene in a Chinese patient with Werner syndrome, Clin. Exp. Dermatol. 33 (2008) 278–281.
[53] L. Zheng, R. Sharma, F. Gaskin, S.M. Fu, S.T. Ju, A novel role of IL-2 in organ-specific autoimmune inflammation beyond regulatory T cell checkpoint: both IL-2 knockout and Fas mutation prolong lifespan of Scurfy mice but by different mechanisms, J. Immunol. 179 (2007) 8035–8041.
[54] A.E. Roux, A. Quissac, P. Chartrand, G. Ferbeyre, L.A. Rokeach, Regulation of chronological aging in Schizosaccharomyces pombe by the protein kinases Pka1 and Sck2, Aging Cell. 5 (2006) 345–357.
[55] P. Fabrizio, L.L. Liou, V.N. Moy, A. Diaspro, J.S. Valentine, E.B. Gralla, et al., SOD2 functions downstream of Sch9 to extend longevity in yeast, Genetics 163 (2003) 35–46.
[56] V.D. Longo, Ras: the other pro-aging pathway, Sci. Aging Knowl. Environ. 2004 (2004) pe36.

[57] W. Wang, X. Yang, V.J. Cristofalo, N.J. Holbrook, M. Gorospe, Loss of HuR is linked to reduced expression of proliferative genes during replicative senescence, Mol. Cell. Biol. 21 (2001) 5889–5898.

[58] C.C. Hsieh, J.H. DeFord, K. Flurkey, D.E. Harrison, J. Papaconstantinou, Effects of the Pit1 mutation on the insulin signaling pathway: implications on the longevity of the long-lived Snell dwarf mouse, Mech. Ageing Dev. 123 (2002) 1245–1255.

[59] J.M. Copeland, J. Cho, T. Lo Jr., J.H. Hur, S. Bahadorani, T. Arabyan, et al., Extension of Drosophila life span by RNAi of the mitochondrial respiratory chain, Curr. Biol. 19 (2009) 1591–1598.

[60] K.W. Wachter, Evolutionary demographic models for mortality plateaus, Proc. Natl. Acad. Sci. U.S.A. 96 (1999) 10544–10547.

[61] S.D. Pletcher, J.W. Curtsinger, Mortality plateaus and the evolution of senescence: why are old-age mortality rates so low? Evolution 52 (1998) 454–464.

[62] L.D. Mueller, M.R. Rose, Evolutionary theory predicts late-life mortality plateaus, Proc. Natl. Acad. Sci. U.S.A. 93 (1996) 15249–15253.

[63] L.A. Gavrilov, N.S. Gavrilova, The Biology of Life Span: A Quantitative Approach, Harwood Academic Publisher, New York, 1991.

[64] L.A. Gavrilov, N.S. Gavrilova, The reliability theory of aging and longevity, J. Theor. Biol. 213 (2001) 527–545.

[65] W.F. Lai, Nucleic acid delivery: roles in biogerontological interventions, Ageing Res. Rev. 12 (2013) 310–315.

[66] M.C. Weinstein, The costs of prevention, J. Gen. Intern. Med. 5 (1990) S89–S92.

[67] S. Kim, X. Bi, M. Czarny-Ratajczak, J. Dai, D.A. Welsh, L. Myers, et al., Telomere maintenance genes SIRT1 and XRCC6 impact age-related decline in telomere length but only SIRT1 is associated with human longevity, Biogerontology 13 (2012) 119–131.

[68] W.R. Swindell, Gene expression profiling of long-lived dwarf mice: longevity-associated genes and relationships with diet, gender and aging, BMC Genomics 8 (2007) 353.

[69] M. Kaeberlein, Longevity genomics across species, Curr. Genomics 8 (2007) 73–78.

[70] E.N. Plyusnina, M.V. Shaposhnikov, A.A. Moskalev, Increase of Drosophila melanogaster lifespan due to D-GADD45 overexpression in the nervous system, Biogerontology 12 (2011) 211–226.

[71] J.N. Sampayo, M.S. Gill, G.J. Lithgow, Oxidative stress and aging—the use of superoxide dismutase/catalase mimetics to extend lifespan, Biochem. Soc. Trans. 31 (2003) 1305–1307.

[72] H. Ohtsuka, Y. Ogawa, H. Mizuno, S. Mita, H. Aiba, Identification of Ecl family genes that extend chronological lifespan in fission yeast, Biosci. Biotechnol. Biochem. 73 (2009) 885–889.

[73] C.J. Kenyon, The genetics of ageing, Nature 464 (2010) 504–512.

[74] L.W. Chen, F. Kuang, L.C. Wei, Y.X. Ding, K.K. Yung, Y.S. Chan, Potential application of induced pluripotent stem cells in cell replacement therapy for Parkinson's disease, CNS Neurol. Disord. Drug Targets 10 (2011) 449–458.

[75] A. Czechowicz, I.L. Weissman, Purified hematopoietic stem cell transplantation: the next generation of blood and immune replacement, Hematol. Oncol. Clin. North Am. 25 (2011) 75–87.

[76] G.B. Fulton, Bioethics and health education: some issues of the biological revolution, J. Sci. Health 47 (1977) 205–211.

[77] M.P. Tenniswood, R.S. Guenette, J. Lakins, M. Mooibroek, P. Wong, J.E. Welsh, Active cell death in hormone-dependent tissues, Cancer Metastasis Rev. 11 (1992) 197–220.

[78] M. Castro-Gago, S. Rodriguez-Segade, F. Camina, A. Bollar, A. Rodriguez-Nunez, Indicators of hypoxia in cerebrospinal fluid of hydrocephalic children with suspected shunt malfunction, Childs Nerv. Syst. 9 (1993) 275–277.

[79] V.G. YSH, B.V. Bhat, P. Chand, K.R. Rao, Hypoxia induced DNA damage in children with isolated septal defect and septal defect with great vessel anomaly of heart, J. Clin. Diagn. Res. 8 (2014) SC01–SC03.

[80] O. Linsell, J.C. Ashton, Cerebral hypoxia-ischemia causes cardiac damage in a rat model, Neuroreport 25 (2014) 796–800.

[81] A. Faa, T. Xanthos, V. Fanos, D. Fanni, C. Gerosa, P. Pampaloni, et al., Hypoxia-induced endothelial damage and microthrombosis in myocardial vessels of newborn landrace/large white piglets, Biomed. Res. Int. 2014 (2014) 619284.

[82] F. Cervellati, C. Cervellati, A. Romani, E. Cremonini, C. Sticozzi, G. Belmonte, et al., Hypoxia induces cell damage via oxidative stress in retinal epithelial cells, Free Radic. Res. 48 (2014) 303–312.

[83] P. van Lommel, R. van Wees, V. Meyers, I. Elfferich, Near-death experience in survivors of cardiac arrest: a prospective study in the Netherlands, Lancet 358 (2001) 2039–2045.

[84] E.W. Cook, B. Greyson, I. Stevenson, Do any near-death experiences provide evidence for the survival of human personality after death? Relevant features and illustrative case reports, J. Sci. Explor. 12 (1998) 377–406.

[85] B. Williams, The Makropulos case: reflections on the tedium of immortality, in: J. Rachels (Ed.), Moral Problems: A Collection of Philosophical Essays, Harper & Row, New York, 1975, pp. 410–428.

Chapter 2

Available delivery technologies for intervention execution

Introduction

To intervene with the aging process, external agents that can elicit physiological changes have to be delivered to the site of action. Delivery technologies, therefore, play important roles in bringing antiaging interventions into reality (Table 2.1). Recently, with advances in delivery technologies, not only has transcytosis of therapeutic agents across tight epithelial and endothelial barriers been facilitated [1], but the efficiency in delivering macromolecular drugs (or those with poor aqueous solubility) has also been improved [1]. The latter has been shown by the case of hydrophobic drugs (e.g., paclitaxel, cyclosporine, and amphotericin B), whose dissolution rate and gastrointestinal absorption efficiency have been significantly improved after being formulated as nanosuspensions [1,2]. Surface modifications of nanoparticles have also enabled particles to adsorb on specific organs and tissues, making targeted delivery of drugs possible at the cellular and tissue levels [1].

TABLE 2.1 Possible roles played by delivery technologies in tackling age-associated problems.

Approach	Problem to be tackled	Possible role
Gerontological approach	Defects in metabolic pathways	To deliver bioactive agents to modulate the expression of various proteins to manipulate metabolic pathways
Engineering approach	Immune system decline	To deliver IL-7, or genes encoding the protein, to specific tissues to stimulate thymopoiesis
	Hormone secretion decline	To deliver hormone-encoding genes or the hormones *per se* to cells and tissues to counteract the effects caused by a decline in hormone secretion
	Accumulation of lysosomal aggregates	To deliver bacterial hydrolase, or genes encoding the protein, to cells and tissues to remove lysosomal aggregates
	Accumulation of extracellular aggregates and cross-links	To deliver genes or bioactive agents to the affected cells to mediate the breakdown or removal of extracellular aggregates and cross-links
	Cell loss	To deliver growth factors to tissues to combat cell loss; or to deliver bioactive agents to stem cells after stem cell transplantation to guide the differentiation of those cells
	Cell senescence	To deliver specific proteins, or genes encoding those proteins, to senescent cells. When those proteins are presented onto the cell surface, phagocytosis of those senescent cells can be stimulated
	Accumulation of mitochondrial mutations	To deliver mtDNA to the cell nucleus for allotopic expression
	Accumulation of nuclear mutations	To deliver bioactive agents to the affected cells and tissues so that apoptosis or angiostasis can be induced
Geriatric approach	Age-associated diseases	To deliver bioactive agents to the affected cells and tissues to treat the disease

Delivery of Therapeutics for Biogerontological Interventions. DOI: https://doi.org/10.1016/B978-0-12-816485-3.00002-7

Over the last several decades, extensive efforts have been devoted to developing carriers for safe and effective transfer of molecules (ranging from nucleic acids to chemical drugs) for clinical applications [3–5]. In this chapter, we will present an overview of major advances in these delivery technologies. Given the huge number of works accumulated in literature, instead of covering all information encyclopedically, only representative examples will be highlighted. More in-depth discussions on the design and engineering of specific delivery systems (e.g., viral vectors, polymeric gene carriers, and hydrogel nanoparticles) for intervention development will be provided in Part II of this book.

Use of delivery technologies in interventions

The roles played by delivery technologies in mediating the execution of interventions have been documented extensively in literature in a variety of medical arenas (e.g., nucleic acid vaccination [6–9] and disease treatment [10–13]). For instance, by using chitosan (CS) nanoparticles as carriers to deliver a plasmid encoding a mite dust allergen to mice, an increase in the level of interferon-γ (IFN-γ) in serum has been induced with subsequent sensitization of T helper 2 (Th2) cell-regulated specific immunoglobulin (Ig) E responses being prevented [14]. With the use of liposome-polycation-hyaluronic acid (LPH) nanoparticles surface functionalized with cyclic RGD, anti-miR-296 has also been successfully delivered to endothelial cells to suppress blood tube formulation and endothelial cell migration [15]. These examples have demonstrated the successful use of delivery technologies both in vitro and in vivo [16].

As far as aging is concerned, tissue deterioration is often involved, and this has attracted efforts from experts in tissue engineering to solve the problem. Advances in tissue engineering have been highly facilitated by the use of delivery technologies. Previously, by using retroviral vectors to overexpress bone morphogenetic protein-2 (BMP2) in murine bone stromal W-20 cells, along with the use of the poly(lactic-co-glycolic acid) (PLGA)/hydroxyapatite composite as a scaffold, heterotopic bone formation has been successfully induced in severe combined immune deficiency (SCID) mice (Fig. 2.1) [17]. Recently, by using an adenoviral vector to deliver vascular endothelial growth factor (VEGF), which is a 40-kDa disulfide-linked dimeric glycoprotein playing an important role in angiogenesis [18], to the muscle layer surrounding the bone defect in a rat model, the number of FVIII-related antigen-positive blood vessels in the defect area has been reported to be increased [19]. Apart from viral vectors, promising results in tissue engineering have been achieved by using nonviral systems. A good example is the work reported by Huang et al. [20], who have used poly(ethylenimine) (PEI) as a carrier to condense therapeutic plasmids and then have incorporated the polyplexes into a PLGA scaffold. A significantly higher level of total bone formation has been observed upon implantation of the scaffold into an in vivo model. All these have evidenced the practical roles played by delivery technologies in tissue regeneration.

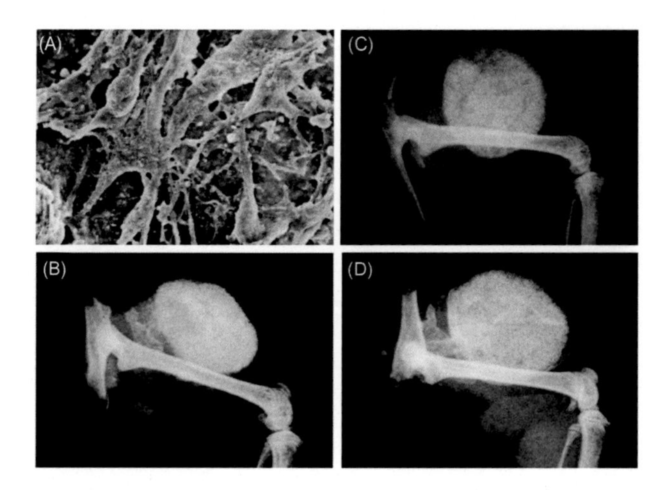

FIGURE 2.1 (A) A scanning electron micrograph of W-20-BMP-2–producing cells grown on the PLAGA/hydroxyapatite composite. (B–D) Radiographs of PLGA/hydroxyapatite composites that have been implanted in SCID mice for 1 month. Before implantation, the composite has either been laden with (B) transfected BMP-2–producing cells or (C) nontransfected W-20 cells, or (D) has been loaded with recombinant human BMP-2. An area of radio-density adjacent to the polymer composite has been found in (B) and (D), indicating the occurrence of heterotopic bone formation. *Reproduced from C.T. Laurencin, M.A. Attawia, L.Q. Lu, M.D. Borden, H.H. Lu, W.J. Gorum, et al., Poly(lactide-co-glycolide)/hydroxyapatite delivery of BMP-2-producing cells: a regional gene therapy approach to bone regeneration, Biomaterials 22 (2001) 1271–1277 with permission from Elsevier B.V, [17].*

In fact, due to the diverse nature of bioactive agents (ranging from small-molecule compounds to nucleic acids and peptides) that possibly can be used for eliciting biomedical changes, over the years different delivery technologies have been developed. Because of the unique yet consistent features shared by DNA and RNA, we will classify the technologies for delivery of nucleic acids as one category, and the technologies for delivering other bioactive agents as another category.

Advances in delivery of nucleic acids

Owing to the rapid development of experimental techniques in molecular genetics, not only has the elucidation of the mechanisms of different genetic pathologies been remarkably facilitated [21–27], but the technical feasibility of tackling diseases with gene therapies has also been enhanced [28–31]. This is one of the reasons explaining the enormous research activities devoted to developing diverse techniques for nucleic acid delivery [26,32]. These techniques can be biological, chemical, or physical in nature.

Nucleic acid delivery mediated by biological means

Nucleic acids can be delivered using diverse biological entities, including bacteria [33], viruses [34–36], virus-like particles [37], erythrocyte ghosts [38], and exosomes [39,40]. Previously, Coller [41] has generated "thrombo-erythrocytes" by conjugating erythrocytes with peptides that possess an Arg-Gly-Asp sequence. The generated erythrocytes selectively bind to activated platelets but not inactivate ones, and can be used as carriers after their intracellular content is replaced with exogenous genetic materials. More recently, a recombinant bacterium derived from species in the enterobacteria family (such as *Salmonella*) has been exploited for nucleic acid delivery [42]. The recombinant bacterium can effectively invade a host cell, but leads to less bacterium-induced programmed cell death in the host as compared to the wild-type counterpart. It shows extensive potential for use in genetic manipulation.

Among different types of biological entities developed for nucleic acid delivery, viral vectors have attracted the most extensive research interests because of their proven record of high delivery efficiency [43]. One example is retroviral vectors. Retroviruses are single-stranded RNA viruses. Their virions are around 80–100 nm in diameter. Retroviruses have three subfamilies: spumaviruses, oncogenic retroviruses, and lentiviruses. In an earlier study, Pensiero and coworkers have adopted cell lines (e.g., TE671, Mv-1-Lu, HT1080, and HOS) that are resistant to human serum–mediated lysis successfully to generate retroviral vectors that can resist complement inactivation mediated by human serum [44]. This has granted retroviral vectors higher potential for intervention execution in human bodies. Later, Choi and Gewirtz have also used the human immunodeficiency-based lentivirus and the murine stem cell retrovirus to generate a hybrid viral vector for genetic manipulation of hematopoietic stem cells and other nondividing cells (such as those in the cardiac and neuronal tissues) [45]. These advances can possibly facilitate the future development of interventions for manipulation of the aging mechanism at the genetic level.

In addition to retroviruses, other viruses have been exploited for use in genetic manipulation. For example, a recombinant human herpesvirus 7 (HHV-7) has been adopted to transport nucleic acids into lymphoid cells [46]. The inactivated Sendai virus (a negative sense, single-stranded RNA virus that belongs to the Paramyxoviridae family [47]) has also been used as a carrier for delivery of exogenous genes [48]. After injection of the Sendai viral vector into the rat cerebrum, transgene expression has been detected in the injection site. Upon intratumoral injection of the vector in vivo, effective expression of the transgene has also been observed. Because the capacity of the vector to achieve persistent systemic expression of the transgene has not yet been examined, the applicability of the vector to manipulate systemic aging has to be further verified. Despite this, for ameliorating age-associated changes, the vector has demonstrated great application prospects. Apart from the aforementioned examples, there are many other viruses that have been used for gene delivery, including adenoviruses and AAVs. Further details of the possible use of viral vectors for the development of biogerontological interventions will be provided in Chapter 3.

Nucleic acid delivery mediated by chemical and physical means

In addition to biological means, delivery of nucleic acids can be mediated by using chemical entities. Since the emergence of the poly(L-lysine) (PLL)-based vector in the 1980s [49], various polymers have been developed as gene carriers [26,32]. For example, an earlier study has adopted alkyl chain–modified low-molecular-weight PEI for delivery of small inferring RNA (siRNA), and has found that hydrophobic modification enhances the ability of PEI to self-assemble into nanoparticles [50]. A β-cyclodextrin (β-CD)-based star copolymer [which consists of an asymmetrically functionalized β-CD core, poly(*N*,*N*-dimethylaminoethyl methacrylate) (PDMA) arms, and covalently conjugated gadolinium (Gd)-1,4,7,10-tetraazacyclododecane-1,4,7,10-tetraacetic acid (DOTA) complexes] has also been developed in recent years [51]. The copolymer is fabricated via azide-alkyne Huisgen cycloaddition, during which atom transfer radical polymerization (ATRP) of *N*,*N*-dimethylaminoethyl methacrylate (DMA) on a β-CD derivative occurs. That derivative possesses 14 α-bromopropionate functionalities at the lower rim for ATRP initiation, and 7 azide moieties at the upper rim for click reactions [51]. Because the copolymer contains polycationic arms, it can interact with plasmids electrostatically for gene delivery. Owing to the presence of Gd-DOTA complexes, the copolymer can mediate contrast enhancement in T1-weighted magnetic resonance imaging (MRI) as well. Among diverse gene carriers developed, some have reached clinical trials. A good example is Genetic Immunity's DermaVir, which is a PEI-formulated vaccine designed for human immunodeficiency virus (HIV)-specific immune reconstitution. In phase II clinical trial, the vaccine has been found to be safe, and has successfully elicited HIV-specific, predominantly central memory T-cell responses [52]. Another example is the BC-819/PEI formulation. No severe adverse events have been reported for the formulation. The formulation has shown high potential to tackle cancers after it has combined with bacillus Calmette-Guérin (BCG) [53]. Over the years, with advances in materials chemistry and engineering, polymers have emerged as promising candidates for the development of carriers to mediate genetic manipulation. Detailed discussions about the design of carriers based on polymers will be provided in Chapter 4.

In addition to polymers, lipid-based carriers, such as liposomes, have received extensive interests as gene carriers. Since the 1970s when liposomes were first reported as carriers for vaccination and drug delivery [54,55], extensive research has followed. In recent years, archaeosomes have been reported as a new class of lipid-based gene carriers, displaying higher stability as compared to conventional liposomes [56]. These carriers can either be generated by using synthetically derived lipids that have structural characteristics in common with those of archaeobacterial ether lipids or can be fabricated by using one or more of the ether lipids found in Archaea [56]. The high transfection efficiency of archaeosomes has previously been reported by Freisleben and colleagues [57], who have chemically modified tetraether archaeal lipids extracted from *Thermoplasma acidophilum* to formulate archaeosomes. Those archaeosomes have displayed transfection efficiency in the same order of magnitude as that of Lipofectin and Lipofectamine. They warrant further exploitation for use in lipofection in the future.

Apart from the aforementioned, gene transfer can be facilitated using physical methods, including laser irradiation [58], sonoporation [59,60], magnetofection [61], microinjection [62], electroporation [63], and particle bombardment [64]. For instance, electroporation has been applied in a previous study to deliver a plasmid encoding interleukin (IL)-12 for treating melanoma. Results have shown that the delivery method is safe, and is able to induce necrosis in melanoma cells in treated tumors [65]. More recently, the potential use of electroporation in DNA vaccination has also been exploited [66]. Nevertheless, applications of physical methods in gene delivery so far have been limited mainly to local transfection. This may not be compatible with the need of combating conditions such as systemic aging. Along with the fact that extensive efforts have been devoted elsewhere to reviewing physical methods in gene delivery [67–71], dwelling into these technologies does not constitute the major focus of this book.

Highlights for experimental design

1. Delivery technologies for genetic materials can be biological, chemical, or physical in nature.
2. Viruses are one of the most important gene delivery systems, though the use of other systems (such as virus-like particles, erythrocyte ghosts, and exosomes) has been reported.
3. Chemical and physical methods can effectively deliver genetic materials into cells without causing concerns on the pathogenicity problem. Examples of these methods include lipofection, magnetofection, and electroporation.

Advances in delivery of nongenetic materials

The use of delivery technologies is not limited to interventions mediated by nucleic acids. Nongenetic materials such as proteins [72] or small-molecule compounds [73,74] can be carried to target sites by using delivery systems. For instance, an injectable drug delivery system comprising thermosensitive biodegradable poly(ether-ester) block copolymers has been reported for transporting various bioactive agents, including thyrotropin releasing hormone (TRH) and chemical drugs (e.g., cisplatin, angiotensins, methotrexate, and 5-fluorouracil) [75]. Biodegradable hydrogels having both pH- and thermo-sensitivity, generated from a solution mixture of hydroxy- and carboxylic acid-terminated PLGA-poly(ethylene glycol) (PEG)-PLGA tri-block copolymers, have also been reported for sustained release of Fc-leptin in vitro [76]. It is true that many of the reported delivery technologies are still in the investigatory stage; however, some of them have already managed to enter clinical trials. One example is liposomes. In 1990, Rahman and coworkers [77] have administered liposome-encapsulated doxorubicin (LED) to cancer patients at different doses. Results have shown that the formulation has not only been well-tolerated at myelosuppressive doses, but it has also produced less venous sclerosis than free doxorubicin. Similar promising effects of LED have later been reported in patients with metastases from primary gastric or colonic tumors [78] and in patients with hepatoma [78]. Apart from LED, other liposomal formulations have been tested in the clinical setting. Examples include paclitaxel liposomes [79,80] and liposome-encapsulated all-trans retinoic acid [81]. These formulations show promise for intervention execution and warrant further exploration for clinical use.

The development of implantable devices for sustained release is also a hotspot in drug delivery research. Previously, an implantable device for sustained drug release has been designed based on progenitor polymers linked via biodegradable linkages [82]. More recently, Anderson and colleagues have reported a retrievable device for release of bioactive agents in patients [83]. The device, on one hand, can deliver bioactive compounds (such as antibiotics and antiproliferative drugs) and, on the other hand, can minimize adverse effects (e.g., infection and hyperplasia) caused by device implantation. As aging is a systemic process, local drug administration might not be a favorable choice for intervention execution in the biogerontological context. For tackling signs of aging in a localized manner (e.g., cutaneous aging), these technologies, however, may play a role.

Finally, delivery of proteins and peptides has been successfully achieved by using genetically modified cells that can synthesize and release the proteins or peptides desired. This strategy has been demonstrated to be viable by an earlier study, which has proposed the possible use of differentiated neural cells (including neural precursor cells or mature neural cells such as glial cells) as protein carriers [84]. The viability of this has later been shown by Tao et al. [85], who have genetically engineered human retinal pigment epithelial cells (ARPE-19) for secretion of the ciliary neurotrophic factor (CNTF). All these works have illustrated the interrelationships between delivery technologies for both genetic and nongenetic materials. In addition, if it is possible to modify autologous cells for in situ production of exogenous agents, those agents will not need to be administered to blood anymore. This can potentially reduce the chance of eliciting immune responses in the host during intervention execution.

Highlights for experimental design

1. Delivery technologies can be designed to deliver both genetic and nongenetic materials to accommodate diverse needs of the intervention development process.
2. Examples of nongenetic materials that can be delivered include, but are not limited to, peptides, proteins, and small-molecule compounds.
3. Delivery of proteins and peptides can be achieved via genetic manipulation of target cells so that the cells can synthesize and release the proteins and peptides desired.

Integration of engineering techniques into therapeutics delivery

While chemical modification is a conventional method to optimize delivery systems (especially chemical systems), in recent years the development of delivery technologies has been promoted by advances in microfluidic technologies, which not only enable precise manipulation of fluids at a small length scale but also allow for automation miniaturization. Over the years, the advent of microfluidic technologies has facilitated the development of various commercial instruments (e.g., the LabChip microfluidic capillary electrophoresis system [86], and the 2100 Electrophoresis Bioanalyzer [87]). It has also enhanced the efficiency of therapeutics delivery. For instance, a microfluidic device containing polyelectrolytic gel electrodes has been reported for continuous low-voltage direct current (DC) electroporation (Fig. 2.2) [88]. To apply the electric potential difference to cells, the device has adopted polydiallyldimethylammonium chloride (pDADMAC) plugs as ionic conductors. An electric field of 0.9 kV/cm across the microchannel has been attained simply by using an input voltage of only 10 V [88]. In in vitro studies in which low DC voltage (5−17 V) has been applied, the transfection efficiency attained in human chronic leukemia K562 cells has been found to be as high as 60% [88], with 80% of cell viability having been maintained after the procedure [88]. This study has shown the potential of using microfluidic technologies to reduce the voltage needed for electroporation and hence to improve the viability of treated cells.

Apart from electroporation, engineering technologies can be used to facilitate the generation of synthetic carriers. One example is liposomes. Traditionally, diverse postprocessing procedures (such as sonication and membrane extrusion) are needed to ensure that the generated liposomes have an appropriate size distribution profile; however, with the incorporation of microfluidic techniques, liposomes having a lower degree of polydispersity

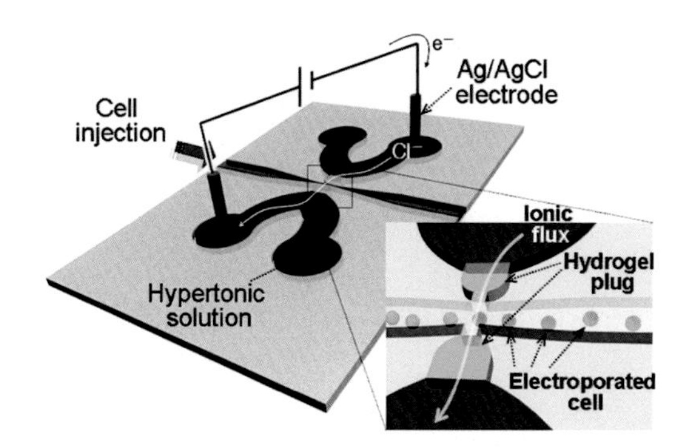

FIGURE 2.2 A schematic diagram of the microelectroporation chip. An electric field gradient is applied to cells when the cells pass through the region between the salt bridges. *Reproduced from S.K. Kim, J.H. Kim, K.P. Kim, T.D. Chung, Continuous low-voltage DC electroporation on a microfluidic chip with polyelectrolytic salt bridges, Anal. Chem. 79 (2007) 7761−7766 with permission from American Chemical Society, [88].*

can be easily produced. This has been documented in an earlier study, which has tuned the size of liposomes by using a microfluidic device that enables hydrodynamic focusing of an ethanol solution of lipids between two sheathed streams of deionized water [89]. The laminar flow in microchannels has facilitated the occurrence of diffusive mixing at the liquid interfaces, leading to self-assembly of lipids into vesicles [89]. Upon incorporation of sonication into the microfluidic fabrication of liposomes, the mean size of the liposomes has even been reduced [90]. More recently, with the use of a touch-and-go lipid wrapping technique in a polydimethylsiloxane (PDMS)/glass microfluidic device [91], multifunctional envelope-type nanodevices (MENDs) have been generated. The MEND is a nanostructure possessing a polycation-condensed nucleic acid core, which is entrapped by a lipid envelope [92]. To facilitate the delivery process, the envelope is, in general, incorporated with different functional moieties [93], including ligands for cell targeting [94], PEG for prolonging the time of blood circulation [95], and protein transduction domains (PTDs) for enhancing intracellular availability [93]. By using the microfluidic method to generate MENDs (which conventionally are fabricated using lipid film hydration [96], thereby requiring a large amount of starting materials and being labor intensive), large-scale production in a shorter time frame is now possible.

Engineering techniques have also been adopted to enhance the fabrication of hydrogel particles. One commonly used technique is electrospray, which produces aerosols via electrostatic dispersion of liquids [97]. This method has previously been used to produce microgels from alginate (Alg) for delivery of paclitaxel-loaded polymeric particles [98]. It allows the size and microstructure of the particles to be fine-tuned and manipulated during the fabrication process (e.g., by altering the magnitude of the electric field strength, the flow rate, and the concentration of the gel-forming polymer solution) (Fig. 2.3). Along with proper design of the electrospray device, hydrogel particles with different microstructures have

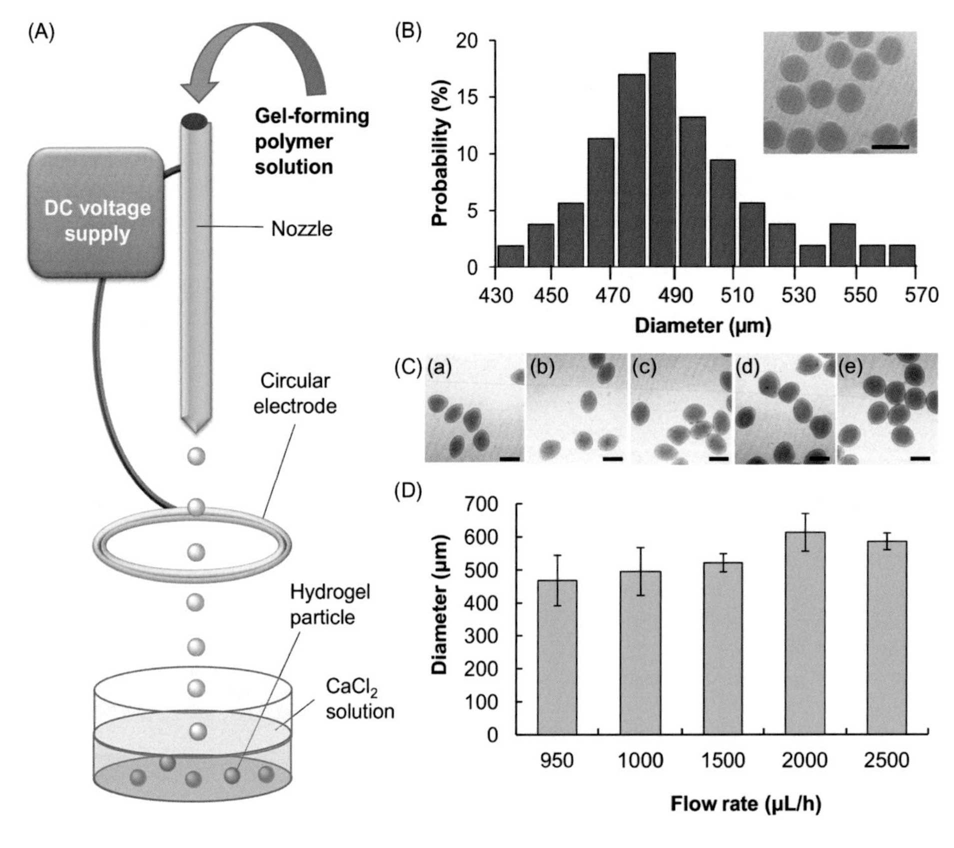

FIGURE 2.3 (A) A schematic diagram of the electrospray setup for the generation of hydrogel particles. (B) Size distribution of the particles generated by using a 4% (w/v) CMC solution, with the flow rate and electric field strength being 1500 μL/h and 5 kV/cm, respectively. An optical image of the particles is shown in the right corner. The scale bar is 500 μm. (C) Optical images of the hydrogel particles generated at different flow rates: (a) 950 μL/h, (b) 1000 μL/h, (c) 1500 μL/h, (d) 2000 μL/h, and (e) 2500 μL/h. The scale bar is 500 μm. (D) A plot of the average diameter of the hydrogel particles as a function of the flow rate. The electric field strength is 5 kV/cm. *Reproduced from W.F. Lai, A.S. Susha, A.L. Rogach, Multicompartment microgel beads for co-delivery of multiple drugs at individual release rates, ACS Appl. Mater. Interfaces 8 (2016) 871–880 with permission from the Royal Society of Chemistry, [99].*

been successfully produced in literature. For instance, by taking advantage of the laminar flow behavior in a capillary tube, multicompartment (MC) microgels have been generated by using Alg and carboxymethylcellulose (CMC) [99]. The average diameter of the microgels has been tuned by changing the electric field strength [99]. During the process of drug delivery, the volume percentage of the drug-loaded compartment of the MC microgel has been increased from 30% to 90% when the flow rate of the dye-load flow has been increased from 50 to 500 μL/h [99]. Intriguingly, with the use of the polymer blending technique, the release rates of individual codelivered drugs has been successfully changed to meet the needs of the treatment [99]. Such high tunability of the microgels can facilitate subsequent optimization for intervention execution. The compartmentalized microstructure of the microgels has also enabled concomitant delivery of multiple therapeutic agents.

Highlights for experimental design

1. Engineering technologies can be adopted to enhance the development and performance of delivery systems.
2. Microfluidic techniques can be used to enhance the performance of various gene delivery technologies (e.g., to increase the efficiency of electroporation, and to facilitate the formation of liposomes).
3. Electrospray methods can enable the generation of monodisperse droplets, especially for the formation of hydrogel particles.

Summary

Development of effective and safe technologies for delivery of exogenous agents is pivotal to the implementation of biogerontological interventions. Over the years, the practical potential of delivery technologies has been reported in diverse fields in literature, ranging from disease treatment [10–13] to DNA vaccination [6–9]. Despite this, many of the reported technologies have been investigatory in nature. Real applications in the clinical setting are highly limited. Moreover, due to the absence of standardized experimental protocols that enable a fair comparison of data from different laboratories, selection of favorable candidates from the pool of existing delivery technologies for further exploitation is highly challenging. Nevertheless, a technical foundation has already been established by the research efforts documented in literature. Based on the foundation laid, in Part II of this book we will explore more deeply the advances in few major categories of delivery technologies that show greater potential for intervention development in antiaging medicine, and will discuss how they can be optimized for interventive biogerontology.

Directions for intervention development

Selecting an appropriate theoretical framework for designing an intervention is the first step in the development of an antiaging therapy. Based on the framework selected, possible routes to manipulate aging mechanisms can be proposed. After that, an intervention can be designed by following the steps below:

1. Choose an agent that can potentially elicit the biological effects desired.
2. Study the physicochemical, structural, and biological properties of the agent.
3. Select a technology that can be used to deliver the agent to target sites.
4. Evaluate the performance of the technology.
5. Optimize the parameters of the delivery process for enhanced efficiency.

References

[1] O.C. Farokhzad, R. Langer, Impact of nanotechnology on drug delivery, ACS Nano 3 (2009) 16–20.

[2] L. Jia, H. Wong, C. Cerna, S.D. Weitman, Effect of nanonization on absorption of 301029: ex vivo and in vivo pharmacokinetic correlations determined by liquid chromatography/mass spectrometry, Pharm. Res. 19 (2002) 1091–1096.

[3] M. Hrynyk, M. Martins-Green, A.E. Barron, R.J. Neufeld, Sustained prolonged topical delivery of bioactive human insulin for potential treatment of cutaneous wounds, Int. J. Pharm. 398 (2010) 146–154.

[4] L. WF, L. MCM, Synthesis and properties of chitosan-PEI graft copolymers as vectors for nucleic acid delivery, J. Mater. Sci. Eng. 4 (2010) 34–40.

[5] N.Y. Yu, A. Schindeler, L. Peacock, K. Mikulec, P.A. Baldock, A. J. Ruys, et al., In vivo local co-delivery of recombinant human bone morphogenetic protein-7 and pamidronate via poly-D, L-lactic acid, Eur. Cell. Mater. 20 (2010) 431–441; discussion 441–432.

[6] M. Bivas-Benita, M.Y. Lin, S.M. Bal, K.E. van Meijgaarden, K.L. Franken, A.H. Friggen, et al., Pulmonary delivery of DNA encoding Mycobacterium tuberculosis latency antigen Rv1733c associated to PLGA-PEI nanoparticles enhances T cell responses in a DNA prime/protein boost vaccination regimen in mice, Vaccine 27 (2009) 4010–4017.

[7] M. Thomas, J.J. Lu, Q. Ge, C. Zhang, J. Chen, A.M. Klibanov, Full deacylation of polyethylenimine dramatically boosts its gene delivery efficiency and specificity to mouse lung, Proc. Natl. Acad. Sci. U.S.A. 102 (2005) 5679–5684.

[8] X. Zhou, B. Liu, X. Yu, X. Zha, X. Zhang, Y. Chen, et al., Controlled release of PEI/DNA complexes from mannose-bearing chitosan microspheres as a potent delivery system to enhance immune response to HBV DNA vaccine, J. Control Release 121 (2007) 200–207.

[9] X. Zhou, B. Liu, X. Yu, X. Zha, X. Zhang, X. Wang, et al., Controlled release of PEI/DNA complexes from PLGA microspheres as a potent delivery system to enhance immune response to HIV vaccine DNA prime/MVA boost regime, Eur. J. Pharm. Biopharm. 68 (2008) 589–595.

[10] K. Aravindaram, N.S. Yang, Gene gun delivery systems for cancer vaccine approaches, Methods Mol. Biol. 542 (2009) 167–178.
[11] S.H. Nezhadi, P.F. Choong, F. Lotfipour, C.R. Dass, Gelatin-based delivery systems for cancer gene therapy, J. Drug. Target. 17 (2009) 731–738.
[12] E. Wagner, R. Kircheis, G.F. Walker, Targeted nucleic acid delivery into tumors: new avenues for cancer therapy, Biomed. Pharmacother. 58 (2004) 152–161.
[13] S.L. Wang, H.H. Yao, Z.H. Qin, Strategies for short hairpin RNA delivery in cancer gene therapy, Expert. Opin. Biol. Ther. 9 (2009) 1357–1368.
[14] G.P. Li, Z.G. Liu, B. Liao, N.S. Zhong, Induction of Th1-type immune response by chitosan nanoparticles containing plasmid DNA encoding house dust mite allergen Der p 2 for oral vaccination in mice, Cell. Mol. Immunol. 6 (2009) 45–50.
[15] X.Q. Liu, W.J. Song, T.M. Sun, P.Z. Zhang, J. Wang, Targeted delivery of antisense inhibitor of miRNA for antiangiogenesis therapy using cRGD-functionalized nanoparticles, Mol. Pharm. 8 (2011) 250–259.
[16] W.F. Lai, Delivery of therapeutics: current status and its relevance to regenerative innovations, Curr. Nanomed. 1 (2011) 7–18.
[17] C.T. Laurencin, M.A. Attawia, L.Q. Lu, M.D. Borden, H.H. Lu, W.J. Gorum, et al., Poly(lactide-co-glycolide)/hydroxyapatite delivery of BMP-2-producing cells: a regional gene therapy approach to bone regeneration, Biomaterials 22 (2001) 1271–1277.
[18] P.J. Keck, S.D. Hauser, G. Krivi, K. Sanzo, T. Warren, J. Feder, et al., Vascular permeability factor, an endothelial cell mitogen related to PDGF, Science 246 (1989) 1309–1312.
[19] T. Tarkka, A. Sipola, T. Jamsa, Y. Soini, S. Yla-Herttuala, J. Tuukkanen, et al., Adenoviral VEGF-A gene transfer induces angiogenesis and promotes bone formation in healing osseous tissues, J. Gene Med. 5 (2003) 560–566.
[20] Y.C. Huang, C. Simmons, D. Kaigler, K.G. Rice, D.J. Mooney, Bone regeneration in a rat cranial defect with delivery of PEI-condensed plasmid DNA encoding for bone morphogenetic protein-4 (BMP-4), Gene Ther. 12 (2005) 418–426.
[21] A. El-Aneed, An overview of current delivery systems in cancer gene therapy, J. Control Release 94 (2004) 1–14.
[22] I. Fajac, P. Briand, M. Monsigny, P. Midoux, Sugar-mediated uptake of glycosylated polylysines and gene transfer into normal and cystic fibrosis airway epithelial cells, Hum. Gene Ther. 10 (1999) 395.
[23] S. Ferrari, A. Pettenazzo, N. Garbati, F. Zacchello, J.P. Behr, M. Scarpa, Polyethylenimine shows properties of interest for cystic fibrosis gene therapy, Biochim. Biophys. Acta 1447 (1999) 219–225.
[24] US Patent, 20100310532, Gene targets in anti-aging therapy and tissue repair, (2010).
[25] WO Patent, 2010059706, Translation factors as anti-aging drug targets, (2010).
[26] W.F. Lai, M.C. Lin, Nucleic acid delivery with chitosan and its derivatives, J. Control Release 134 (2009) 158–168.
[27] J.M. Olefsky, Gene therapy for rats and mice, Nature 408 (2000) 420–421.
[28] B. Perez, L. Rodriguez-Pascau, L. Vilageliu, D. Grinberg, M. Ugarte, L.R. Desviat, Present and future of antisense therapy for splicing modulation in inherited metabolic disease, J. Inherit. Metab. Dis. 33 (2010) 397–403.
[29] P.S. Sharp, H. Bye-a-Jee, D.J. Wells, Physiological characterization of muscle strength with variable levels of dystrophin restoration in mdx mice following local antisense therapy, Mol. Ther. 19 (2011) 165–171.
[30] X. Sun, M. Vale, X. Jiang, R. Gupta, G. Krissansen, Antisense HIF-1α prevents acquired tumor resistance to angiostatin gene therapy, Cancer Gene. Ther. 17 (2010) 532.
[31] I. Vaneckova, Z. Dobesova, J. Kunes, J. Zicha, The effect of repeated antisense therapy on various vasoactive systems in young ren-2 transgenic rats, J. Hypertens. 28 (2010) e492.
[32] W.F. Lai, In vivo nucleic acid delivery with PEI and its derivatives: current status and perspectives, Expert. Rev. Med. Devices 8 (2011) 173–185.
[33] M. Tangney, Gene therapy for cancer: dairy bacteria as delivery vectors, Discov. Med. 10 (2010) 195–200.
[34] C. Ma, Z. Fan, Z. Gao, S. Wang, Z. Shan, Delivery of human erythropoietin gene with an adeno-associated virus vector through parotid glands to treat renal anaemia in a swine model, Gene Ther. 24 (2017) 692–698.
[35] M.G. Katz, A.S. Fargnoli, T. Weber, R.J. Hajjar, C.R. Bridges, Use of adeno-associated virus vector for cardiac gene delivery in large-animal surgical models of heart failure, Hum. Gene Ther. Clin. Dev. 28 (2017) 157–164.
[36] L. Xu, Y. Gao, Y.S. Lau, R. Han, Adeno-associated virus-mediated delivery of CRISPR for cardiac gene editing in mice, J. Vis. Exp. (2018). Available from: https://doi.org/10.3791/57560.
[37] Y.-F. Xu, Y.-Q. Zhang, X.-M. Xu, G.-X. Song, Papillomavirus virus-like particles as vehicles for the delivery of epitopes or genes, Arch. Virol. 151 (2006) 2133–2148.
[38] S.H. Kim, E.J. Kim, J.H. Hou, J.M. Kim, H.G. Choi, C.K. Shim, et al., Opsonized erythrocyte ghosts for liver-targeted delivery of antisense oligodeoxynucleotides, Biomaterials 30 (2009) 959–967.
[39] S.H. Kim, N. Bianco, R. Menon, E.R. Lechman, W.J. Shufesky, A.E. Morelli, et al., Exosomes derived from genetically modified DC expressing FasL are anti-inflammatory and immunosuppressive, Mol. Ther. 13 (2006) 289–300.
[40] Y. Seow, M.J. Wood, Biological gene delivery vehicles: beyond viral vectors, Mol. Ther. 17 (2009) 767–777.
[41] US Patent, 5328840, Method for preparing targeted carrier erythrocytes, (1994).
[42] US Patent, WO2010135563, Recombinant bacterium and methods of antigen and nucleic acid delivery, (2010).
[43] X. Zhang, W.T. Godbey, Viral vectors for gene delivery in tissue engineering, Adv. Drug Deliv. Rev. 58 (2006) 515–534.
[44] US Patent, 5952225, Retroviral vectors produced by producer cell lines resistant to lysis by human serum, (1999).
[45] US Patent, 6218186, HIV-MSCV hybrid viral vector for gene transfer, (2001).
[46] US Patent, 7820436, Recombinant viral vector for gene transfer into lymphoid cells, (2010).
[47] P. Faisca, D. Desmecht, Sendai virus, the mouse parainfluenza type 1: a longstanding pathogen that remains up-to-date, Res. Vet. Sci. 82 (2007) 115–125.
[48] US Patent, 7803621, Virus envelope vector for gene transfer, (2010).
[49] G.Y. Wu, C.H. Wu, Receptor-mediated gene delivery and expression in vivo, J. Biol. Chem. 263 (1988) 14621–14624.

[50] G. Guo, L. Zhou, Z. Chen, W. Chi, X. Yang, W. Wang, et al., Alkane-modified low-molecular-weight polyethylenimine with enhanced gene silencing for siRNA delivery, Int. J. Pharm. 450 (2013) 44–52.

[51] Y. Li, Y.F. Qian, T. Liu, G.Y. Zhang, J.M. Hu, S.Y. Liu, Asymmetrically functionalized beta-cyclodextrin-based star copolymers for integrated gene delivery and magnetic resonance imaging contrast enhancement, Polym. Chem.-UK. 5 (2014) 1743–1750.

[52] Genetic Immunity, Dermavir. <http://www.geneticimmunity.com/dermavir.html> (accessed 18.09.18).

[53] Anchiano Therapeutics, BioCanCell Advances Phase III Clinical Program for Bladder Cancer. <https://www.anchiano.com/2015oct1/> (accessed 19.09.18).

[54] A.D. Bangham, Liposome Letters, Academic Press, London, 1983.

[55] S.X. Straub, R.F. Garry, W.E. Magee, Interferon induction by poly (I): poly (C) enclosed in phospholipid particles, Infect. Immun. 10 (1974) 783–792.

[56] T. Benvegnu, L. Lemiegre, S. Cammas-Marion, New generation of liposomes called archaeosomes based on natural or synthetic archaeal lipids as innovative formulations for drug delivery, Recent Pat. Drug Deliv. Formul. 3 (2009) 206–220.

[57] US Patent, 6316260, Tetraether lipid derivatives and liposomes and lipid agglomerates containing tetraether lipid derivatives, and use thereof (2001).

[58] US Patent, 7223600, Photochemical internalization for delivery of molecules into the cytosol, (2007).

[59] US Patent, 7141044, Alternate site gene therapy, (2006).

[60] US Patent, 7211248, Enhancement of transfection of DNA into the liver, (2007).

[61] US Patent, 7547473, Magnetic nanoparticles and method for producing the same, (2009).

[62] US Patent, 7550650, Production of a transgenic avian by cytoplasmic injection, (2009).

[63] US Patent, 7547551, Transfection of eukaryontic cells with linear polynucleotides by electroporation, (2009).

[64] US Patent, 7638332, Low pressure gas accelerated gene gun, (2009).

[65] L.C. Heller, R. Heller, Electroporation gene therapy preclinical and clinical trials for melanoma, Curr. Gene Ther. 10 (2010) 312–317.

[66] L. Lambricht, A. Lopes, S. Kos, G. Sersa, V. Preat, G. Vandermeulen, Clinical potential of electroporation for gene therapy and DNA vaccine delivery, Expert. Opin. Drug Deliv. 13 (2016) 295–310.

[67] T.H. Chou, S. Biswas, S. Lu, Gene delivery using physical methods: an overview, Methods Mol. Biol. 245 (2004) 147–166.

[68] K.E. Matthews, A. Keating, Gene therapy with physical methods of gene transfer, Transfus. Sci. 17 (1996) 29–34.

[69] S. Mehier-Humbert, R.H. Guy, Physical methods for gene transfer: improving the kinetics of gene delivery into cells, Adv. Drug Deliv. Rev. 57 (2005) 733–753.

[70] S. Naqvi, M. Samim, A.K. Dinda, Z. Iqbal, S. Telagoanker, F.J. Ahmed, et al., Impact of magnetic nanoparticles in biomedical applications, Recent Pat. Drug Deliv. Formul. 3 (2009) 153–161.

[71] D.J. Wells, Gene therapy progress and prospects: electroporation and other physical methods, Gene Ther. 11 (2004) 1363–1369.

[72] Y. Zhang, W. Wei, P. Lv, L. Wang, G. Ma, Preparation and evaluation of alginate-chitosan microspheres for oral delivery of insulin, Eur. J. Pharm. Biopharm. 77 (2011) 11–19.

[73] E.K. Hui, R.J. Boado, W.M. Pardridge, Tumor necrosis factor receptor-IgG fusion protein for targeted drug delivery across the human blood-brain barrier, Mol. Pharm. 6 (2009) 1536–1543.

[74] P. Xitian, L. Hongying, W. Kang, L. Yulin, Z. Xiaolin, W. Zhiyu, A novel remote controlled capsule for site-specific drug delivery in human GI tract, Int. J. Pharm. 382 (2009) 160–164.

[75] US Patent, 5702717, Thermosensitive biodegradable polymers based on poly(ether-ester) block copolymers, (2007).

[76] US Patent, 6451346, Biodegradable pH/thermosensitive hydrogels for sustained delivery of biologically active agents, (2002).

[77] A. Rahman, J. Treat, J.K. Roh, L.A. Potkul, W.G. Alvord, D. Forst, et al., A phase I clinical trial and pharmacokinetic evaluation of liposome-encapsulated doxorubicin, J. Clin. Oncol. 8 (1990) 1093–1100.

[78] R.R. Owen, R.A. Sells, I.T. Gilmore, R.R. New, R.E. Stringer, A phase I clinical evaluation of liposome-entrapped doxorubicin (Lip-Dox) in patients with primary and metastatic hepatic malignancy, Anticancer Drugs 3 (1992) 101–107.

[79] X. Wang, J. Zhou, Y. Wang, Z. Zhu, Y. Lu, Y. Wei, et al., A phase I clinical and pharmacokinetic study of paclitaxel liposome infused in non-small cell lung cancer patients with malignant pleural effusions, Eur. J. Cancer 46 (2010) 1474–1480.

[80] Q. Zhang, X.E. Huang, L.L. Gao, A clinical study on the premedication of paclitaxel liposome in the treatment of solid tumors, Biomed. Pharmacother. 63 (2009) 603–607.

[81] S.A. Boorjian, M.I. Milowsky, J. Kaplan, M. Albert, M.V. Cobham, D.M. Coll, et al., Phase 1/2 clinical trial of interferon alpha2b and weekly liposome-encapsulated all-trans retinoic acid in patients with advanced renal cell carcinoma, J. Immunother. 30 (2007) 655–662.

[82] US Patent, 7517914, Controlled degradation materials for therapeutic agent delivery, (2009).

[83] US Patent, 7824704, Controlled release bioactive agent delivery device, (2010).

[84] WO Patent, 2005118789, Therapeutic delivery of adenosine into a tissue, (2005).

[85] US Patent, 7115257, ARPE-19 as a platform cell line for encapsulated cell-based delivery, (2006).

[86] L. Birch, C.L. Archard, H.C. Parkes, D.G. McDowell, Evaluation of LabChip™ technology for GMO analysis in food, Food Control 12 (2001) 535–540.

[87] D. Janasek, J. Franzke, A. Manz, Scaling and the design of miniaturized chemical-analysis systems, Nature 442 (2006) 374–380.

[88] S.K. Kim, J.H. Kim, K.P. Kim, T.D. Chung, Continuous low-voltage DC electroporation on a microfluidic chip with polyelectrolytic salt bridges, Anal. Chem. 79 (2007) 7761–7766.

[89] A. Jahn, W.N. Vreeland, D.L. DeVoe, L.E. Locascio, M. Gaitan, Microfluidic directed formation of liposomes of controlled size, Langmuir 23 (2007) 6289–6293.

[90] X. Huang, R. Caddell, B. Yu, S. Xu, B. Theobald, L.J. Lee, et al., Ultrasound-enhanced microfluidic synthesis of liposomes, Anticancer Res. 30 (2010) 463–466.

[91] K. Kitazoe, J. Wang, N. Kaji, Y. Okamoto, M. Tokeshi, K. Kogure, et al., A touch-and-go lipid wrapping technique in microfluidic channels for rapid fabrication of multifunctional envelope-type gene delivery nanodevices, Lab. Chip. 11 (2011) 3256–3262.
[92] K. Kogure, H. Akita, Y. Yamada, H. Harashima, Multifunctional envelope-type nano device (MEND) as a non-viral gene delivery system, Adv. Drug Deliver. Rev. 60 (2008) 559–571.
[93] I.A. Khalil, K. Kogure, S. Futaki, S. Hama, H. Akita, M. Ueno, et al., Octaarginine-modified multifunctional envelope-type nanoparticles for gene delivery, Gene Ther. 14 (2007) 682–689.
[94] T. Ishitsuka, H. Akita, H. Harashima, Functional improvement of an IRQ-PEG-MEND for delivering genes to the lung, J. Control Release 154 (2011) 77–83.
[95] H. Hatakeyama, H. Akita, H. Harashima, A multifunctional envelope type nano device (MEND) for gene delivery to tumours based on the EPR effect: a strategy for overcoming the PEG dilemma, Adv. Drug Deliver. Rev 63 (2011) 152–160.
[96] K. Kogure, R. Moriguchi, K. Sasaki, M. Ueno, S. Futaki, H. Harashima, Development of a non-viral multifunctional envelope-type nano device by a novel lipid film hydration method, J. Control Release 98 (2004) 317–323.
[97] A.K. Stark, M. Schilling, D. Janasek, J. Franzke, Characterization of dielectric barrier electrospray ionization for mass spectrometric detection, Anal. Bioanal. Chem. 397 (2010) 1767–1772.
[98] S.H. Ranganath, I. Kee, W.B. Krantz, P.K. Chow, C.H. Wang, Hydrogel matrix entrapping PLGA-paclitaxel microspheres: drug delivery with near zero-order release and implantability advantages for malignant brain tumour chemotherapy, Pharma. Res. 26 (2009) 2101–2114.
[99] W.F. Lai, A.S. Susha, A.L. Rogach, Multicompartment microgel beads for co-delivery of multiple drugs at individual release rates, ACS Appl. Mater. Interfaces 8 (2016) 871–880.

Part II

From Plans to Technologies

Chapter 3

Design of viral vectors for genetic manipulation

Introduction

Accompanying with the gradual elucidation of the underlying biology of aging, genetic manipulation has emerged as a viable strategy to manipulate the aging process. Viral vectors are by far the most extensively studied gene carriers. Owing to their high efficiency in nucleic acid delivery, copious viral vectors have progressed into clinical trials in as early as the 1990s [1,2]. Since the turn of the last century, increasing efforts have been devoted to developing viral vectors capable of sensing specific environmental inputs because these vectors can potentially enable more precise control of the gene delivery process. In fact, many viruses have already naturally evolved to be able to respond to different endogenous stimuli (e.g., pH and redox) [3–6]. Efforts in the production of viruses that can respond to exogenous stimuli (e.g., temperature, optical light, and magnetic field) have also been reported in recent years [7–10]. To apply a viral vector to genetic manipulation, we often need to remove most of the genes coding for viral proteins from the viral genome and to enable the viral proteins required for viral replication to be expressed. The latter can be achieved by using a transiently transfected plasmid that encodes the proteins, or by infecting the packaging cells concomitantly with a helper virus in which the protein-encoding genes are present. Until now, applications of members of the *Retroviridae* family (gammaretroviruses and lentiviruses), adenoviruses, AAVs, and herpesviruses have constituted more than 50% of clinical trials of gene therapy mediated by virus-mediated gene delivery [11].

When a biogerontological intervention is developed, the viral vector to be adopted should ideally be safe. The possibility of offering stable and prolonged gene expression is also desired as this can enable one-off treatment for antiaging purposes. This need can be partially fulfilled by using retroviruses. After cellular internalization, the RNA genome of this virus can be reverse-transcribed into double-stranded DNA, which then undergoes nuclear localization and finally gets integrated into the host genome. Stable transgene expression brought by genomic integration, however, is often accompanied by insertional mutagenesis. The problem of insertional mutagenesis was brought to particular attention in the early 2000s when two patients suffering from SCID-X1 developed acute lymphoblastic leukemia after gammaretroviral gene therapy of hematopoietic stem cells. This has later been shown to be caused by the retroviral vector, which has been inserted into the LMO2 proto-oncogene [12–14]. Instances of preneoplastic or truly neoplastic cell expansion led by insertional mutagenesis have also been reported in gene therapy of X-linked chronic granulomatous diseases [15] and Wiskott–Aldrich syndrome (WAS) [16].

Example protocols for experimental design

The method below is an example protocol for preparing a lentiviral vector for transduction:

1. Plate HEK 293T cells in a 10-cm plastic cell culture dish
2. Incubate the dish at 37°C under a humidified atmosphere with 5% CO_2 until a confluence of 30%–60% is obtained
3. Dissolve PEI in 1 mL of the serum-free medium to reach a concentration of 30 μg/mL
4. Prepare four lentiviral plasmid solutions (pRSV-Rev, pMD2.G, pMDLg/pRRE, and the lentiviral transfer vector, which encodes the gene to be packaged), each with a concentration of 1 mg/mL
5. Add the four plasmid solutions into the PEI solution
6. Vortex the mixture for 30 seconds
7. Incubate the mixture at ambient conditions for 15 minutes to generate polyplexes
8. Replace the cell culture medium in the dish with 9 mL of the fresh cell culture medium that contains no antibiotics
9. Add the polyplexes into the dish
10. Swirl the dish gently to mix well

(*Continued*)

Delivery of Therapeutics for Biogerontological Interventions. DOI: https://doi.org/10.1016/B978-0-12-816485-3.00003-9

(Continued)

11. Add 10 μL of a chloroquine solution into the dish to reach a concentration of 20 μM
12. Incubate the dish at 37°C under a humidified atmosphere with 5% CO_2 for 5 hours
13. Replace the cell culture medium with 10 mL of the fresh medium containing 10% fetal bovine serum (FBS) and 20 mM 4-(2-hydroxyethyl)-1-piperazineethanesulfonic acid (HEPES)
14. Incubate the dish at 37°C under a humidified atmosphere with 5% CO_2 for 2–3 days
15. Collect the lentiviral solution and filter it through a 0.45-μm cellulose acetate membrane filter
16. Add 1 M HEPES (pH 7.4) into the filtered lentiviral solution to reach a concentration of 10 mM
17. Add the solution to cells for transduction

To avoid the occurrence of insertional mutagenesis, adenoviral vectors can be used as an alternative. The adenovirus belongs to the family *Adenoviridae* and the genus *Mastadenovirus*. It is a nonenveloped virus possessing a double-stranded DNA genome. Over 50 viral proteins are encoded by the viral genome. Among them, hexon, fiber, and penton, along with few other structural proteins needed for capsid stabilization and genome association, constitute the exterior of the icosahedral viral capsid. Until now, diverse serotypes of the adenovirus have been exploited as carriers to deliver nucleic acids to diverse tissues, ranging from the hepatic tissue [17] to the inner ear [18]. Favorable properties of the adenovirus include its large cloning capacity and its ability to transduce both dividing and nondividing cells [19]. In addition, it can be easily prepared as a high-titer stock [19]. Yet, the virus can elicit host innate inflammatory responses after intravascular administration [20]. This is a major obstacle to effective use of the vector as a gene carrier.

Example protocols for experimental design

The method below is an example protocol for determining the 50% tissue culture infectious dose (TCID50) for evaluating the infectability of an adenoviral stock:

1. Plate HEK 293T cells in a 10 cm plastic cell culture dish
2. Incubate the dish at 37°C under a humidified atmosphere with 5% CO_2 until a confluence of 80% is obtained
3. Add 60 μL of the serum-containing cell culture medium to each well of a 96-well flat-bottom cell culture plate
4. Dilute 1 μL of the adenoviral stock to 1 mL
5. Add 20 μL of the diluted stock to the first row
6. Mix well
7. Transfer 20 μL from each well in the first row of the plate to the corresponding well in the next row

(Continued)

(Continued)

8. Repeat steps 6 and 7 until reaching the second last row of the plate, with the last row of the plate being used as the control
9. Discard 20 μL from each well in the second last row of the plate
10. Triturate the cells cultured in the dish into 10 mL of the fresh cell culture medium
11. Add 50 μL of the cell solution into each well of the plate
12. Incubate the plate at 37°C under a humidified atmosphere with 5% CO_2 for 4 days
13. Add 50 μL of the fresh cell culture medium into each well of the plate
14. Repeat Steps 12 and 13
15. Incubate the dish at 37°C under a humidified atmosphere with 5% CO_2 for another 4 days
16. Measure the cytopathic effects in each well of the plate
17. Determine TCID50 using the following formula, with S being the sum of the ratios of positive wells per row, D_{1st} being the dilution factor for the first row of the plate (in the protocol above, D_{1st} is 4×10^3), and D_R being the dilution value from one row to the next row (in the protocol above, D_R is 4)

$$\mathrm{TCID50} = D_{1st} + D_R^{(S-0.5)} \quad (3.1)$$

The AAV is another type of viral vectors that do not raise the concerns of insertional mutagenesis. The AAV belongs to the family *Parvoviridae* and the genus *Dependovirus*. It is a nonenveloped virus having a single-stranded genome. To complete the lifecycle of this virus, the presence of proteins from a helper virus (e.g., adenovirus) is needed. As, at the moment, this virus has not been found to be associated with any human diseases, together with its capacity of mediating long-term transgene expression in both dividing and nondividing cells [21–23], the AAV has emerged as one of the favorable candidates for virus-mediated gene delivery. In the viral genome, there are two inverted terminal repeats (ITRs) that flank two genes, namely, *cap* and *rep*. These two genes can produce diverse proteins with partially overlapping sequences [23]. In brief, the *cap* gene encodes three proteins (VP1, VP2, and VP3) that form the viral capsid [23]; whereas the *rep* gene can lead to the generation of four proteins (Rep78, Rep68, Rep52, and Rep40) that are needed for various viral functions, ranging from virion assembly to transcription regulation [21]. Although the application potential of the AAV has been significantly improved by technological advances that enable scalable virus production and purification [24,25], the clinical use of the AAV has been hindered by the limited genome packaging size and the possible infection of off-target cells [21].

Finally, while the adenoviral vector and AAV can avoid the risk of insertional mutagenesis, because of the lack of integration of the transgene into the host genome, transgene expression is less persistent as compared to that achieved by lentiviral vectors. This problem can be partially addressed by using the prototype foamy virus (PFV), which shows genome integration during transduction and tends to integrate outside of the active transcription units of the host genome [26]. In addition, the PFV possesses limited seroprevalence in humans [27], broad cell tropism [28], and importantly, a large packaging capacity that enables it to carry payloads greater than 9 kb [29]. All these make PFV vectors one of the candidates that warrant further exploitation for applications in genetic manipulation. The biodistribution profile of PFV vectors has previously been examined in neonatal mice. On Day 11 after an intravenous (i.v.) injection of PFV-CMV-EGFP, which is a PFV vector packaged with enhanced green fluorescent protein (EGFP) driven by the cytomegalovirus (CMV) promoter, EGFP fluorescence has been detected in the heart, lung, liver, and spleen [28], whereas an intraperitoneal (i.p.) injection of the vector has led to transgene expression in the liver, small intestine, xiphisternum, pancreas, and gut (Fig. 3.1) [28]. Moreover, after an intracranial injection of the vector into the anterior horn of the lateral ventricle on the left side of the brain, expression of the transgene has been detected mainly at the hippocampal architecture, with dense expression being localized to the dentate gyrus [28]. This is in contrast to the AAV vector and lentiviral vector, which can give a broad and nonspecific spread through the brain of neonatal mice without displaying regional localization [28]. The localized intracranial expression profile given by the PFV vector might be desirable if genetic manipulation of hippocampal neurons is desired, but if more widespread genetic manipulation of the brain is required, the AAV vector and lentiviral vector might be a more favorable choice (Fig. 3.2).

Engineering the properties of a viral vector

With technical advances, it is now possible to generate a complex viral vector by combining components from multiple viral variants into one single virus. One strategy to achieve this is to swap viral attachment protein (VAP) domains among different serotypes or to replace a VAP from one serotype with that of the other. This has been documented in an earlier study [30], which has

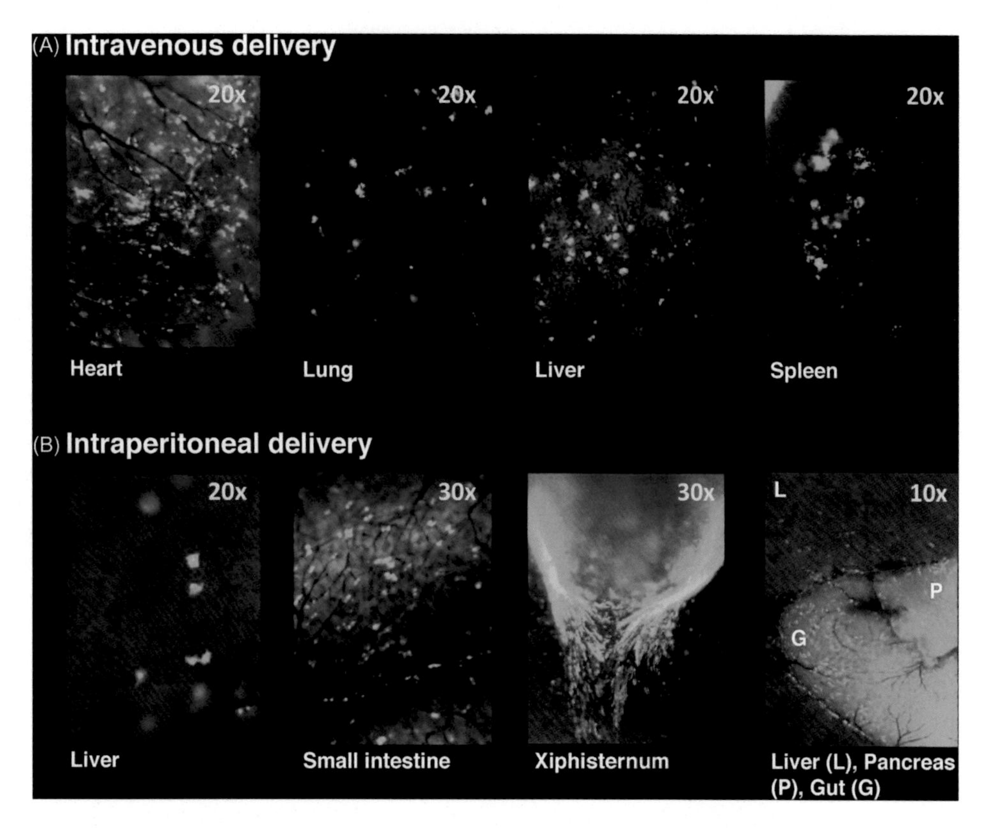

FIGURE 3.1 EGFP fluorescence imaging of mice on Day 11 after administration of PFV-CMV-EGFP via the (A) i.v. and (B) i.p. routes. *CMV*, cytomegalovirus; *EGFP*, enhanced green fluorescent protein; *i.p.*, intraperitoneal; *i.v.*, intravenous; *PFV*, prototype foamy virus. *Reproduced from J.R. Counsell, R. Karda, J.A. Diaz, L. Carey, T. Wiktorowicz, S.M.K. Buckley, et al., Foamy virus vectors transduce visceral organs and hippocampal structures following in vivo delivery to neonatal mice, Mol. Ther. Nucleic Acids 12 (2018) 626–634, with permission from Elsevier B.V. [28].*

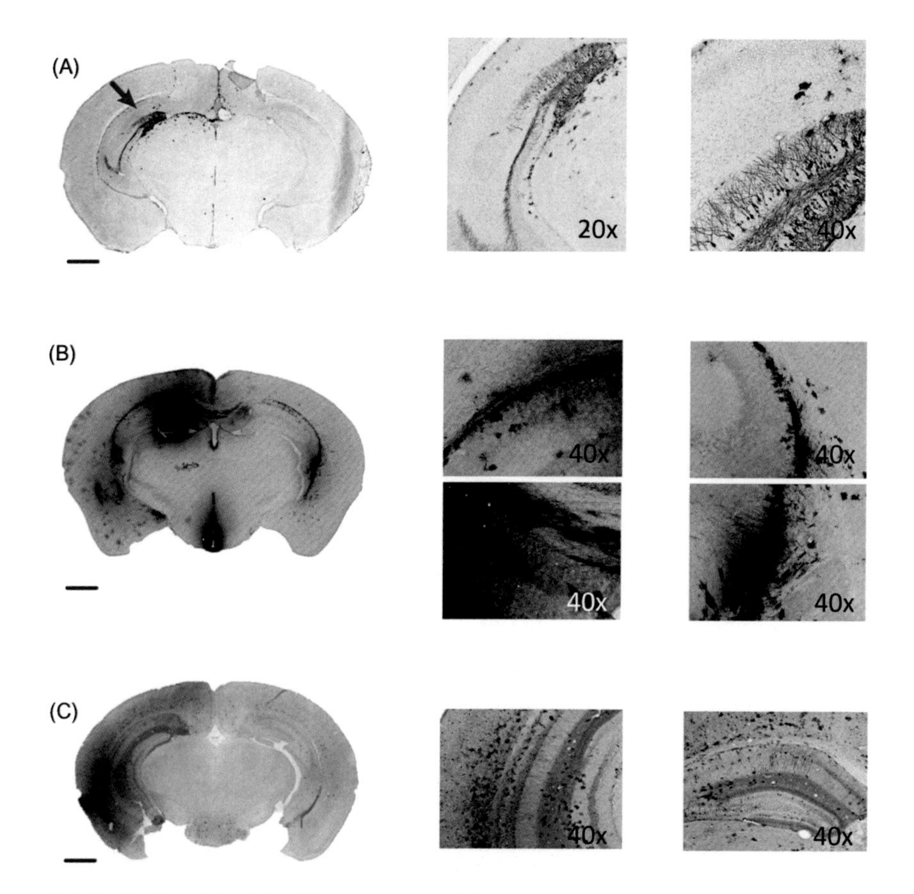

FIGURE 3.2 Intracranial biodistribution of the (A) PFV vector, (B) lentiviral vector, and (C) AAV vector packaged with EGFP, on Day 11 after intracranial administration of the vector into neonatal mice. The EGFP immunostaining localized to the hippocampal architecture is indicated by a *blue arrow*. The dose of the PFV vector, lentiviral vector, and AAV vector is 2.5×10^7, 4.2×10^6, and 2.5×10^6 genome copies, respectively. The scale bar is 5 μm. *EGFP*, enhanced green fluorescent protein; *PFV*, prototype foamy virus. *Reproduced from J.R. Counsell, R. Karda, J.A. Diaz, L. Carey, T. Wiktorowicz, S.M.K. Buckley, et al., Foamy virus vectors transduce visceral organs and hippocampal structures following in vivo delivery to neonatal mice, Mol. Ther. Nucleic Acids 12 (2018) 626–634, with permission from Elsevier B.V. [28].*

constructed chimeric replication-incompetent adenovirus serotype 5 (rAd5) vectors by replacing the seven short hypervariable regions (HVR) on the surface of the Ad5 hexon protein with the corresponding HVRs from the rare adenovirus serotype Ad48. By removing the neutralizing epitopes on the surface of viral capsid proteins, the preexisting antivector immunity in the host has been circumvented. Apart from the adenoviral vector, the success in generating chimeric recombinant viral vectors has been demonstrated in other viruses, such as the AAV. As far as AAV-based gene delivery is concerned, AAV2 has been extensively used and can be easily purified by using the heparin column [31]. Yet, when transduction to muscle tissues is involved, AAV1 often shows higher efficiency [31]. To combine the strengths of both serotypes into one vector, Hauck and coworkers have cotransfected AAV packaging cells with a mixture of AAV helper plasmids encoding both serotypes, generating packaged virions containing capsid proteins from AAV1 and AAV2 [31]. Results have shown that the chimeric AAV vector cannot only be purified readily using the heparin column but also exhibits the transduction characteristics of both parent vectors, resulting in a level of transgene expression similar to that of AAV1 in muscles or AAV2 in the liver. Despite this encouraging result, the cotransfection strategy often results in a heterogeneous mixture of chimeric vectors. Attainment of the desired phenotype is stochastic.

As aging is a systemic phenomenon, genetic manipulation in different tissues is needed for antiaging purposes. Since many cells (e.g., neurons, cardiomyocytes, skeletal muscle cells, endothelial cells, and the vast majority of peripheral blood lymphocytes) seldom divide or do not divide at all, viral vectors (e.g., gammaretroviral vectors) that fail to transduce nonreplicating cells may not be a favorable choice. With the use of pseudotyping, the cell tropism of a virus can be engineered. This has been illustrated by the case of lentiviral vectors. Studies have revealed the phenomenon that, while the lentiviral vector pseudotyped with the vesicular stomatitis virus (VSV) glycoprotein gives little transgene expression upon apical application to polarized well-differentiated human airway epithelial cell cultures [32], the vector pseudotyped with envelope proteins from filoviruses (e.g., respiratory syncytial virus, fowl plaque virus, influenza virus, and Ebola virus) enables effective transduction of the airway epithelial cells from the apical side [33]. Cell targeting can also be achieved by incorporating surface proteins with domains that are sensitive to specific cell properties. For example, some age-associated diseases (such as Alzheimer's disease [34] and Parkinson's disease [35]) are known to be associated with a higher level of expression of matrix metalloproteinases (MMPs) such as MMP-2, MMP-7, and MMP-9. Incorporating viral vectors with protease-sensitive domains can, therefore, enable controlled delivery of transgenes into pathological sites that show an elevated protease level. This has already been achieved by Szecsi and colleagues [36], who have generated a retroviral vector that harbors engineered surface glycoproteins derived from avian influenza virus hemagglutinins (HAs). The engineered glycoproteins have been inserted with MMP-2-cleavable peptide sequences, which block the cell entry of the virus when in place. Using this strategy, the virus can preferentially infect MMP-2-expressing diseased cell lines but not healthy cells [36]. In addition to protein engineering, directed evolution can be used to select mutant viruses displaying MMP-activatable properties. It has been used to generate retroviral vectors showing MMP-dependent tropisms [37].

Chemical manipulation for controlled delivery

One of the common approaches to chemical modification of a viral vector is PEGylation. For instance, a previous study has PEGylated a helper-dependent adenoviral vector encoding the human apolipoprotein A-I [38]. Upon administration to a mouse model of familial hypercholesterolemia, the PEGylated vector has been found to induce less secretion of proinflammatory cytokines, thereby eliciting less acute toxicity, than the unmodified vector. Despite this, PEGylation of a viral vector requires proper optimization of the degree of modification, or the infectivity of the virus will be jeopardized [39]. Another strategy to modify a viral vector is to incorporate the vector with various functional moieties. For example, incorporation with targeting ligands can enhance the internalization of the vector by target cells, whereas attachment of a stimuli-responsive element to a viral vector can enable more precise control of transgene expression.

In fact, viruses sometimes have been naturally evolved to respond to stimuli for infection purposes. One example is reoviruses, which can undergo conformational changes when exposed to changes in pH, resulting in exposure of the protein μ1 that subsequently undergoes autocleavage and finally permeabilizes the endosomal membrane for endolysosomal escape [40,41]. Another example is the Epstein–Barr virus (EBV), which shows redox-responsiveness that plays an important role in regulating viral replication during host cell infection [42]. This regulation mechanism is mediated by the protein Zta, whose redox-sensitive cysteine is oxidized under oxidative conditions, leading to the formation of a disulfide bond between Zta proteins and preventing the binding of the protein to DNA [42]; however, when the cysteine is reduced, binding of Zta to DNA is allowed, causing activation of transcription [42]. In fact, changes in parameters such as pH and redox have always been observed during the aging process. For example, alternations in the redox state, as manifested by the disturbance of the cellular redox homeostasis, have always been associated with cardiovascular aging and the declined functional capacity of immune cells [43,44]. Moreover, a difference in pH between the normal tissue and the diseased tissue has been identified in age-associated disorders such as cancers [45]. Because of these reasons, naturally evolved stimuli-responsiveness has always been welcomed because it may endow the corresponding viral vector favorable possibilities for more controlled and targeted delivery of transgenes during genetic manipulation.

Example protocols for experimental design

The method below is an example protocol for determining the transduction efficiency of a viral vector:

1. Seed the cells into a plastic cell culture dish
2. Incubate the dish at 37°C under a humidified atmosphere with 5% CO_2 until a confluence of 80% is obtained
3. Replace the cell culture medium with the fresh medium that contains no antibiotics
4. Add an appropriate amount of a filtered solution of a viral vector (which encodes the gene of interest) to the dish
5. Swirl the dish gently to mix well
6. Incubate the dish at 37°C under a humidified atmosphere with 5% CO_2 for 6 hours
7. Replace the cell culture medium with the fresh medium
8. Incubate the dish at 37°C under a humidified atmosphere with 5% CO_2 for 3 days
9. Use an appropriate method [e.g., Western blotting, fluorescence microscopy, luciferase activity assay, or immunocytochemistry (ICH)] to determine the expression of the transgene in the cells transduced with the viral vector

Besides relying on natural evolution, stimuli-responsive viral vectors can be generated artificially by using chemical methods. A good example is the photoactivatable retroviral vector reported by Pandori and Sano [46], who have modified a murine leukemia virus (MLV)–derived vector, which bears the envelope glycoprotein of the 4070A amphotropic murine retrovirus, with a biotin derivative [which has an *N*-hydroxysuccinimide (NHS) ester moiety that enables easy attachment of the biotin moiety to viral proteins and possesses a photocleavable 1-(2-nitrophenyl) ethyl moiety that allows for the release of the biotin moieties from proteins upon UV irradiation]. Results have shown that, after conjugation, the viral envelop glycoproteins fail to properly bind to cellular receptors, partly due to the steric hindrance caused by the presence of the biotin moieties. This leads to the loss of viral infectivity in the absence of light, but the infectivity can be easily restored simply by using UV irradiation.

In fact, to modify a viral vector chemically, the availability of exposed amino acids on the surface of the capsid is one feature of which we may take advantage. By incorporating some amine-reactive groups into a functional moiety, the attachment of the moiety to a virus can be achieved via conventional protein biology methods. For instance, by taking advantage of the fact that around 300 of the 1080 lysine residues in the capsid of each AAV are surface-exposed (Fig. 3.3), taxol-NHS ester has been conjugated to an AAV for codelivery of genes and chemical drugs [47], with the structural integrity of the virus having been retained after chemical conjugation. Apart from NHS-esters, other amine-reactive groups that can be exploited include isocyanates, isothiocyanates, sulfonyl chlorides, aldehydes, carbodiimides, acyl azides, anhydrides, carbonates, fluorobenzenes, imidoester, epoxides, and fluorophenyl esters. These groups can allow the functional moieties to be attached to a viral capsid easily.

Applications in genetic manipulation

During aging, a decrease in the striatal dopamine D_2 receptor (D_2R) expression, which is one of the notable features in aged mammalian brains, leads to changes in the dopamine-acetylcholine (ACh) interaction in the striatum, leading to a reduction in the striatal ACh release [48]. Based on this observation, a previous study has perfused adenoviral vectors encoding D_2R into the striatum of an aged rat and has successfully restored the functional decline [49]. In fact, to deliver genes to the central nervous system (CNS), adenoviral vectors are a favorable choice because of their capacity of transducing postmitotic neurons [50–53]. Conventional (E1-deleted) recombinant adenoviral vectors, however, have been found to induce host tissue inflammatory and immune responses [20,54]. As the leaky expression of viral genes accounts for most episodes of toxicity and immunogenicity [55,56], the viral genetic material has been proposed to be completely removed from a virus before use. This leads to the advent of virus-like particles.

Virus-like particles resemble viruses but are noninfectious. Their possible use in combating systemic biological problems, such as aging, has been supported by an earlier study, in which Alexa Fluor 680 (AF680)-labeled adenoviral dodecahedrons have been administered intravenously into a xenograft murine model of subcutaneous melanoma. Five hours after injection, strong signals have been detected in the liver and skin, but weak signals have also been observed in diverse organs (e.g., bladder, spleen, kidneys, lungs, uterus, and guts) (Fig. 3.4) [57]. Such a widespread distribution of virus-like particles after systemic administration may be exploited for the execution of antiaging interventions. Unfortunately, the fluorescence

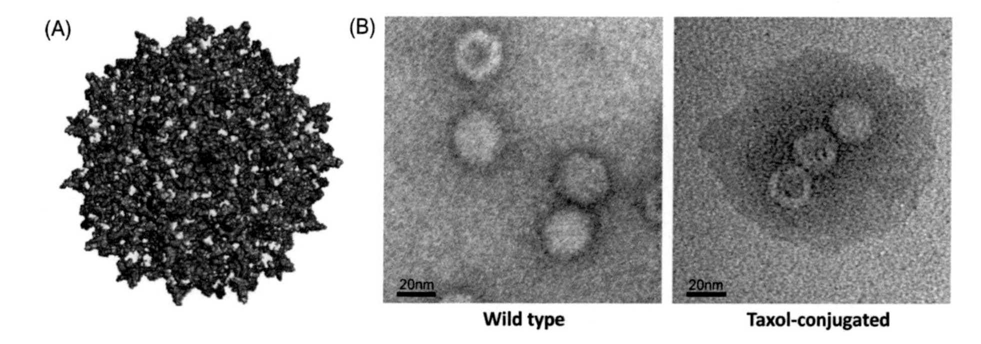

FIGURE 3.3 Structural analysis of the taxol-conjugated AAV. (A) An image of the capsid of an AAV. The surface-exposed lysines are shown in blue. The image is produced by using Pymol. (B) Transmission electron micrographs of the wild-type AAV and the conjugated AAV. Both samples were negatively stained with the uranyl formate solution. *Reproduced from F. Wei, K.I. McConnell, T.K. Yu, J. Suh, Conjugation of paclitaxel on adeno-associated virus (AAV) nanoparticles for co-delivery of genes and drugs, Eur. J. Pharm. Sci., 46 (2012) 167–172, with permission from Elsevier B.V. [47].*

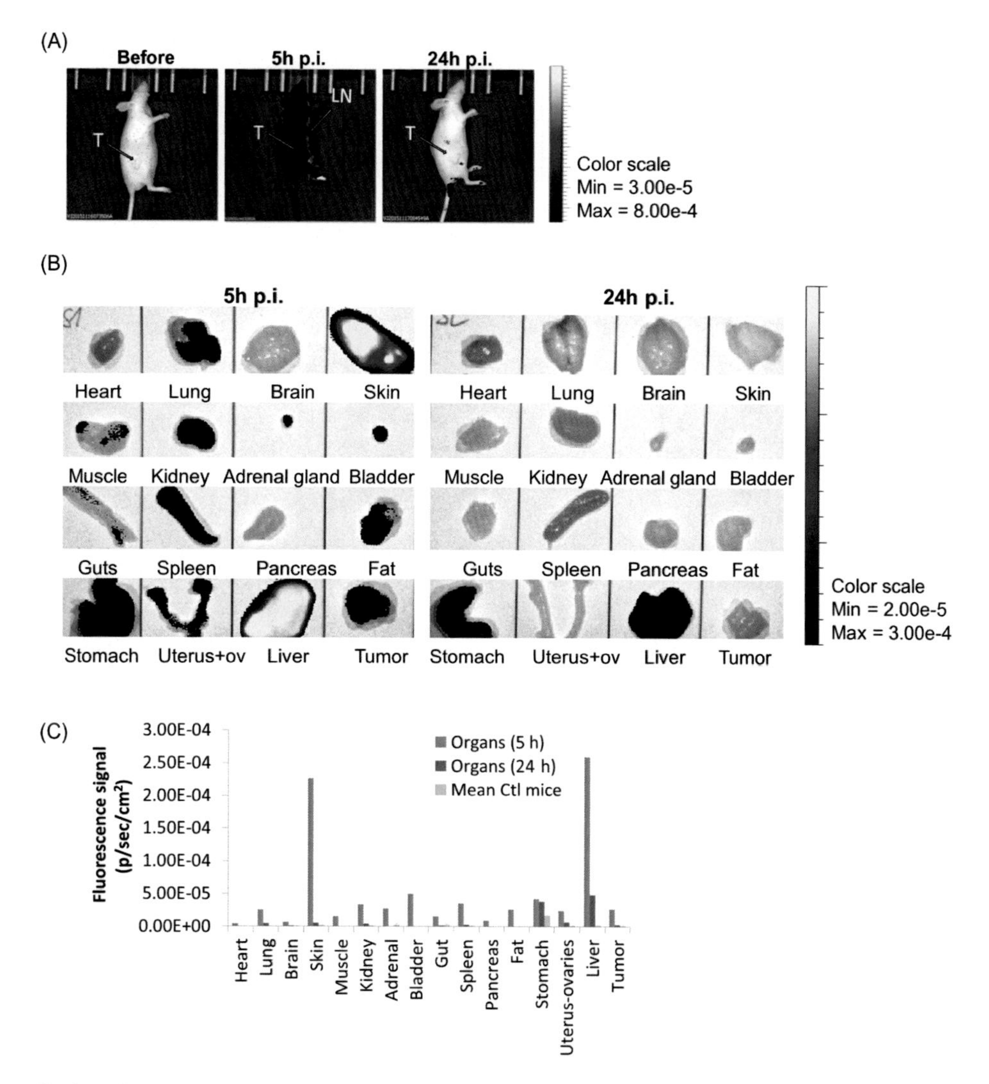

FIGURE 3.4 Localization of the AF680-labeled adenoviral dodecahedrons after i.v. injection to nude mice. (A) In vivo fluorescence imaging of the labeled adenoviral dodecahedrons before injection, 5 h postinjection, and 24 h postinjection (*T*, tumor; *LN*, lymph nodes; *p.i.*, postinjection); (B) Ex vivo fluorescence imaging; and (C) fluorescence analysis of isolated organs and tumors obtained at indicated time points from both the control (Ctl, noninjected) mice and the treated mice ($p/s/cm^2$, number of photons per second per cm^2). *Reproduced from M. Jedynak, D. Laurin, P. Dolega, M. Podsiadla-Bialoskorska, I. Szurgot, J. Chroboczek, et al., Leukocytes and drug-resistant cancer cells are targets for intracellular delivery by adenoviral dodecahedron, Nanomedicine 14 (2018) 1853–1865, with permission from Elsevier B.V. [57].*

signals in most of the organs have disappeared 24 hours after administration, with fluorescence signals having been able to be detected only in the liver (and in the stomach that has displayed nonspecific autofluorescence during the experiment) [57]. If the adenoviral dodecahedrons are used for the execution of a biogerontological intervention, the persistence of transgene expression may be too short for persistent manipulation of the aging network. In addition to virus-like particles, gutless or helper-dependent adenoviral vectors have been generated by removing the entire viral genome (except the genes for replication and packaging) [58]. Compared to the conventional vector that gives transgene expression only for around 30–60 days after intraventricular and intrahippocampal inoculation in the aged rat brain, the helper-dependent vector enables transgene expression to last for over 180 days. As the helper-dependent vector gives long-term transgene expression without eliciting chronic toxicity [59], it offers encouraging potential for gene therapy, though large-scale production of the helper-dependent virus, at the moment, is technically demanding and time-consuming.

In addition to adenoviral vectors, the recombinant AAV has been applied to gene transfer to combat cognitive decline. For instance, by transducing mitral cells of the olfactory bulb and pyramidal cells of CA1, overexpression of the neurotrophic factor has been induced, resulting in the transport of the glial cell-line-derived neurotrophic factor to the piriform cortex and the contralateral CA1 region, respectively [60]. More recently, the AAV has been adopted to deliver two antioxidant enzymes (viz., superoxide dismutase (SOD) 1 and catalase) to the hippocampus of an aged rat. Results have shown that by increasing the expression of both enzymes, cognitive decline in advanced age has been ameliorated [61]. As brain aging is related to an imbalance between intracellular concentrations of ROS and antioxidant defenses [62,63], virus-mediated overexpression of antioxidant enzymes may help to combat the functional decline caused by oxidative stress in other tissues as well. This possibility has been verified by Li et al. [64], who have used a recombinant AAV to overexpress catalase in the mitochondria, which is the site of free radical production during muscle contraction, and have successfully improved the treadmill performance of mice during exhaustive exercises. This work has paved the platform for ameliorating the damage caused by free radicals in tissues via virus-mediated ectopic catalase expression in mitochondria.

Apart from cognitive decline, the innate and adaptive immune functions of an individual are jeopardized during aging, leading to a reduction in the ability to resist infections and to respond to vaccinations. The success in using virus-mediated gene delivery to enhance the immunity of the aged population has been demonstrated by Vemula et al. [65]. By using an adenoviral vector expressing murine β-defensin 2 (Mbd2), which can help to recruit and activate immature dendritic cells as well as professional antigen-presenting cells (APC) to the site of immunization, the efficacy of an adenoviral vector-based H5N1 influenza vaccine in aged mice has been enhanced [65].

Finally, aging associates with cell senescence, which is partly attributed by the mitotic clock, leading to age-associated tissue degeneration (especially in tissues that show poor efficiency in regeneration). To tackle these problems, guided differentiation of stem cells may provide a cell source for cell replacement; however, sources of stem cells are not easily accessible, with the differentiation procedures sometimes being too time-consuming to be practical. As an alternative to the use of stem cells, cellular reprogramming of somatic cells (e.g., fibroblasts) via direct genetic manipulation has been proposed. The feasibility of this has been demonstrated by an earlier study, in which adult ear tip fibroblasts (ETFs), which have been isolated from C57/B6 mice of 6–8 weeks old, have been converted into retinal ganglion-like cells upon transduction with adenoviral vectors carrying *Ascl1*, *Brn3b*, and *Ngn2* [66]. Those cells have been confirmed to be positive for fibronectin but negative for Tuj1, Nestin, GFAP, RPF-1, Ath5, Brn3a, and Thy1.2 before transduction; however, after transduction, the cells have become positive for Tuj1 and have exhibited neuronal morphologies [66] (Fig. 3.5). In addition, they have coexpressed multiple retinal ganglion cell (RGC)–related markers and have shown the membrane properties of functional neurons [66]. This has shed light on the feasibility of manipulating cell types by genetic manipulation for possible cell replacement and tissue regeneration in the future.

Opportunities and challenges

Viruses are highly complex biological entities, in which overlapping reading frames and alternative splicing are adopted in the genome to enable economical use of the coding space. Such high complexity has made the engineering of viral particles challenging because manipulation of a virus for a novel property sometimes may disrupt key properties (e.g., capsid assembly and cargo packaging) needed for viral functions. The problem has been made more complicated by the insufficient knowledge of viral structure–function relationships and virus–cell interactions in practice. In addition, as the modified vector has to be stable in the biological environment for practical use, proper evaluation of the performance of the chemically modified virus, as well as fine-tuning of parameters in the modification process, is often required.

At this moment, safety concerns on the use of viruses are still one of the major hurdles jeopardizing wide applications of viral vectors. Such concerns have been stemmed partly from the development of leukemia in SCID-X1 patients that have been treated with a gammaretroviral vector [67] and also from the death of a relatively fit 18-year-old patient whose inherited enzyme deficiency has been attempted to be treated with an adenoviral vector [68]. Here it is worth mentioning that the potentially higher resistance of aged tissues to transduction is a phenomenon that should always be noted when a biogerontological intervention is developed, as it makes virus-mediated gene delivery for genetic manipulation less effective in aged tissues than in young tissues. This phenomenon has been evidenced by an earlier study [69], which has examined the effect of aging on viral transduction by injecting the AAV expressing the green fluorescent protein (GFP) to the rat midbrain. After vector administration, the midbrain of the aged rats has displayed 40% fewer GFP-expressing cells when compared to the young counterparts. This has demonstrated that the age *per se* is a factor imposing hurdles to the execution of biogerontological interventions.

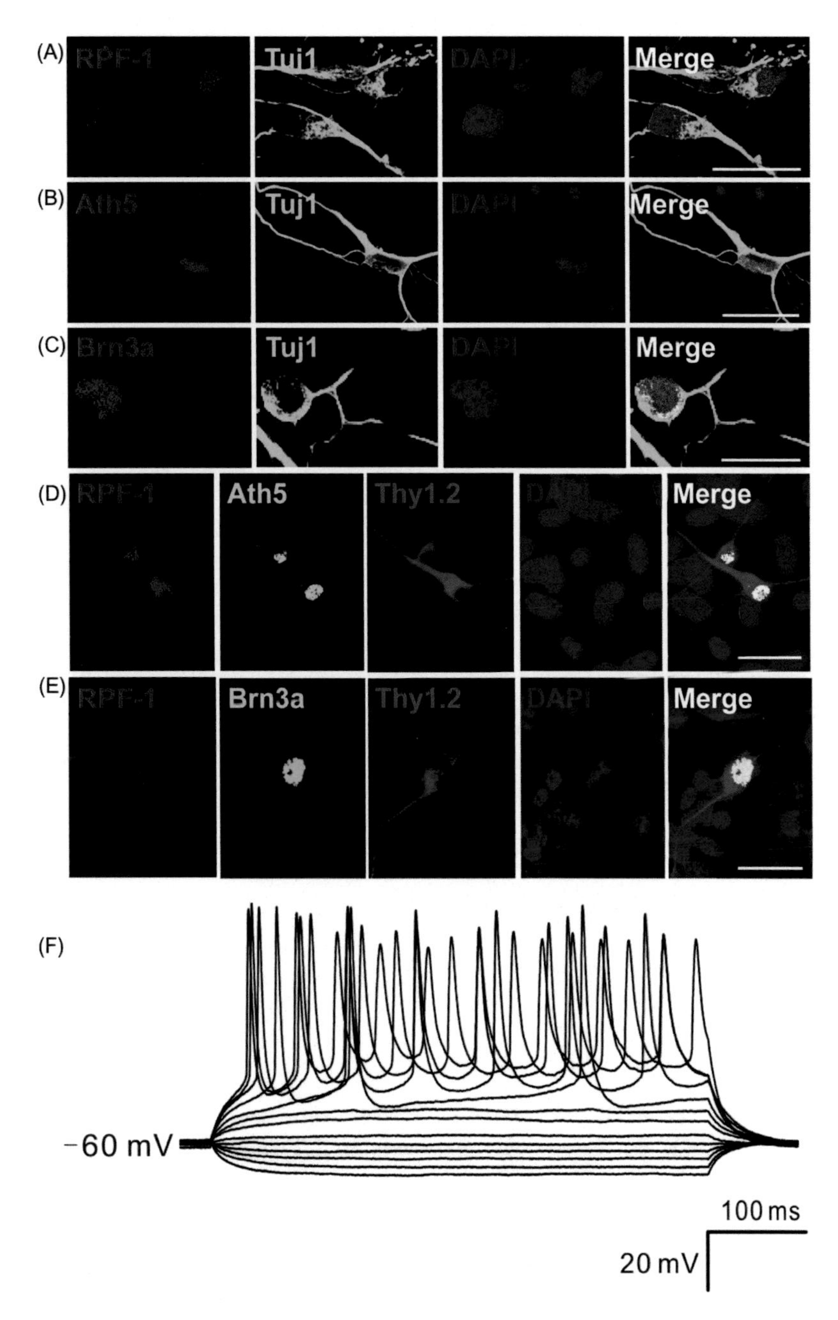

FIGURE 3.5 The induced retinal ganglion-like cells derived from ETFs. (A–E) The induced retinal ganglion-like cells derived from ETFs expressed (A, B, C) nuclear markers of RPF-1/Ath5/Brn3a together with Tuj1, as well as (D, E) multiple RGC-related markers. The scale bar is 20 μm. (F) Action potentials for the induced retinal ganglion-like cells. *ETFs*, ear tip fibroblasts; *RGC*, retinal ganglion cell. *Reproduced from F. Meng, X. Wang, P. Gu, Z. Wang, W. Guo, Induction of retinal ganglion-like cells from fibroblasts by adenoviral gene delivery, Neuroscience 250 (2013) 381–393, with permission from Elsevier B.V. [66].*

Summary

The concept to use genes as therapeutic agents was originally conceived in the late 1900s as a logical consequence of the increasing knowledge of the close relationship between gene regulation and biological processes ranging from development to angiogenesis. Aging is a complicated degenerative process caused by both extrinsic (e.g., oxidative damage) and intrinsic factors (e.g., anomalous epigenetic signaling) [70,71]. It can undoubtedly be modulated by genetic manipulation. This, therefore, links viruses and antiaging medicine together, as viral vectors have been recognized as by far the most effective systems for gene delivery. In this chapter, we have reviewed some major techniques, ranging from protein engineering to chemical modification, to manipulate the properties of a viral vector for controlled gene delivery. Although several challenges (e.g., immunogenicity and insertional mutagenesis) may have to be resolved before viruses can be ideal for human use, taking the high versatility and high transduction efficiency into account, it is not difficult to see that viral vectors will continue to play an important role in gene delivery and gene therapy in the forthcoming decade.

Directions for intervention development

When genetic manipulation is involved in a proposed biogerontological intervention, the proposed intervention can be designed by following the steps below:

1. Choose an appropriate virus with desired properties for intervention development
2. Evaluate the limitations of the virus
3. Choose a suitable strategy to engineer the properties of the virus
4. Engineer the virus for improving its performance
5. Evaluate and optimize the performance of the engineered virus in vitro and in vivo

References

[1] Y. Itoh, N. Maruyama, M. Kitamura, T. Shirasawa, K. Shigemoto, T. Koike, Induction of endogenous retroviral gene product (SU) as an acute-phase protein by IL-6 in murine hepatocytes, Clin. Exp. Immunol. 88 (1992) 356–359.

[2] C. Hesdorffer, J. Ayello, M. Ward, A. Kaubisch, L. Vahdat, C. Balmaceda, et al., Phase I trial of retroviral-mediated transfer of the human MDR1 gene as marrow chemoprotection in patients undergoing high-dose chemotherapy and autologous stem-cell transplantation, J. Clin. Oncol. 16 (1998) 165–172.

[3] H. Petrs-Silva, R. Linden, Advances in recombinant adeno-associated viral vectors for gene delivery, Curr. Gene Ther. 13 (2013) 335–345.

[4] Y. Lu, C.O. Madu, Viral-based gene delivery and regulated gene expression for targeted cancer therapy, Expert Opin. Drug Deliv. 7 (2010) 19–35.

[5] Y. Gan, Z. Jing, R.A. Stetler, G. Cao, Gene delivery with viral vectors for cerebrovascular diseases, Front. Biosci. 5 (2013) 188–203.

[6] C. Capasso, M. Hirvinen, V. Cerullo, Beyond gene delivery: strategies to engineer the surfaces of viral vectors, Biomedicines 1 (2013) 3–16.

[7] X. Zhang, W.T. Godbey, Viral vectors for gene delivery in tissue engineering, Adv. Drug Deliv. Rev. 58 (2006) 515–534.

[8] T.L. Wu, D. Zhou, Viral delivery for gene therapy against cell movement in cancer, Adv. Drug Deliv. Rev. 63 (2011) 671–677.

[9] S.J. Gray, K.T. Woodard, R.J. Samulski, Viral vectors and delivery strategies for CNS gene therapy, Ther. Deliv. 1 (2010) 517–534.

[10] B.L. Davidson, X.O. Breakefield, Viral vectors for gene delivery to the nervous system, Nat. Rev. Neurosci. 4 (2003) 353–364.

[11] M. Giacca, S. Zacchigna, Virus-mediated gene delivery for human gene therapy, J. Controlled Release 161 (2012) 377–388.

[12] M.P. McCormack, T.H. Rabbitts, Activation of the T-cell oncogene LMO2 after gene therapy for X-linked severe combined immunodeficiency, N. Engl. J. Med. 350 (2004) 913–922.

[13] D.B. Kohn, M. Sadelain, J.C. Glorioso, Occurrence of leukaemia following gene therapy of X-linked SCID, Nat. Rev. Cancer 3 (2003) 477–488.

[14] S. Hacein-Bey-Abina, A. Garrigue, G.P. Wang, J. Soulier, A. Lim, E. Morillon, et al., Insertional oncogenesis in 4 patients after retrovirus-mediated gene therapy of SCID-X1, J. Clin. Invest. 118 (2008) 3132–3142.

[15] M.G. Ott, M. Schmidt, K. Schwarzwaelder, S. Stein, U. Siler, U. Koehl, et al., Correction of X-linked chronic granulomatous disease by gene therapy, augmented by insertional activation of MDS1-EVI1, PRDM16 or SETBP1, Nat. Med. 12 (2006) 401–409.

[16] K. Boztug, M. Schmidt, A. Schwarzer, P.P. Banerjee, I.A. Diez, R.A. Dewey, et al., Stem-cell gene therapy for the Wiskott-Aldrich syndrome, N. Engl. J. Med. 363 (2010) 1918–1927.

[17] N. Brunetti-Pierri, T. Ng, D.A. Iannitti, D.J. Palmer, A.L. Beaudet, M.J. Finegold, et al., Improved hepatic transduction, reduced systemic vector dissemination, and long-term transgene expression by delivering helper-dependent adenoviral vectors into the surgically isolated liver of nonhuman primates, Hum. Gene Ther. 17 (2006) 391–404.

[18] A. Taura, K. Taura, Y.H. Choung, M. Masuda, K. Pak, E. Chavez, et al., Histone deacetylase inhibition enhances adenoviral vector transduction in inner ear tissue, Neuroscience 166 (2010) 1185–1193.

[19] C. Zhang, D. Zhou, Adenoviral vector-based strategies against infectious disease and cancer, Hum. Vaccines 12 (2016) 2064–2074.

[20] P. Piccolo, N. Brunetti-Pierri, Challenges and prospects for helper-dependent adenoviralvector-mediated gene therapy, Biomedicines 2 (2014) 132–148.

[21] M.F. Naso, B. Tomkowicz, W.L. Perry 3rd, W.R. Strohl, Adeno-associated virus (AAV) as a vector for gene therapy, BioDrugs 31 (2017) 317–334.

[22] S. Bass-Stringer, B.C. Bernardo, C.N. May, C.J. Thomas, K.L. Weeks, J.R. McMullen, Adeno-associated virus gene therapy: translational progress and future prospects in the treatment of heart failure, Heart Lung Circ. 27 (2018) 1285–1300.

[23] J.J. Aponte-Ubillus, D. Barajas, J. Peltier, C. Bardliving, P. Shamlou, D. Gold, Molecular design for recombinant adeno-associated virus (rAAV) vector production, Appl. Microbiol. Biotechnol. 102 (2018) 1045–1054.

[24] W. Qu, M. Wang, Y. Wu, R. Xu, Scalable downstream strategies for purification of recombinant adeno-associated virus vectors in light of the properties, Curr. Pharm. Biotechnol. 16 (2015) 684–695.

[25] P.O. Buclez, G. Dias Florencio, K. Relizani, C. Beley, L. Garcia, R. Benchaouir, Rapid, scalable, and low-cost purification of recombinant adeno-associated virus produced by baculovirus expression vector system, Mol. Ther. Methods Clin. Dev. 3 (2016) 16035.

[26] G.D. Trobridge, D.G. Miller, M.A. Jacobs, J.M. Allen, H.P. Kiem, R. Kaul, et al., Foamy virus vector integration sites in normal human cells, Proc. Natl. Acad. Sci. U.S.A. 103 (2006) 1498–1503.

[27] A.S. Khan, Simian foamy virus infection in humans: prevalence and management, Expert Rev. Anti-infect. Ther. 7 (2009) 569–580.

[28] J.R. Counsell, R. Karda, J.A. Diaz, L. Carey, T. Wiktorowicz, S. M.K. Buckley, et al., Foamy virus vectors transduce visceral organs and hippocampal structures following in vivo delivery to neonatal mice, Mol. Ther. Nucleic Acids 12 (2018) 626–634.

[29] G. Trobridge, N. Josephson, G. Vassilopoulos, J. Mac, D.W. Russell, Improved foamy virus vectors with minimal viral sequences, Mol. Ther. 6 (2002) 321–328.

[30] D.M. Roberts, A. Nanda, M.J. Havenga, P. Abbink, D.M. Lynch, B.A. Ewald, et al., Hexon-chimaeric adenovirus serotype 5 vectors circumvent pre-existing anti-vector immunity, Nature 441 (2006) 239–243.

[31] B. Hauck, L. Chen, W. Xiao, Generation and characterization of chimeric recombinant AAV vectors, Mol. Ther. 7 (2003) 419–425.

[32] L.G. Johnson, J.C. Olsen, L. Naldini, R.C. Boucher, Pseudotyped human lentiviral vector-mediated gene transfer to airway epithelia in vivo, Gene Ther. 7 (2000) 568–574.

[33] G.P. Kobinger, D.J. Weiner, Q.C. Yu, J.M. Wilson, Filovirus-pseudotyped lentiviral vector can efficiently and stably transduce airway epithelia in vivo, Nat. Biotechnol. 19 (2001) 225–230.

[34] N.A. Py, A.E. Bonnet, A. Bernard, Y. Marchalant, E. Charrat, F. Checler, et al., Differential spatio-temporal regulation of MMPs in the 5xFAD mouse model of Alzheimer's disease: evidence for a pro-amyloidogenic role of MT1-MMP, Front. Aging Neurosci. 6 (2014) 247.

[35] S. Lorenzl, D.S. Albers, S. Narr, J. Chirichigno, M.F. Beal, Expression of MMP-2, MMP-9, and MMP-1 and their endogenous counterregulators TIMP-1 and TIMP-2 in postmortem brain tissue of Parkinson's disease, Exp. Neurol. 178 (2002) 13–20.

[36] J. Szecsi, R. Drury, V. Josserand, M.P. Grange, B. Boson, I. Hartl, et al., Targeted retroviral vectors displaying a cleavage site-engineered hemagglutinin (HA) through HA-protease interactions, Mol. Ther. 14 (2006) 735–744.

[37] R.M. Schneider, Y. Medvedovska, I. Hartl, B. Voelker, M.P. Chadwick, S.J. Russell, et al., Directed evolution of retroviruses activatable by tumour-associated matrix metalloproteases, Gene Ther. 10 (2003) 1370–1380.

[38] E. Leggiero, D. Astone, V. Cerullo, B. Lombardo, C. Mazzaccara, G. Labruna, et al., PEGylated helper-dependent adenoviral vector expressing human Apo A-I for gene therapy in LDLR-deficient mice, Gene Ther. 20 (2013) 1124–1130.

[39] D.V. Schaffer, J.T. Koerber, K.I. Lim, Molecular engineering of viral gene delivery vehicles, Annu. Rev. Biomed. Eng. 10 (2008) 169–194.

[40] C.L. Moyer, G.R. Nemerow, Viral weapons of membrane destruction: variable modes of membrane penetration by non-enveloped viruses, Curr. Opin. Virol. 1 (2011) 44–49.

[41] M.L. Nibert, A.L. Odegard, M.A. Agosto, K. Chandran, L.A. Schiff, Putative autocleavage of reovirus mu1 protein in concert with outer-capsid disassembly and activation for membrane permeabilization, J. Mol. Biol. 345 (2005) 461–474.

[42] P. Wang, L. Day, J. Dheekollu, P.M. Lieberman, A redox-sensitive cysteine in Zta is required for Epstein-Barr virus lytic cycle DNA replication, J. Virol. 79 (2005) 13298–13309.

[43] M.H. Jacob, R. Janner Dda, A.S. Araujo, M.P. Jahn, L.C. Kucharski, T.B. Moraes, et al., Redox imbalance influence in the myocardial Akt activation in aged rats treated with DHEA, Exp. Gerontol. 45 (2010) 957–963.

[44] M. De la Fuente, J. Cruces, O. Hernandez, E. Ortega, Strategies to improve the functions and redox state of the immune system in aged subjects, Curr. Pharm. Des. 17 (2011) 3966–3993.

[45] M. Sharma, M. Astekar, S. Soi, B.S. Manjunatha, D.C. Shetty, R. Radhakrishnan, pH gradient reversal: an emerging hallmark of cancers, Recent Pat. Anticancer Drug Discov. 10 (2015) 244–258.

[46] M.W. Pandori, T. Sano, Photoactivatable retroviral vectors: a strategy for targeted gene delivery, Gene Ther. 7 (2000) 1999–2006.

[47] F. Wei, K.I. McConnell, T.K. Yu, J. Suh, Conjugation of paclitaxel on adeno-associated virus (AAV) nanoparticles for co-delivery of genes and drugs, Eur. J. Pharm. Sci. 46 (2012) 167–172.

[48] S. Kurotani, H. Umegaki, K. Ishiwata, Y. Suzuki, A. Iguchi, The age-associated changes of dopamine-acetylcholine interaction in the striatum, Exp. Gerontol. 38 (2003) 1009–1013.

[49] H. Umegaki, Y. Yamaguchi, K. Ishiwata, D.K. Ingram, G.S. Roth, A. Iguchi, Functional recovery of the striatal cholinergic system in aged rats by adenoviral vector-mediated gene transfer of dopamine D2 receptor, Mech. Ageing Dev. 127 (2006) 813–815.

[50] T.S. Watanabe, S. Ohtori, M. Koda, Y. Aoki, H. Doya, H. Shirasawa, et al., Adenoviral gene transfer in the peripheral nervous system, J. Orthop. Sci. 11 (2006) 64–69.

[51] R.J. Parks, J.L. Bramson, Adenoviral vectors: prospects for gene delivery to the central nervous system, Gene Ther. 6 (1999) 1349–1350.

[52] R. Pedersini, E. Vattemi, P.P. Claudio, Adenoviral gene therapy in high-grade malignant glioma, Drug News Perspect. 23 (2010) 368–379.

[53] M.G. Castro, M. Candolfi, T.J. Wilson, A. Calinescu, C. Paran, N. Kamran, et al., Adenoviral vector-mediated gene therapy for gliomas: coming of age, Expert Opin. Biol. Ther. 14 (2014) 1241–1257.

[54] C.A. Davies, H. Gollins, N. Stevens, A.P. Fotheringham, I. Davies, The glial cell response to a viral vector in the aged brain, Neuropathol. Appl. Neurobiol. 30 (2004) 30–38.

[55] K. Kajiwara, A.P. Byrnes, H.M. Charlton, M.J. Wood, K.J. Wood, Immune responses to adenoviral vectors during gene transfer in the brain, Hum. Gene Ther. 8 (1997) 253–265.

[56] M.J. Parr, P.Y. Wen, M. Schaub, S.J. Khoury, M.H. Sayegh, H.A. Fine, Immune parameters affecting adenoviral vector gene therapy in the brain, J. Neurovirol. 4 (1998) 194–203.

[57] M. Jedynak, D. Laurin, P. Dolega, M. Podsiadla-Bialoskorska, I. Szurgot, J. Chroboczek, et al., Leukocytes and drug-resistant cancer cells are targets for intracellular delivery by adenoviral dodecahedron, Nanomedicine 14 (2018) 1853–1865.

[58] L. Zou, X. Yuan, H. Zhou, H. Lu, K. Yang, Helper-dependent adenoviral vector-mediated gene transfer in aged rat brain, Hum. Gene Ther. 12 (2001) 181–191.

[59] N. Brunetti-Pierri, P. Ng, Gene therapy with helper-dependent adenoviral vectors: lessons from studies in large animal models, Virus Genes 53 (2017) 684–691.

[60] I. Kanter-Schlifke, B. Georgievska, D. Kirik, M. Kokaia, Brain area, age and viral vector-specific glial cell-line-derived neurotrophic factor expression and transport in rat, Neuroreport 18 (2007) 845–850.

[61] W.H. Lee, A. Kumar, A. Rani, J. Herrera, J. Xu, S. Someya, et al., Influence of viral vector-mediated delivery of superoxide dismutase and catalase to the hippocampus on spatial learning and memory during aging, Antioxid. Redox Signal. 16 (2012) 339–350.
[62] I. Hajjar, S.S. Hayek, F.C. Goldstein, G. Martin, D.P. Jones, A. Quyyumi, Oxidative stress predicts cognitive decline with aging in healthy adults: an observational study, J. Neuroinflamm. 15 (2018) 17.
[63] W. Droge, H.M. Schipper, Oxidative stress and aberrant signaling in aging and cognitive decline, Aging Cell 6 (2007) 361–370.
[64] D. Li, Y. Lai, Y. Yue, P.S. Rabinovitch, C. Hakim, D. Duan, Ectopic catalase expression in mitochondria by adeno-associated virus enhances exercise performance in mice, PLoS One 4 (2009) e6673.
[65] S.V. Vemula, A. Pandey, N. Singh, J.M. Katz, R. Donis, S. Sambhara, et al., Adenoviral vector expressing murine beta-defensin 2 enhances immunogenicity of an adenoviral vector based H5N1 influenza vaccine in aged mice, Virus. Res. 177 (2013) 55–61.
[66] F. Meng, X. Wang, P. Gu, Z. Wang, W. Guo, Induction of retinal ganglion-like cells from fibroblasts by adenoviral gene delivery, Neuroscience 250 (2013) 381–393.
[67] R.H. Buckley, Gene therapy for SCID-a complication after remarkable progress, Lancet 360 (2002) 1185–1186.
[68] E. Marshall, Gene therapy death prompts review of adenovirus vector, Science 286 (1999) 2244–2245.
[69] N.K. Polinski, S.E. Gombash, F.P. Manfredsson, J.W. Lipton, C.J. Kemp, A. Cole-Strauss, et al., Recombinant adenoassociated virus 2/5-mediated gene transfer is reduced in the aged rat midbrain, Neurobiol. Aging 36 (2015) 1110–1120.
[70] M. Holzenberger, J. Dupont, B. Ducos, P. Leneuve, A. Geloen, P. C. Even, et al., IGF-1 receptor regulates lifespan and resistance to oxidative stress in mice, Nature 421 (2003) 182–187.
[71] S.F. Gilbert, Ageing and cancer as diseases of epigenesis, J. Biosci. 34 (2009) 601–604.

Chapter 4

Design of polymeric vectors for genetic manipulation

Introduction

With the unraveling of the mechanisms of various genetic pathologies [1,2], genetic manipulation has emerged as a viable strategy to tackle diseases that are hitherto incurable [3–6]. Genetic manipulation involves two components to work together. One component is the carrier, and the other one is the therapeutic nucleic acid, which can be plasmids that encode therapeutic genes/proteins (e.g., costimulatory molecules [7], cytokines [8,9], suicide genes [10], and tumor suppressor genes [11,12]) or small RNA [such as microRNA (miRNA), antisense RNA, and siRNA] that confronts disease conditions. As mentioned in Chapter 3, viral vectors have been the most extensively studied carriers thus far. Owing to their high transduction efficiency, copious viral vectors have progressed into clinical trials since the 1990s [13,14]. Several years ago, a retroviral construct containing a B domain-deleted human factor VIII (hFVIII) gene was administered, via peripheral i.v. infusion, to 13 hemophilia A patients in a phase I clinical trial. The retroviral vector has been reported to be tolerated well by patients, and has remained in some subjects' peripheral blood mononuclear cells for more than 1 year [15].

Notwithstanding the high transduction efficiency, the safety concerns involved have limited the broad use of viral vectors [16–19]. For instance, although retroviruses show broad cell tropism of infectivity and enable stable integration of the exogenous transgene into the host genome, they may lead to insertional mutagenesis [20]. Furthermore, AAVs can transduce both dividing and nondividing cells; however, the preparation of AAV vectors with high virus titers is technically challenging. Although adenoviral vectors can be easily prepared as a high-titer stock, they are highly immunogenic [20]. The immunogenicity of adenoviral vectors has been shown by a recent study, which has tested an adenoviral vector encoding interferon-β (IFN-β) on patients having malignant pleural effusions (MPE) or malignant pleural mesothelioma (MPM) [21]. Though, in most of the patients, antibody responses against tumor antigens have been successfully elicited, the induction of neutralizing Ad antibodies has obviated the possibility of effective adenovirus-mediated gene delivery after the second dose. Such findings have illustrated the inherent immunogenicity of viruses and hence the fundamental limitations of treatment modalities utilizing viral vectors. This, together with the pathogenicity and toxicity of viruses, has stimulated the development of nonviral alternatives. In 1988, a polymeric DNA vehicle [which consists of a galactose-terminal (asialo-)glycoprotein covalently linked to PLL] was reported [22]. After that, other polymeric gene carriers, ranging from polypropylenimine dendrimers [23] to PLGA [24], have been developed. In this chapter, we will introduce the working principles underlying the development and derivatization of polymeric vectors. The objective of this chapter is to lay a foundation of knowledge from which different polymeric vectors can be designed for the execution of genetic manipulation.

Selection of polymeric vectors

The mechanism of nucleic acid delivery mediated by nonviral vectors has been modeled earlier by Kopatz et al. [25]. According to the model, gradual electrostatic zippering of the cell membrane onto the vector is sustained by the lateral diffusion of syndecan molecules, which cluster into cholesterol-rich rafts, after the initial binding of the vector to cells [25]. With the clustering of syndecan molecules, the protein kinase C (PKC) activity is triggered. Actin binding to the syndecans' cytoplasmic tail is also induced. Polyplexes of the nonviral vector are finally engulfed [25]. This model has offered a theoretical perspective on cellular internalization of nonviral vectors, and has enabled researchers to improve the physical and chemical properties of the vectors for nucleic acid delivery. In general, to select a polymer to be developed as a gene carrier, the most fundamental guiding principle is that the polymer should possess positive charges. These

Delivery of Therapeutics for Biogerontological Interventions. DOI: https://doi.org/10.1016/B978-0-12-816485-3.00004-0

charges can enable the polymer to complex with nucleic acids, which usually possess negative charges.

Taking this into consideration, it is not difficult to understand that cationic polymers have often been exploited directly for gene delivery applications. One example is PEI [26–29], which is an aziridine polymer that exists as a polycation with a high proton buffering capacity over a broad range of pH [30]. Over the years, PEI and its derivatives have been widely studied for delivering diverse nucleic acid materials, ranging from plasmids [31] and oligonucleotides [32] to ribozymes [33], for reagent consuming animal studies. Structurally, PEI may assume either a branched configuration (containing primary, secondary, and tertiary amine groups) or a linear configuration (containing secondary amine groups only). Compared to branched PEI (bPEI), linear PEI (lPEI) is less commercially available, and is generally harder to be modified because the reactivity of primary amines is higher than that of the secondary ones [34]. This partially explains why bPEI has been more intensively studied than lPEI for gene delivery. Practically, the transfection efficiency of bPEI depends largely on the physical–chemical features (e.g., cationic charge density, degrees of branching, and molecular weights) of the PEI molecules [35–37], although the polyplex properties (e.g., zeta potential, and particle size) [38] and the experimental conditions (e.g., polyplex concentration, and incubation time) [38] have also played a significant determining role.

Another example of cationic polymers that has been directly adopted for gene delivery is CS, which is a linear polysaccharide consisting of β(1→4) linked D-glucosamine residues with a variable number of randomly distributed *N*-acetyl-D-glucosamine units (Fig. 4.1) [6,39]. CS can be obtained from chitin after alkaline hydrolysis with an inorganic base [40]. It is digestible by lysozymes or some bacterial enzymes produced by the intestinal flora [41,42]. Mumper et al. [43] are one of the first groups to exploit CS as a gene carrier. During the process of gene delivery, the carbohydrate backbone of CS generates both ionic and nonionic interactions with cell surface proteins [44]. CS contains primary amine groups with a p*K*a value of around 6.5 [45]. It can form positively charged single helicoidal stiff chains in an acidic aqueous solution [46]. The positive surface charges render CS with the capacity to interact with the negatively charged entitles (e.g., the plasma membrane, and exogenous nucleic acids) [47]. Other than its free base, CS in its salt form (or in coformulation with other agents) has been studied. CS salts, such as CS lactate, CS aspartate, CS hydrochloride, CS glutamate, and CS acetate, have been examined in vitro by Weecharangsan et al. [48], and have been found to exhibit transfection efficiency superior to that of standard CS. As shown by Fang et al. [49], who have examined the effects of pH and the molecular weight of CS on 1,2-dipalmitoyl-*sn*-glycero-3-phosphocholine (DPPC) bilayer–CS interactions, CS possesses the ability to swirl across the membrane lipid bilayer. Together with the fact that the configuration of CS can be fully displaced under acidic conditions (with this uncoiled configuration being able to disrupt the tight junctions and to facilitate the paracellular transport of hydrophilic compounds [50]), CS is a worthwhile candidate for vector development. Apart from PEI and CS, two other representative examples of cationic polymers that have been exploited for gene delivery are PLL and poly(amidoamine) (PAMAM). The use of these four polymers has been reviewed elsewhere [51–54]; readers interested in going deeply and specifically into any of these cationic polymers are referred to relevant articles for details.

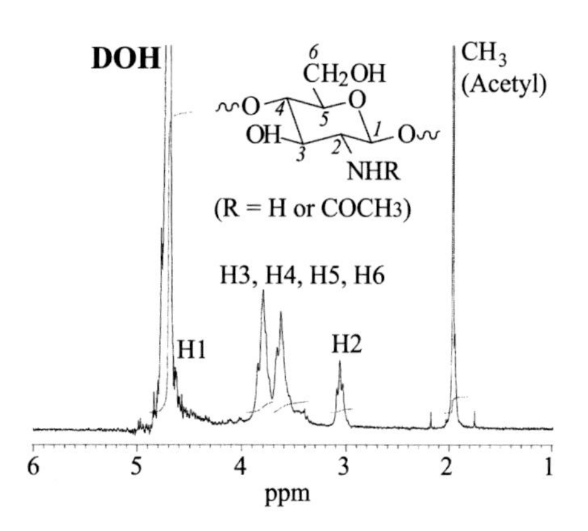

FIGURE 4.1 The structure and proton nuclear magnetic resonance (NMR) spectrum of CS. D_2O/HCl (100/1 v/v) is used as the solvent for NMR. *Reproduced from W.F. Lai, M.C. Lin, Nucleic acid delivery with chitosan and its derivatives, J. Control Release 134 (2009) 158–168, with permission from Elsevier B.V. [6].*

Example protocols for experimental design

The method below is an example protocol for depolymerizing CS:

1. Dissolve 1 g of CS in a 1% (w/w) acetic acid solution to reach a final concentration of approximately 1% (w/v).
2. Dissolve KNO_2 in distilled water to reach a concentration of 0.1 M.
3. Add the KNO_2 solution into the CS solution dropwise under magnetic stirring until the CS/KNO_2 mole ratio reaches 0.01.
4. Shake the solution mixture vigorously for 2 hours.
5. Centrifuge at 2600 $\times g$ for 5 minutes.
6. Adjust the pH to 8.0.
7. Centrifuge at 2600 $\times g$ for 15 minutes.
8. Remove the supernatant.
9. Resuspend the pellet in distilled water.
10. Centrifuge at 2600 $\times g$ for 15 minutes.
11. Repeat Steps 8–10 twice.
12. Lyophilize the pellet for 3 days.

Structural modification of polymeric vectors

When a polymer is developed as a gene carrier, its capacity to condense nucleic acids into polyplexes has to be evaluated by using the gel retardation assay because this capacity is needed for effective in vitro transfection. A polymer, however, may not necessarily be effective in transfection in the in vivo context even if it can condense nucleic acids. This is because, upon in vivo administration, polyplexes may interact with erythrocytes, transthyretin, apolipoproteins, IgM, IgD, fibrinogen, albumin, and components of the complement system [55–58]. This makes the polyplexes prone to be accumulated in capillary beds and be removed by the reticuloendothelial system [59–61]. By reason of this, different derivatization strategies have been adopted to enhance the in vivo transfection efficiency. The pros and cons of different strategies have been summarized in Table 4.1.

Copolymerization

Copolymerization is a method of creating a polymer that possesses functional properties different from those of the homopolymer components. Among different polymeric modifiers used for copolymerization purposes, poly(ethylene glycol) (PEG) is one of the popular choices. PEGylation actually is a widely used approach to enhance the solubility of the polymer [62], to mitigate the aggregation of polyplexes [63], and to increase the retention of the polyplexes in tumors. The latter is made possible by the enhanced permeability and retention (EPR) effect [64–66] caused by the leaky tumor vasculature and the impaired lymphatic drainage of malignant tissues. Moreover, owing to the hydrophilic nature and the brush-type polymer crowding of PEG [67], PEGylated vectors, in general, are less prone to opsonization, and have a more extended blood circulation time [67,68]. This has

TABLE 4.1 The pros and cons of major strategies for vector modification [6].

Pros	Cons
Structural modification	
Copolymerization	
• The practice is well-developed in literature on polymer chemistry	• A significant increase in the size of the carrier often results after the modification process • The number of conventionally used polymeric modifiers is limited
Functional group modification	
• Compared to copolymerization, functional group modification can cause a less significant (or even negligible) increase in the size of the carrier	• Sophisticated consideration of the structure–activity relationship (SAR) is needed; however, the knowledge of SAR in vector design is currently limited • Side reactions may occur, imposing challenges to the removal of by-products
Ligand conjugation	
With proteinaceous ligand(s)	
• Proteins and peptides have diverse functionalities to match different biological needs • A large variety of possible ligand candidates is available • Owing to the availability of techniques for bioconjugation, design of the ligand conjugation process generally demands less chemical rigor	• The generated product may be chemically unstable • The ligand may cause the carrier immunogenic • Proteins and ligands are expensive to be purchased
With nonproteinaceous ligand(s)	
• Nonproteinaceous ligands are cheaper to be purchased than proteinaceous ones • Owing to the availability of techniques for bioconjugation, design of the ligand conjugation process generally demands less chemical rigor	• The variety of conventionally used ligands for selection is comparatively low

been demonstrated by a previous study, which has found that, on day 1 after bile duct infusion and day 3 after portal vein infusion, PEGylated CS has produced a much higher level of transgene expression than unmodified CS in the rat liver [69]. A more recent study has also shown that upon caudal-vein infusion into a murine model of hepatic cancer, the PEGylated polyplexes have mediated a higher level of transgene expression in hepatoma tissues as compared to the non-PEGylated counterparts [70]. Every coin, however, has two sides. While PEGylation can extend the blood circulation time of polyplexes, it may reduce the efficiency of the vector in DNA condensation and endosomal escape, thereby impairing the performance in transfection [58,63,71,72]. The latter has been revealed by the observation that PEGylated polyplexes may experience premature unpackaging in blood, thereby resulting in less effective nucleic acid transfer to the liver [73]. Mishra et al. have also reported that PEGylation, though conferring salt stability to polyplexes, has hampered the cellular uptake and intracellular trafficking of the vector during transfection [74]. In this regard and the fact that properties (e.g., density, conformation, molecular weight, and flexibility) of the PEG moiety are important factors determining the outcome of PEGylation [67], optimizing the degree of PEG grafting as well as the coating size is indubitably imperative for the construction of effective PEGylated vectors.

The possible drawbacks of PEGylation may also be mitigated by connecting the PEG shield to the polyplex core via an acid labile linkage (such as vinyl ethers [75], acetals [76,77], and hydrazones [78]). In the acidic milieu of the endosomal compartment, the linkage can be hydrolyzed, leading to shielding destabilization [79]. Incorporation of targeting ligands to PEG is another means to reduce the nonspecific cellular uptake of polyplexes [80]. The effectiveness of this strategy has been supported not only by Chen et al.'s galactose-PEG-bPEI copolymers that, under optimal conditions, have exhibited a 4.5-fold and 11.6-fold increas in transfection efficiency in the A549 cell line and in the lung, respectively, as compared to the conventional bPEI/DNA formulation [81], but also by Kleemann et al.'s peptide-conjugated bPEI vector, in which an oligopeptide TAT has been attached via a PEG linker to bPEI 25 kDa [82]. In vivo studies have shown that, upon intratracheal instillation, the polyplexes (containing pGL3-Alexa Fluor as well as a polymeric carrier that has been labeled with Oregon Green 488 carboxyl acid) have successfully entered bronchial epithelia cells and the alveolar region, in which transgene expression has been detected 48 hours after administration (Fig. 4.2) [82]. Judging from all evidence offered above, incorporation of acid labile linkages or targeting ligands is a possible direction for refinement of PEGylated vectors.

In addition to PEG, another example of polymeric modifiers is stearic acid. Upon incorporation of stearic acid, it is possible to render the polymeric vector capable of undergoing self-assembly, in aqueous solutions, into micelles [83]. Sometimes cationic polymers or other polymeric vectors can be copolymerized to generate a graft copolymer. This has been shown by a previous study, which has reported that PLL-graft-CS (PLL grafting ratio = 14.0, polymer/DNA weight ratio = 10:1) has displayed higher transfection efficiency in 293 T cells than CS, PLL, and PEI [84]. Owing to its relatively high cytotoxicity as demonstrated in L929 cells, this copolymer, however, requires further optimization before it can be possibly used for intervention development.

Functional group modification

Although copolymerization has been extensively adopted for polymer modification, the repertoire of commonly used polymeric modifiers is relatively small in nucleic acid delivery research. Along with an inevitable increase in the vector size after copolymerization, functional group modification becomes a favorable alternative for polymeric vector derivatization. The rationale of functional group modification is to modify the physical and chemical properties of an existing cationic polymer so as to augment the performance in transfection. One example of derivatives prepared by using this method is alkylated CS (ACS). It has been reported by Yao's group, which has modified CS ($M_v = 50\ 000$, D.D. = 99%) with dodecyl bromide [85,86]. In vitro data have suggested that, in general, the larger the alkyl chain is, the more efficient the ACS is in C_2C_{12} cell transfection. This is presumably because longer alkyl side chains can allow ACS to have a stronger hydrophobic interaction with DPPC (a phospholipid commonly found in the plasma membrane of eukaryotic cells) and to be more effective in membrane perturbation [87]. More recently, Ercelen et al. [88] have attempted to prepare another set of ACS by grafting low-molecular-weight CS with *N*-/2(3)-(dodec-2-enyl)succinoyl groups. Despite the observed interactions between their ACS vectors and the model membrane (which consists of dimyristoylphosphatidylglycerol and dimyristoylphosphatidylcholine), robust data on the transfection efficiency of those vectors are lacking and have to be gathered before the potency of those vectors as nucleic acid carriers can be confirmed.

Another example of derivatives prepared by using functional group modification is thiolated CS (CSH), which can be prepared by first activating the carboxyl groups of thioglycolic acid with 1-ethyl-3-(3-dimethylaminopropyl)-carbodiimide hydrochloride (EDCI), followed by a reaction between the activated groups with free amine groups of CS. CSH has been reported to

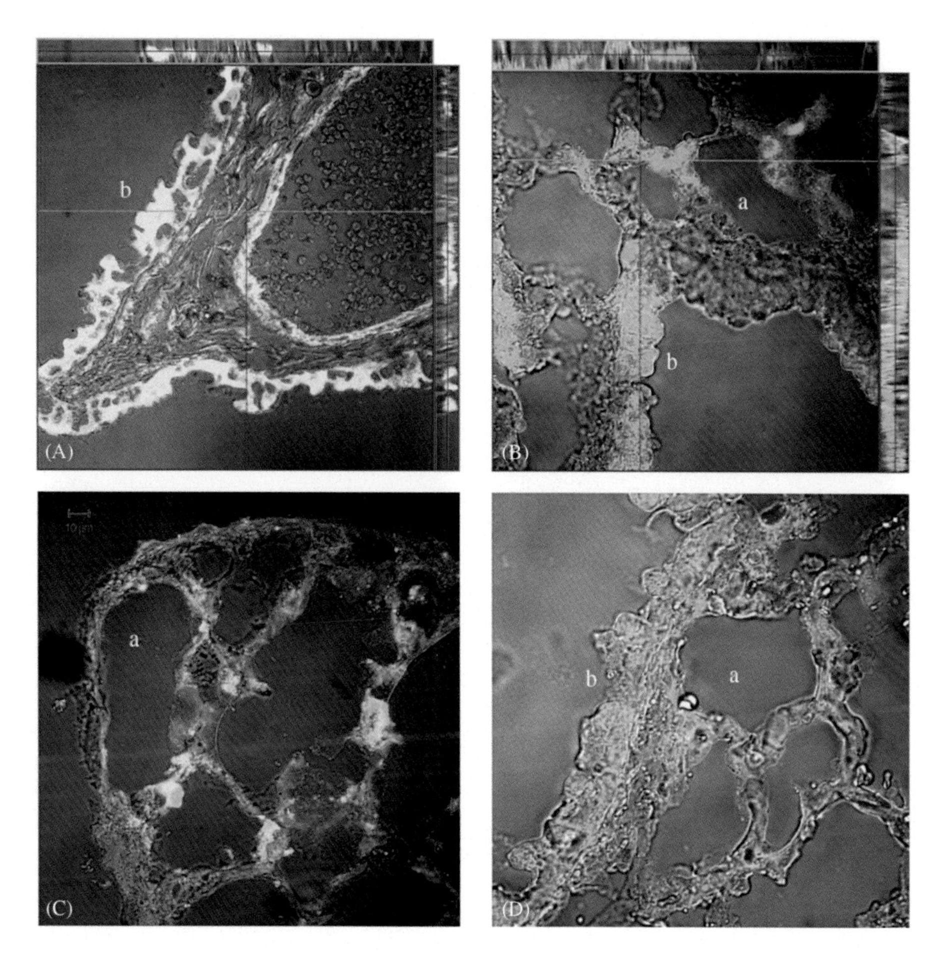

FIGURE 4.2 The localization of (A, C) the double-labeled polyplexes and (B, D) transgene expression in (A and B) bronchial epithelia endothelial cells, and (C and D) the alveolar region in mice, as detected at 4 and 48 hours after intratracheal instillation of the polyplexes, respectively. In the figure, "a" indicates the alveolar region; and "b" indicates the bronchial region. *Reproduced from E. Kleemann, M. Neu, N. Jekel, L. Fink, T. Schmehl, T. Gessler, et al., Nano-carriers for DNA delivery to the lung based upon a TAT-derived peptide covalently coupled to PEG-PEI, J. Control Release 109 (2005) 299–316, with permission from Elsevier B.V. [82].*

display good mucoadhesive properties and permeation enhancing effects in oral drug delivery [89–91]. Recently, Lee et al. [92] have noted that CSH with a higher thiol group content exhibits higher efficiency in transfection in HEK 293 cells. This has been ascribed to the attached thiol groups that facilitate CS to undergo cell permeation and hence cellular uptake. At 60 hours after transfection, a substantial increase ($P < .01$) in transgene expression has been observed in HEK 293 cells transfected with CSH but not with Lipofectin or unmodified CS. This may be because thiolation has attenuated the positive charge density of CS, leading to a more rapid release of transgenes from polyplexes of CSH; however, the actual mechanism has yet to be fully elucidated.

Ligand conjugation

Contrary to functional group modification, which necessitates a thorough understanding of the SAR for the design of the procedure for chemical modification, the most important issue to be considered in ligand conjugation is only the biological properties of the ligand adopted. This simplifies the process of vector design. In fact, ligand conjugation is at the moment the most prevailing and direct strategy for offering target specificity to a polymeric carrier. Among different classes of ligands used for conjugation, proteins/peptides are one of the favorable choices. One example of proteinaceous ligands is transferrin, an iron-binding and iron-transport protein that can function as a targeting moiety toward various cancer cell lines (including those of colon cancer, ovarian cancer, and glioblastoma) [93]. The efficiency of transferrin conjugation in enhancing transfection has been partially supported by the case of CS, whose transfection efficiency in HEK 293 and HeLa cells has been substantially increased after conjugation [94]. Other examples of proteinaceous ligands include antibodies [95], the KNOB protein (which is the C-terminal globular domain of the fiber protein on

TABLE 4.2 Examples of proteins and peptides that may potentially function as ligands for vector modification [6].

Potential use	Mechanisms of action	Examples
Tumor targeting	To allow the carrier to bind to specific antigens on cancer cells to enhance tumor targeting	OV-TL16, PSMA-specific monoclonal antibody J591, HER-2 antibody, anti-CD3 antibody, GRP-78 targeting peptide,
Membrane fusion and/or destabilization	To facilitate the endolysosomal escape of the carrier by disrupting the endolysosomal membrane	Melittin, ppTG1, ppTG20, penetratin, transportan, KALA
Nuclear localization signaling	To localize the carrier to the nucleus and allow the carrier to be transported across the nuclear pore complex (NPC)	SV40 T antigen, adenovirus E1a, PARP, SV40 Vp3, M9-ScT conjugate
Passing through the blood–brain barrier (BBB)	To enhance the efficiency of the carrier to pass across the BBB via low-density lipoprotein receptor-mediated endocytosis	Tandem dimer sequence of apoprotein-E (141–150)

the adenoviral capsid [94]), and the TAT peptide [96]. Some more proteins and peptides that may potentially function as ligands for vector modification are listed in Table 4.2. Although proteins and peptides have versatile functional properties (e.g., nuclear localization and cell targeting) favorable to nucleic acid delivery [6], when these ligands are adopted, the immunogenicity of the generated vector should be examined. In addition, due to the large size of proteins and peptides, changes in the size of the polyplexes may be significant. This may also affect the performance in delivery.

Example protocols for experimental design

The method below is an example protocol for conjugating the TAT peptide to PEI:

1. Dissolve 120 mg of PEI in 5 mL of dimethyl sulfoxide (DMSO).
2. Dissolve 3 mg of *N*-succinimidyl 3-(2-pyridyldithio)propionate (SPDP) in 1.5 mL of DMSO.
3. Add the two solutions together.
4. Incubate the solution mixture at room temperature for 1 hour under an inert atmosphere with constant stirring.
5. Dissolve 21 mg of a 13-amino acid peptide YGRKKRRQRRRPC derived from TAT in 1.5 mL of DMSO.
6. Add the peptide solution into the solution in Step 4.
7. React at ambient conditions under constant stirring for 3 hours.
8. Dialyze the reaction mixture (molecular weight cut-off = 12 kDa) against doubly deionized water for 2 days.
9. Lyophilize to obtain TAT-conjugated PEI.

Other than proteins and peptides, various nonproteinaceous ligands have been adopted. One example is hexoses (including galactose and lactose), which have been used to fabricate CS derivatives such as galactosylated CS-graft-PEG (GCP) [97], galactosylated low-molecular-weight CS (Gal-LMWCS) [98], and lactosylated CS [99]. Lately, mannosylated CS (MCS), which can induce mannose receptor-mediated endocytosis and hence can deliver a plasmid directly to dendritic cells [100], has been produced. In vivo studies have shown that, upon intratumoral administration, the polyplexes encoding murine IL-12 have suppressed angiogenesis and tumor growth in CT-26 carcinoma-bearing mice. The expression of Apaf-1, Bad, Bax, and G_1-S checkpoints molecules (e.g., p21 and p27) have been elevated in these mice; whereas the expression of the early G_1 marker, proliferating cell nuclear antigens, cyclin D1, cyclin-dependent kinase 4, and antiapoptotic Bcl-xL have been reduced. These alternations in protein expression have illustrated that, after intratumoral administration of the polyplexes, cell cycle arrest and apoptosis have been induced [100]. Deoxycholic acid (DCA) and folic acid (FA) are also some of the important nonproteinaceous ligands used in vector modification. Their use in gene delivery has been previously reviewed by Liu and Yao [101] and Mansouri et al. [102]. Though nonproteinaceous ligands are generally more stable and less immunogenic than the proteinaceous counterparts, the variety of conventionally used ligands for selection is comparatively low.

Example protocols for experimental design

The method below is an example protocol for conjugating FA to PEI:

1. Dissolve 4.5 mg of FA in 10 mL of DMSO.
2. Dissolve 4 mg of 1,1′-carbonyldimidazole (CDI) in 1 mL of DMSO.
3. Mix the two solutions together under magnetic stirring.
4. React for 3 hours at ambient conditions.
5. Dissolve 5 g of 25 kDa PEI in 15 mL of DMSO.
6. Add the solution in Step 4 to the PEI solution dropwise under magnetic stirring.
7. React for 24 hours at ambient conditions.
8. Dialyze the reaction mixture (molecular weight cut-off = 12 kDa) against doubly deionized water for 2 days.
9. Lyophilize to obtain FA-conjugated PEI.

Optimization for intervention execution

In the in vitro context, the efficiency of nucleic acid delivery mediated by polymeric vectors is affected mainly by the vector structure, the method of vector derivatization, and the transfection conditions. The reality, however, is much more complex. Many more factors may determine the usability and efficiency of the polymeric system. One of these factors is the route of administration. Actually, systemic i.v. administration has some advantages over other routes for nucleic acid transfer. For example, PEI polyplexes can target and accumulate in fenestrated tissues (e.g., bone marrow, liver, spleen, and certain tumors) via passive diffusion, even though the polymer has not undergone much derivatization beforehand [38]. Unfortunately, because of their interactions with blood constituents, polyplexes can be cleared rapidly from the blood circulation. To enhance the stability of polyplexes in vivo, previously PEI polyplexes have been coated with an Alg-based hydrogel to generate Alg/PEI/DNA nanoparticles (Fig. 4.3). These nanoparticles have demonstrated enhanced stability in vitro, and have shown a much longer blood circulation time as compared to unmodified PEI polyplexes [103]. In the case of cancers, the longer retention time in the blood circulation may allow for a higher level of accumulation of the nanoparticles at the tumor site via the EPR effect [103], thereby enhancing the potential of the nanoparticles to be used as carriers for tackling cancers.

Apart from i.v. administration, other routes can be used to enhance recipients' compliance (e.g., nasal and oral administration), or to increase the dose of the administered agent in the target site (e.g., intratumoral administration). The feasibility of the latter has been suggested by the observation that the weight of a tumor could be reduced by over 80% (on day 21) after a direct injection of lPEI

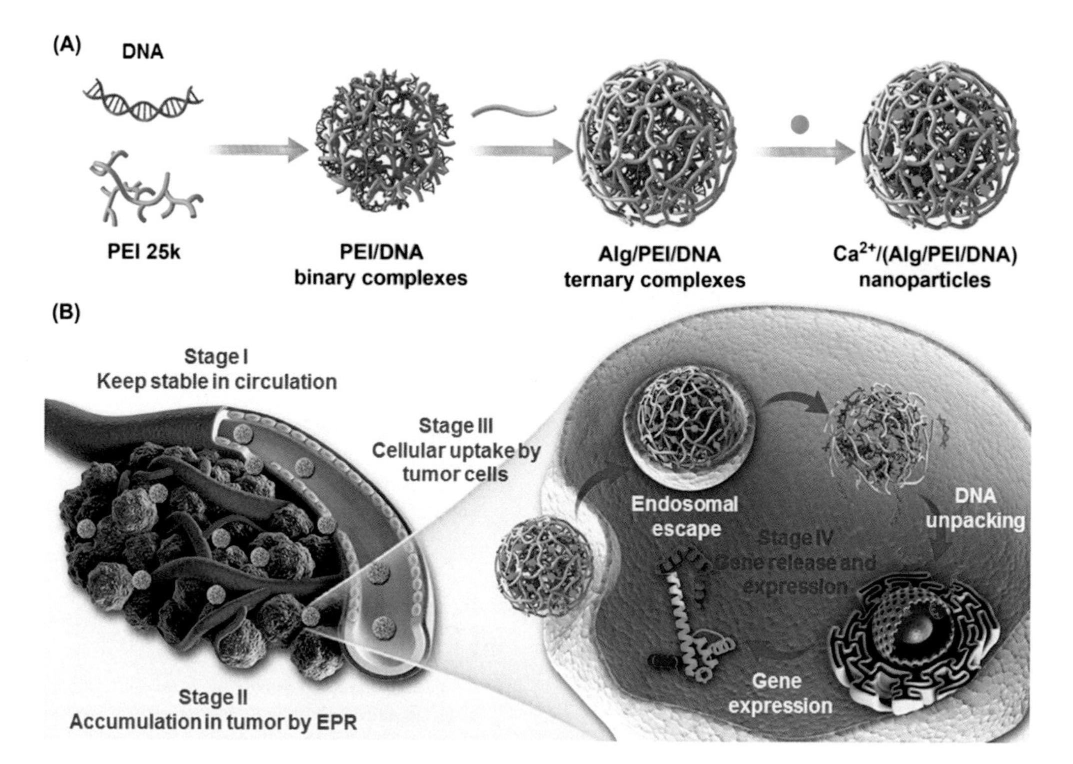

FIGURE 4.3 Use of the ionic-cross-linked Alg/PEI/DNA nanoparticle as a gene carrier. (A) Steps for the fabrication of the nanoparticles, and (B) the subsequent transportation process in vivo. *Reproduced from Y. Zhang, L. Lin, L. Liu, F. Liu, A. Maruyama, H. Tian, et al., Ionic-crosslinked polysaccharide/PEI/DNA nanoparticles for stabilized gene delivery, Carbohydr. Polym. 201 (2018) 246–256, with permission from Elsevier B.V. [103].*

polyplexes (containing pPB/TK and pPBase plasmids) to the tumor site in a SKOV3 xenograft murine model [104]. Although intratumoral administration is a simple and direct method to target therapeutic agents to tumor sites, its practicality is severely compromised in the case of metastatic tumor involvement. To tackle this problem, immunotherapy is a possible solution. At present, efforts devoted to immunotherapy mediated by polymeric gene carriers are still highly limited, but the feasibility of it has already been suggested in vivo by the use of gene delivery technologies in DNA vaccination [105–108]. It is hoped that with more follow-up research, polymeric vectors can contribute more significantly to treatments for age-related diseases, including but not limited to cancers, in the future.

Another important factor determining the performance in in vivo transfection is the persistence of transgene expression. Unfortunately, at the moment most of the polymeric vectors reported in literature also fail to confer persistent transgene expression [109,110]. The shortest reported half-life does not exceed minutes [73]; whereas subcutaneous implantation of DNA-incorporated scaffolds in rats has not extended the expression duration for more than weeks, either [111]. Such poor persistence of transgene expression is not only due to immune removal of the vector, but is also caused by the transcription silencing of the transgene promoter and by the loss of transfected cells through cellular turnover. Because of the short duration of transgene expression, repeated dosing may be required [112]. Owing to the toxicity of some polymeric vectors, such as PEI, repeated systemic dosing, however, may lead to adverse biological effects [113].

Other factors that may affect the efficiency in polymer-mediated nucleic acid delivery include the biodistribution of polyplexes and the physiological status of the individual. In general, after i.v. administration, polyplexes tend to localize in the liver, kidney, lung, or even spleen [114]. But after covalent attachment of ligands, such as transferrin [56] and the CNGRC peptide [115], the expression level of the transgene can be elevated in specific tissues, such as tumor tissues. Regarding the relevance of the physiological status of an individual to transgene expression, Sasaki's group [116] has examined the association between the proliferative state of the liver and the hepatic expression of transgenes, and has demonstrated that liver resection can be a possible factor modifying gene expression mediated by polymeric vectors in mice. Furthermore, by using a murine hepatitis model established by subcutaneous administration of tetrachlorocarbon (CCl_4), the expression of transgenes delivered by PEI (both linear and branched isomers) has been found to be much lower at the severe hepatitis stage (18 hours after CCl_4 injection) than at the liver regeneration stage (48 hours after CCl_4 injection) [117]. All these have pointed to the fact that careful timing of genetic manipulation is necessary.

Summary

When the safety risks of viral vectors are of concern, the use of polymeric vectors as nonviral alternatives is a sensible choice. Although the transfection efficiency of polymeric vectors is generally lower than that of viral vectors, polymeric vectors are flexible in structures, with the physiochemical properties and targeting capacity easily fine-tuned to meet the actual needs during intervention execution. Various strategies applicable to the design and optimization of polymeric vectors have already been discussed in this chapter. With this as the foundation, in the next chapter we will discuss how CDs can be used to further enhance the flexibility and tunability of polymeric vectors for nucleic acid delivery.

Directions for intervention development

When the safety risks of viral vectors are of concern but genetic manipulation is needed for executing the biogerontological intervention, polymeric vectors can be considered to be adopted. By following the steps below, one can streamline the process of polymer selection and vector development.

1. Either select an existing polymer that has the potential to be applied as a gene carrier or design a new polymer from scratch.
2. When a new polymer is designed, try to incorporate the polymer with positively charged functional groups so that its capacity to complex with nucleic acids can be enhanced.
3. Evaluate the performance of the polymer in nucleic acid delivery in vitro and in vivo.
4. Modify and optimize the structure, and hence the physiochemical properties and delivery performance, of the polymer.
5. Re-evaluate the performance of the modified polymeric vector as a gene carrier.
6. Repeats Steps 4–5 until the polymeric vector can reach satisfactory performance for execution of the proposed intervention.

References

[1] L. O'Connor, B. Glynn, Recent advances in the development of nucleic acid diagnostics, Expert. Rev. Med. Devices 7 (2010) 529–539.

[2] P. Sebastiani, N. Solovieff, S.W. Hartley, J.N. Milton, A. Riva, D. A. Dworkis, et al., Genetic modifiers of the severity of sickle cell anemia identified through a genome-wide association study, Am. J. Hematol. 85 (2010) 29–35.

[3] I. Fajac, P. Briand, M. Monsigny, P. Midoux, Sugar-mediated uptake of glycosylated polylysines and gene transfer into normal and cystic fibrosis airway epithelial cells, Hum. Gene Ther. 10 (1999) 395.

[4] S. Ferrari, A. Pettenazzo, N. Garbati, F. Zacchello, J.P. Behr, M. Scarpa, Polyethylenimine shows properties of interest for cystic fibrosis gene therapy, Biochim. Biophys. Acta 1447 (1999) 219–225.

[5] A. El-Aneed, An overview of current delivery systems in cancer gene therapy, J. Control Release 94 (2004) 1–14.
[6] W.F. Lai, M.C. Lin, Nucleic acid delivery with chitosan and its derivatives, J. Control Release 134 (2009) 158–168.
[7] A. Porgador, K.R. Irvine, A. Iwasaki, B.H. Barber, N.P. Restifo, R.N. Germain, Predominant role for directly transfected dendritic cells in antigen presentation to CD8+ T cells after gene gun immunization, J. Exp. Med. 188 (1998) 1075–1082.
[8] S. Lee, M. Gierynska, S.K. Eo, N. Kuklin, B.T. Rouse, Influence of DNA encoding cytokines on systemic and mucosal immunity following genetic vaccination against herpes simplex virus, Microbes Infect. 5 (2003) 571–578.
[9] L. Dupre, L. Kremer, I. Wolowczuk, G. Riveau, A. Capron, C. Locht, Immunostimulatory effect of IL-18-encoding plasmid in DNA vaccination against murine Schistosoma mansoni infection, Vaccine 19 (2001) 1373–1380.
[10] M. Mesnil, H. Yamasaki, Bystander effect in herpes simplex virus-thymidine kinase/ganciclovir cancer gene therapy: role of gap-junctional intercellular communication, Cancer Res. 60 (2000) 3989–3999.
[11] D. Lee, J.W. Kim, T. Seo, S.G. Hwang, E.J. Choi, J. Choe, SWI/SNF complex interacts with tumor suppressor p53 and is necessary for the activation of p53-mediated transcription, J. Biol. Chem. 277 (2002) 22330–22337.
[12] D.M. Nguyen, S.A. Wiehle, P.E. Koch, C. Branch, N. Yen, J.A. Roth, et al., Delivery of the p53 tumor suppressor gene into lung cancer cells by an adenovirus/DNA complex, Cancer Gene Ther. 4 (1997) 191–198.
[13] Y. Itoh, N. Maruyama, M. Kitamura, T. Shirasawa, K. Shigemoto, T. Koike, Induction of endogenous retroviral gene product (SU) as an acute-phase protein by IL-6 in murine hepatocytes, Clin. Exp. Immunol. 88 (1992) 356–359.
[14] C. Hesdorffer, J. Ayello, M. Ward, A. Kaubisch, L. Vahdat, C. Balmaceda, et al., Phase I trial of retroviral-mediated transfer of the human MDR1 gene as marrow chemoprotection in patients undergoing high-dose chemotherapy and autologous stem-cell transplantation, J. Clin. Oncol. 16 (1998) 165–172.
[15] J.S. Powell, M.V. Ragni, G.C. White 2nd, J.M. Lusher, C. Hillman-Wiseman, T.E. Moon, et al., Phase 1 trial of FVIII gene transfer for severe hemophilia A using a retroviral construct administered by peripheral intravenous infusion, Blood 102 (2003) 2038–2045.
[16] S. Lehrman, Virus treatment questioned after gene therapy death, Nature 401 (1999) 517–518.
[17] Q. Liu, D.A. Muruve, Molecular basis of the inflammatory response to adenovirus vectors, Gene Ther. 10 (2003) 935–940.
[18] S. Hargreaves, Rules on gene therapy are tightened after-leukaemia alert, Br. Med. J 325 (2002) 791.
[19] A. Cole, Child in gene therapy programme develops leukaemia, Br. Med. J. 336 (2008) 13.
[20] X. Zhang, W.T. Godbey, Viral vectors for gene delivery in tissue engineering, Adv. Drug Deliv. Rev. 58 (2006) 515–534.
[21] D.H. Sterman, A. Recio, A.R. Haas, A. Vachani, S.I. Katz, C.T. Gillespie, et al., A phase I trial of repeated intrapleural adenoviral-mediated interferon-beta gene transfer for mesothelioma and metastatic pleural effusions, Mol. Ther. 18 (2010) 852–860.
[22] G.Y. Wu, C.H. Wu, Receptor-mediated gene delivery and expression in vivo, J. Biol. Chem. 263 (1988) 14621–14624.
[23] A.G. Schatzlein, B.H. Zinselmeyer, A. Elouzi, C. Dufes, Y.T. Chim, C.J. Roberts, et al., Preferential liver gene expression with polypropylenimine dendrimers, J. Control Release 101 (2005) 247–258.
[24] K. Tahara, H. Yamamoto, H. Takeuchi, Y. Kawashima, Development of gene delivery system using PLGA nanospheres, Yakugaku Zasshi 127 (2007) 1541–1548.
[25] I. Kopatz, J.S. Remy, J.P. Behr, A model for non-viral gene delivery: through syndecan adhesion molecules and powered by actin, J. Gene Med. 6 (2004) 769–776.
[26] I. Chemin, D. Moradpour, S. Wieland, W.B. Offensperger, E. Walter, J.P. Behr, et al., Liver-directed gene transfer: a linear polyethlenimine derivative mediates highly efficient DNA delivery to primary hepatocytes in vitro and in vivo, J. Viral. Hepat. 5 (1998) 369–375.
[27] M.E. Davis, Non-viral gene delivery systems, Curr. Opin. Biotechnol. 13 (2002) 128–131.
[28] O. Boussif, F. Lezoualc'h, M.A. Zanta, M.D. Mergny, D. Scherman, B. Demeneix, et al., A versatile vector for gene and oligonucleotide transfer into cells in culture and in vivo: polyethylenimine, Proc. Natl. Acad. Sci. U.S.A. 92 (1995) 7297–7301.
[29] D. Goula, J.S. Remy, P. Erbacher, M. Wasowicz, G. Levi, B. Abdallah, et al., Size, diffusibility and transfection performance of linear PEI/DNA complexes in the mouse central nervous system, Gene Ther. 5 (1998) 712–717.
[30] M. Neu, D. Fischer, T. Kissel, Recent advances in rational gene transfer vector design based on poly(ethylene imine) and its derivatives, J. Gene Med. 7 (2005) 992–1009.
[31] Y. Liu, J. Nguyen, T. Steele, O. Merkel, T. Kissel, A new synthesis method and degradation of hyper-branched polyethylenimine grafted polycaprolactone block mono-methoxyl poly(ethylene glycol) copolymers (hy-PEI-g-PCL-b-mPEG) as potential DNA delivery vectors, Polymer. (Guildf). 50 (2009) 3895–3904.
[32] P. Bandyopadhyay, X.M. Ma, C. Linehan-Stieers, B.T. Kren, C.J. Steer, Nucleotide exchange in genomic DNA of rat hepatocytes using RNA/DNA oligonucleotides: targeted delivery of liposomes and polyethyleneimine to the asialoglycoprotein receptor, J. Biol. Chem. 274 (1999) 10163–10172.
[33] A. Aigner, D. Fischer, T. Merdan, C. Brus, T. Kissel, F. Czubayko, Delivery of unmodified bioactive ribozymes by an RNA-stabilizing polyethylenimine (LMW-PEI) efficiently downregulates gene expression, Gene Ther. 9 (2002) 1700–1707.
[34] G. Creusat, G. Zuber, Self-assembling polyethylenimine derivatives mediate efficient siRNA delivery in mammalian cells, Chembiochem 9 (2008) 2787–2789.
[35] A. von Harpe, H. Petersen, Y.X. Li, T. Kissel, Characterization of commercially available and synthesized polyethylenimines for gene delivery, J. Control Release 69 (2000) 309–322.
[36] K. Kunath, A. von Harpe, D. Fischer, H. Peterson, U. Bickel, K. Voigt, et al., Low-molecular-weight polyethylenimine as a non-viral vector for DNA delivery: comparison of physicochemical properties, transfection efficiency and in vivo distribution with high-molecular-weight polyethylenimine, J. Control Release 89 (2003) 113–125.
[37] D. Fischer, T. Bieber, Y.X. Li, H.P. Elsasser, T. Kissel, A novel non-viral vector for DNA delivery based on low molecular weight, branched polyethylenimine: effect of molecular weight on transfection efficiency and cytotoxicity, Pharm. Res. 16 (1999) 1273–1279.

[38] U. Lungwitz, M. Breunig, T. Blunk, A. Gopferich, Polyethylenimine-based non-viral gene delivery systems, Eur. J. Pharm. Biopharm. 60 (2005) 247–266.
[39] J.K. Suh, H.W. Matthew, Application of chitosan-based polysaccharide biomaterials in cartilage tissue engineering: a review, Biomaterials. 21 (2000) 2589–2598.
[40] W. Paul, C.P. Garside, Chitosan, a drug carrier for the 21st century: a review, STP Pharma. Sci 10 (2000) 5–22.
[41] S. Aiba, Lysozymic hydrolysis of partially N-acetylated chitosans, Int. J. Biol. Macromol. 14 (1992) 225–228.
[42] H. Zhang, S.H. Neau, In vitro degradation of chitosan by bacterial enzymes from rat cecal and colonic contents, Biomaterials. 23 (2002) 2761–2766.
[43] R.J. Mumper, J. Wang, J.M. Claspell, A.P. Rolland, Novel polymeric condensing carriers for gene delivery, Proc. Int. Symp. Control. Release Bioact. Mater. 22 (1995) 178–179.
[44] S. Venkatesh, T.J. Smith, Chitosan-membrane interactions and their probable role in chitosan-mediated transfection, Biotechnol. Appl. Biochem. 27 (1998) 265–267.
[45] M. Anthonsen, O. Smidsrod, Hydrogen ion titration of chitosans with varying degrees of N-acetylation by monitoring induced 1H-NMR chemical shifts, Carbohydr. Polym. 26 (1995) 303–305.
[46] G. Berth, H. Dautzenberg, M.G. Peter, Physico-chemical characterization of chitosans varying in degree of acetylation, Carbohydr. Polym. 36 (1998) 205–216.
[47] R. Hejazi, M. Amiji, Chitosan-based gastrointestinal delivery systems, J. Control Release 89 (2003) 151–165.
[48] W. Weecharangsan, P. Opanasopit, T. Ngawhirunpat, A. Apirakaramwong, T. Rojanarata, U. Ruktanonchai, et al., Evaluation of chitosan salts as non-viral gene vectors in CHO-K1 cells, Int. J. Pharm. 348 (2008) 161–168.
[49] N. Fang, V. Chan, H.Q. Mao, K.W. Leong, Interactions of phospholipid bilayer with chitosan: effect of molecular weight and pH, Biomacromolecules 2 (2001) 1161–1168.
[50] M. Thanou, J.C. Verhoef, H.E. Junginger, Chitosan and its derivatives as intestinal absorption enhancers, Adv. Drug Deliv. Rev. 50 (2001) S91–S101.
[51] Q. Xu, C.H. Wang, D.W. Pack, Polymeric carriers for gene delivery: chitosan and poly(amidoamine) dendrimers, Curr. Pharm. Des. 16 (2010) 2350–2368.
[52] D. Shcharbin, A. Shakhbazau, M. Bryszewska, Poly(amidoamine) dendrimer complexes as a platform for gene delivery, Expert. Opin. Drug. Deliv. 10 (2013) 1687–1698.
[53] X. Sun, N. Zhang, Cationic polymer optimization for efficient gene delivery, Mini Rev. Med. Chem. 10 (2010) 108–125.
[54] S.C. De Smedt, J. Demeester, W.E. Hennink, Cationic polymer based gene delivery systems, Pharm. Res. 17 (2000) 113–126.
[55] M. Kursa, G.F. Walker, V. Roessler, M. Ogris, W. Roedl, R. Kircheis, et al., Novel shielded transferrin-polyethylene glycol-polyethylenimine/DNA complexes for systemic tumor-targeted gene transfer, Bioconjug. Chem. 14 (2003) 222–231.
[56] R. Kircheis, L. Wightman, A. Schreiber, B. Robitza, V. Rossler, M. Kursa, et al., Polyethylenimine/DNA complexes shielded by transferrin target gene expression to tumors after systemic application, Gene Ther. 8 (2001) 28–40.
[57] H. Petersen, P.M. Fechner, A.L. Martin, K. Kunath, S. Stolnik, C. J. Roberts, et al., Polyethylenimine-graft-poly(ethylene glycol) copolymers: influence of copolymer block structure on DNA complexation and biological activities as gene delivery system, Bioconjug. Chem. 13 (2002) 845–854.
[58] M. Ogris, G. Walker, T. Blessing, R. Kircheis, M. Wolschek, E. Wagner, Tumor-targeted gene therapy: strategies for the preparation of ligand-polyethylene glycol-polyethylenimine/DNA complexes, J. Control Release 91 (2003) 173–181.
[59] C.H. Ahn, S.Y. Chae, Y.H. Bae, S.W. Kim, Biodegradable poly (ethylenimine) for plasmid DNA delivery, J. Control Release 80 (2002) 273–282.
[60] G.P. Tang, J.M. Zeng, S.J. Gao, Y.X. Ma, L. Shi, Y. Li, et al., Polyethylene glycol modified polyethylenimine for improved CNS gene transfer: effects of PEGylation extent, Biomaterials 24 (2003) 2351–2362.
[61] H.K. Nguyen, P. Lemieux, S.V. Vinogradov, C.L. Gebhart, N. Guerin, G. Paradis, et al., Evaluation of polyether-polyethyleneimine graft copolymers as gene transfer agents, Gene Ther. 7 (2000) 126–138.
[62] A. Kichler, M. Chillon, C. Leborgne, O. Danos, B. Frisch, Intranasal gene delivery with a polyethylenimine-PEG conjugate, J. Control Release 81 (2002) 379–388.
[63] S.J. Sung, S.H. Min, K.Y. Cho, S. Lee, Y.J. Min, Y.I. Yeom, et al., Effect of polyethylene glycol on gene delivery of polyethylenimine, Biol. Pharm. Bull. 26 (2003) 492–500.
[64] X. Gao, L. Huang, Potentiation of cationic liposome-mediated gene delivery by polycations, Biochemistry 35 (1996) 1027–1036.
[65] L. Vitiello, A. Chonn, J.D. Wasserman, C. Duff, R.G. Worton, Condensation of plasmid DNA with polylysine improves liposome-mediated gene transfer into established and primary muscle cells, Gene Ther. 3 (1996) 396–404.
[66] K. Hong, W. Zheng, A. Baker, D. Papahadjopoulos, Stabilization of cationic liposome-plasmid DNA complexes by polyamines and poly(ethylene glycol)-phospholipid conjugates for efficient in vivo gene delivery, FEBS Lett. 400 (1997) 233–237.
[67] A. Vonarbourg, C. Passirani, P. Saulnier, J.P. Benoit, Parameters influencing the stealthiness of colloidal drug delivery systems, Biomaterials 27 (2006) 4356–4373.
[68] J.Y. Wong, T.L. Kuhl, J.N. Israelachvili, N. Mullah, S. Zalipsky, Direct measurement of a tethered ligand-receptor interaction potential, Science 275 (1997) 820–822.
[69] X. Jiang, H. Dai, K.W. Leong, S.H. Goh, H.Q. Mao, Y.Y. Yang, Chitosan-g-PEG/DNA complexes deliver gene to the rat liver via intrabiliary and intraportal infusions, J. Gene Med. 8 (2006) 477–487.
[70] Y. Zhang, J. Chen, Y. Pan, J. Zhao, L. Ren, M. Liao, et al., A novel PEGylation of chitosan nanoparticles for gene delivery, Biotechnol. Appl. Biochem. 46 (2007) 197–204.
[71] A.K. Pannier, J.A. Wieland, L.D. Shea, Surface polyethylene glycol enhances substrate-mediated gene delivery by nonspecifically immobilized complexes, Acta Biomater. 4 (2008) 26–39.
[72] P. Erbacher, T. Bettinger, P. Belguise-Valladier, S. Zou, J.L. Coll, J.P. Behr, et al., Transfection and physical properties of various saccharide, poly(ethylene glycol), and antibody-derivatized polyethylenimines (PEI), J. Gene Med. 1 (1999) 210–222.

[73] R.S. Burke, S.H. Pun, Extracellular barriers to in vivo PEI and PEGylated PEI polyplex-mediated gene delivery to the liver, Bioconjug. Chem. 19 (2008) 693–704.
[74] S. Mishra, P. Webster, M.E. Davis, PEGylation significantly affects cellular uptake and intracellular trafficking of non-viral gene delivery particles, Eur. J. Cell Biol. 83 (2004) 97–111.
[75] J. Shin, P. Shum, D.H. Thompson, Acid-triggered release via dePEGylation of DOPE liposomes containing acid-labile vinyl ether PEG-lipids, J. Control Release 91 (2003) 187–200.
[76] R. Tomlinson, J. Heller, S. Brocchini, R. Duncan, Polyacetal-doxorubicin conjugates designed for pH-dependent degradation, Bioconjug. Chem. 14 (2003) 1096–1106.
[77] N. Murthy, J. Campbell, N. Fausto, A.S. Hoffman, P.S. Stayton, Design and synthesis of pH-responsive polymeric carriers that target uptake and enhance the intracellular delivery of oligonucleotides, J. Control Release 89 (2003) 365–374.
[78] R.S. Greenfield, T. Kaneko, A. Daues, M.A. Edson, K.A. Fitzgerald, L.J. Olech, et al., Evaluation in vitro of adriamycin immunoconjugates synthesized using an acid-sensitive hydrazone linker, Cancer Res. 50 (1990) 6600–6607.
[79] M. Morille, C. Passirani, A. Vonarbourg, A. Clavreul, J.P. Benoit, Progress in developing cationic vectors for non-viral systemic gene therapy against cancer, Biomaterials 29 (2008) 3477–3496.
[80] M. Lee, S.W. Kim, Polyethylene glycol-conjugated copolymers for plasmid DNA delivery, Pharm. Res. 22 (2005) 1–10.
[81] J. Chen, X.L. Gao, K.L. Hu, Z.Q. Pang, J. Cai, J.W. Li, et al., Galactose-poly(ethylene glycol)-polyethylenimine for improved lung gene transfer, Biochem. Biophys. Res. Commun. 375 (2008) 378–383.
[82] E. Kleemann, M. Neu, N. Jekel, L. Fink, T. Schmehl, T. Gessler, et al., Nano-carriers for DNA delivery to the lung based upon a TAT-derived peptide covalently coupled to PEG-PEI, J. Control Release 109 (2005) 299–316.
[83] F.Q. Hu, M.D. Zhao, H. Yuan, J. You, Y.Z. Du, S. Zeng, A novel chitosan oligosaccharide-stearic acid micelles for gene delivery: properties and in vitro transfection studies, Int. J. Pharm. 315 (2006) 158–166.
[84] Y. Xiang, Q. Yu, Z. Qi, Z. Du, S. Xu, H. Zhang, Enhancement of immunological activity of CpG ODN by chitosan gene carrier, J. Huazhong Univ. Sci. Technol. Med. Sci. 27 (2007) 128–130.
[85] F. Li, W.G. Liu, K.D. Yao, Preparation of oxidized glucose-crosslinked N-alkylated chitosan membrane and in vitro studies of pH-sensitive drug delivery behaviour, Biomaterials 23 (2002) 343–347.
[86] W.G. Liu, K.D. Yao, Q.G. Liu, Formation of a DNA/N-dodecylated chitosan complex and salt-induced gene delivery, J. Appl. Polym. Sci. 82 (2001) 3391–3395.
[87] W.G. Liu, X. Zhang, S.J. Sun, G.J. Sun, K.D. Yao, D.C. Liang, et al., N-alkylated chitosan as a potential nonviral vector for gene transfection, Bioconjug. Chem. 14 (2003) 782–789.
[88] S. Ercelen, X. Zhang, G. Duportail, C. Grandfils, J. Desbrieres, S. Karaeva, et al., Physicochemical properties of low molecular weight alkylated chitosans: a new class of potential nonviral vectors for gene delivery, Colloids Surf. B Biointerfaces 51 (2006) 140–148.
[89] A. Bernkop-Schnurch, M. Hornof, T. Zoidl, Thiolated polymers—thiomers: synthesis and in vitro evaluation of chitosan-2-iminothiolane conjugates, Int. J. Pharm. 260 (2003) 229–237.
[90] A. Bernkop-Schnurch, D. Guggi, Y. Pinter, Thiolated chitosans: development and in vitro evaluation of a mucoadhesive, permeation enhancing oral drug delivery system, J. Control Release 94 (2004) 177–186.
[91] M. Roldo, M. Hornof, P. Caliceti, A. Bernkop-Schnurch, Mucoadhesive thiolated chitosans as platforms for oral controlled drug delivery: synthesis and in vitro evaluation, Eur. J. Pharm. Biopharm. 57 (2004) 115–121.
[92] D. Lee, W. Zhang, S.A. Shirley, X. Kong, G.R. Hellermann, R.F. Lockey, et al., Thiolated chitosan/DNA nanocomplexes exhibit enhanced and sustained gene delivery, Pharm. Res. 24 (2007) 157–167.
[93] A. Calzolari, I. Oliviero, S. Deaglio, G. Mariani, M. Biffoni, N.M. Sposi, et al., Transferrin receptor 2 is frequently expressed in human cancer cell lines, Blood Cells Mol. Dis. 39 (2007) 82–91.
[94] H.Q. Mao, K. Roy, V.L. Troung-Le, K.A. Janes, K.Y. Lin, Y. Wang, et al., Chitosan-DNA nanoparticles as gene carriers: synthesis, characterization and transfection efficiency, J. Control Release 70 (2001) 399–421.
[95] Y. Aktas, M. Yemisci, K. Andrieux, R.N. Gursoy, M.J. Alonso, E. Fernandez-Megia, et al., Development and brain delivery of chitosan-PEG nanoparticles functionalized with the monoclonal antibody OX26, Bioconjug. Chem. 16 (2005) 1503–1511.
[96] W.F. Lai, G.P. Tang, X. Wang, G. Li, H. Yao, Z. Shen, et al., Cyclodextrin-PEI-tat polymer as a vector for plasmid DNA delivery to placenta mesenchymal stem cells, BioNanoScience 1 (2011) 89–96.
[97] I.K. Park, T.H. Kim, Y.H. Park, B.A. Shin, E.S. Choi, E.H. Chowdhury, et al., Galactosylated chitosan-graft-poly(ethylene glycol) as hepatocyte-targeting DNA carrier, J. Control Release 76 (2001) 349–362.
[98] S. Gao, J. Chen, X. Xu, Z. Ding, Y.H. Yang, Z. Hua, et al., Galactosylated low molecular weight chitosan as DNA carrier for hepatocyte-targeting, Int. J. Pharm. 255 (2003) 57–68.
[99] P. Erbacher, S. Zou, T. Bettinger, A.M. Steffan, J.S. Remy, Chitosan-based vector/DNA complexes for gene delivery: biophysical characteristics and transfection ability, Pharm. Res. 15 (1998) 1332–1339.
[100] T.H. Kim, H. Jin, H.W. Kim, M.H. Cho, C.S. Cho, Mannosylated chitosan nanoparticle-based cytokine gene therapy suppressed cancer growth in BALB/c mice bearing CT-26 carcinoma cells, Mol. Cancer Ther. 5 (2006) 1723–1732.
[101] W.G. Liu, K.D. Yao, Chitosan and its derivatives—a promising non-viral vector for gene transfection, J. Control Release 83 (2002) 1–11.
[102] S. Mansouri, P. Lavigne, K. Corsi, M. Benderdour, E. Beaumont, J.C. Fernandes, Chitosan-DNA nanoparticles as non-viral vectors in gene therapy: strategies to improve transfection efficacy, Eur. J. Pharm. Biopharm. 57 (2004) 1–8.
[103] Y. Zhang, L. Lin, L. Liu, F. Liu, A. Maruyama, H. Tian, et al., Ionic-crosslinked polysaccharide/PEI/DNA nanoparticles for stabilized gene delivery, Carbohydr. Polym. 201 (2018) 246–256.
[104] Y. Kang, X.Y. Zhang, W. Jiang, C.Q. Wu, C.M. Chen, Y.F. Zheng, et al., Tumor-directed gene therapy in mice using a composite nonviral gene delivery system consisting of the piggyBac transposon and polyethylenimine, BMC. Cancer 9 (2009).
[105] X. Zhou, B. Liu, X. Yu, X. Zha, X. Zhang, Y. Chen, et al., Controlled release of PEI/DNA complexes from mannose-bearing chitosan microspheres as a potent delivery system to

enhance immune response to HBV DNA vaccine, J. Control Release 121 (2007) 200–207.

[106] M. Thomas, J.J. Lu, Q. Ge, C.C. Zhang, J.Z. Chen, A.M. Klibanov, Full deacylation of polyethylenimine dramatically boosts its gene delivery efficiency and specificity to mouse lung, Proc. Natl. Acad. Sci. U.S.A. 102 (2005) 5679–5684.

[107] X.F. Zhou, B. Liu, X.H. Yu, X. Zha, X.Z. Zhang, X.Y. Wang, et al., Controlled release of PEI/DNA complexes from PLGA microspheres as a potent delivery system to enhance immune response to HIV vaccine DNA prime/MVA boost regime, Eur. J. Pharm. Biopharm. 68 (2008) 589–595.

[108] M. Bivas-Benita, M.Y. Lin, S.M. Bal, K.E. van Meijgaarden, K. L. Franken, A.H. Friggen, et al., Pulmonary delivery of DNA encoding Mycobacterium tuberculosis latency antigen Rv1733c associated to PLGA-PEI nanoparticles enhances T cell responses in a DNA prime/protein boost vaccination regimen in mice, Vaccine 27 (2009) 4010–4017.

[109] S. Wang, N. Ma, S.J. Gao, H. Yu, K.W. Leong, Transgene expression in the brain stem effected by intramuscular injection of polyethylenimine/DNA complexes, Mol. Ther. 3 (2001) 658–664.

[110] C. Rudolph, J. Lausier, S. Naundorf, R.H. Muller, J. Rosenecker, In vivo gene delivery to the lung using polyethylenimine and fractured polyamidoamine dendrimers, J. Gene Med. 2 (2000) 269–278.

[111] Y.C. Huang, K. Riddle, K.G. Rice, D.J. Mooney, Long-term in vivo gene expression via delivery of PEI-DNA condensates from porous polymer scaffolds, Hum. Gene Ther. 16 (2005) 609–617.

[112] P. Dames, A. Ortiz, U. Schillinger, E. Lesina, C. Plank, J. Rosenecker, et al., Aerosol gene delivery to the murine lung is mouse strain dependent, J. Mol. Med. 85 (2007) 371–378.

[113] P. Chollet, M.C. Favrot, A. Hurbin, J.L. Coll, Side-effects of a systemic injection of linear polyethylenimine-DNA complexes, J. Gene Med. 4 (2002) 84–91.

[114] W.F. Lai, In vivo nucleic acid delivery with PEI and its derivatives: current status and perspectives, Expert. Rev. Med. Devices 8 (2011) 173–185.

[115] S. Moffatt, S. Wiehle, R.J. Cristiano, Tumor-specific gene delivery mediated by a novel peptide-polyethylenimine-DNA polyplex targeting aminopeptidase N/CD13, Hum. Gene Ther. 16 (2005) 57–67.

[116] Y. Tada, T. Kitahara, T. Yoshioka, T. Nakamura, N. Ichikawa, M. Nakashima, et al., Partial hepatectomy enhances polyethyleniminemediated plasmid DNA delivery, Biol. Pharm. Bull. 29 (2006) 1712–1716.

[117] H. Sasaki, S. Yoshida, T. Kitahara, T. Yoshioka, H. Nakagawa, T. Nakamura, et al., Influence of disease stage on polyethylenimine-mediated plasmid DNA delivery in murine hepatitis, Int. J. Pharm. 318 (2006) 139–145.

Chapter 5

Design of cyclodextrin-based systems for intervention execution

Introduction

Cyclodextrins (CDs) are cyclic (α-1,4)-linked oligosaccharides of α-D-glucopyranose [1]. The most common forms of CDs are α-, β-, and γ-CDs. They are made up of six, seven, and eight α-D-glucopyranose units, respectively (Fig. 5.1) [2,3]. The high content of hydroxyl groups renders CDs soluble in water. Yet, the aqueous solubility of CDs in general is lower than that of the comparable linear dextrins, owing to the relatively high crystal energy of CDs [4]. CDs were documented first by Villiers, who isolated a crystalline substance, namely "cellulosine," from a bacterial digest of starch. That substance has been found to resist acid hydrolysis [5]. It has later been known as "cyclodextrin." The basic physicochemical features of CDs (including chemical structure, reactivity, cavity size, solubility, and inclusion complexation capacity) have also been subsequently documented by Cramer in his book *Einschlussverbindungen* [6].

Over the last several decades, CDs have exhibited promising practical potential in diverse areas, ranging from controlled drug release [7–12] to chiral separation of basic drugs [13]. These applications are largely mediated by the ability of CDs to form host–guest complexes. In fact, compared to other conventional host molecules (such as cucurbiturils, pillararenes, crown ethers, and calixarenes), CDs display distinctive features that make them attractive in structural design and engineering of polymeric vectors [14]. For instance, while many other host molecules have to be synthesized via multistep synthetic procedures before use, CDs are commercially available and "ready-made" molecular entities. In addition, the native forms of many host molecules display

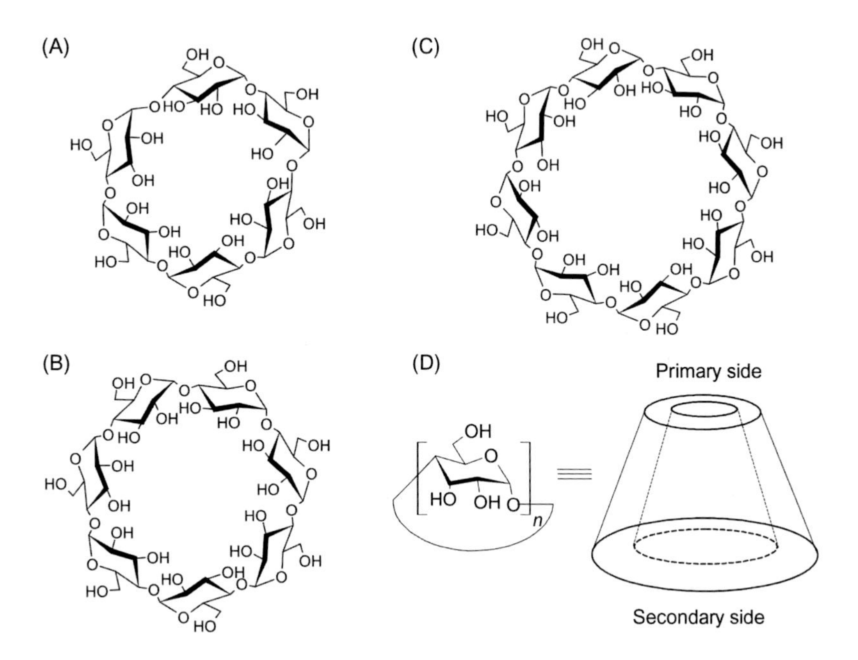

FIGURE 5.1 The structures of (A) α-CD, (B) β-CD, and (C) γ-CD, as well as (D) the torus shape of the CD molecule. The shape of the molecule is not drawn to scale. *Reproduced from W.F. Lai, A.L. Rogach, W.T. Wong, Chemistry and engineering of cyclodextrins for molecular imaging, Chem. Soc. Rev. 46 (2017) 6379–6419 with permission from RSC, [3].*

Delivery of Therapeutics for Biogerontological Interventions. DOI: https://doi.org/10.1016/B978-0-12-816485-3.00005-2

poor aqueous solubility, whereas native CDs are highly water soluble. This makes the direct use of CDs more convenient. Because of these, CD-based "guest–host chemistry" has gained extensive research interests in the development and modification of polymeric gene carriers. In this chapter, we will first have an overview of the basic properties of CDs, followed by a discussion of incorporation of CDs into the design and modification of polymeric vectors for the execution of biogerontological interventions that involve genetic manipulation.

Basic properties of cyclodextrins

CDs have the apolar cavity interiors and the hydrophilic cavity exteriors. Because of this unique structure, CD molecules can provide a microenvironment for encapsulation and solubilization of hydrophobic guest molecules [15,16]. This makes CDs a favorable candidate to be exploited as excipients in pharmaceutical formulation. In addition to delivering chemical drugs, since the turn of the last century more and more efforts have been directed to exploiting the use of CDs in delivering nucleic acids [17,18]. For instance, Agrawal's group has examined the possible use of CDs (and their analogs) in enhancing cellular internalization of oligonucleotides [19,20]. Abdou and coworkers have also evaluated the capacity of various native and derivatized CDs to enhance the activity of an 18-mer phosphodiester oligodeoxynucleotide (ODN) (which is complementary to the initiation region of a messenger RNA (mRNA) molecule coding for the spike protein, and possesses the intergenic consensus sequence of an enteric coronavirus) against viral growth in human adenocarcinoma cells [21]. They have found that upon complexation with an β-CD derivative, namely 6-deoxy-6-S-β-D-galactopyranosyl-6-thio-cyclomalto-heptaose, in a molar ratio of 1:100, the ODN has induced up to 90% of viral inhibition. This is much higher than that achieved (12%–34%) by using the naked ODN [22–26]. This, along with other studies [26,27], has established the foundation on which subsequent research on CD-mediated gene delivery can be built.

CDs exhibit a binding affinity with nucleic acids [17,28], and can also attenuate the cytotoxicity of other nucleic acid carriers. The latter has been revealed by an earlier study [29], which has linked diamino-CD monomers with diimidate comonomers to generate a number of linear cationic β-CD-based polymers. In BHK-21 cells, the IC_{50} of the polymers produced has been found to be much lower than that of the CD-lacking polyamidines [29]. This has suggested that, by incorporating CD moieties into the backbone of a cationic polymer, the cytotoxicity of the polymer can possibly be reduced. Apart from this, CDs can facilitate the efficiency of virus-mediated nucleic acid delivery by enhancing viral binding and internalization into a host cell. This has been illustrated by the successful use of CDs to improve adenoviral-mediated gene transfer to the rat jejunum [30].

Strategies for structural engineering of cyclodextrins

Until now, a wide diversity of CD derivatives (whose OH groups have undergone alkylation, esterification, or even random derivatization) have been reported. Through chemical modification of the OH groups, charged groups can usually be incorporated into the molecular structure, causing changes in the solubility of CDs in water or organic solvents [31]. Over the years, different methods have been adopted to functionalize CDs. These methods generally take place at the OH groups located either in the upper rim (primary side) or in the lower rim (secondary side). The reactivity of hydroxyl groups in the CD ring varies with the positions of those groups. In general, there are secondary hydroxyl groups at 2- and 3-positions, and primary hydroxyl groups at the 6-position of the glucopyranose ring [32]. Among them, 3-OHs are the least accessible, whereas 2-OHs are the most acidic. Compared to the hydroxyl groups at the other two positions, 6-OHs are the most nucleophilic, and can be readily converted into other functional groups. These reactivity differences among hydroxyl groups at different locations have been exploited extensively for the development of strategies for selective chemical functionalization of CDs [32,33].

At the moment, most of the functionalization approaches involve the primary hydroxyl group at the 6-position. A typical functionalization method is to use a nucleophilic substitution reaction, during which mono-6-(*p*-toluenesulfonyl)-6-deoxy-CD (Ts-CD) is first synthesized as a precursor, whose tosyl group then undergoes nucleophilic displacement by selected nucleophiles (amines, azide, carboxylate, hydroxylamine, iodide, polyamines, and thiols) to form a monofunctionalized CD [34]. The process of tosylation can be performed by using tosyl chloride in dry pyridine [35], or in water along with either NaOH [36] or $CuSO_4$ [37]. Either 6-monotosylate or a mixture of 6-polytosylates can be formed via this process, and the direction of the process depends mainly on the molar ratio between CDs and tosyl chloride. Furthermore, the formation of regioisomers during the process may increase the difficulties of subsequently isolating pure tosylates. This problem can be alleviated by using mesitylenesulfonyl chloride or other oversized sulfonyl chlorides. This method has been shown to be able to limit the number of regioisomers formed [35].

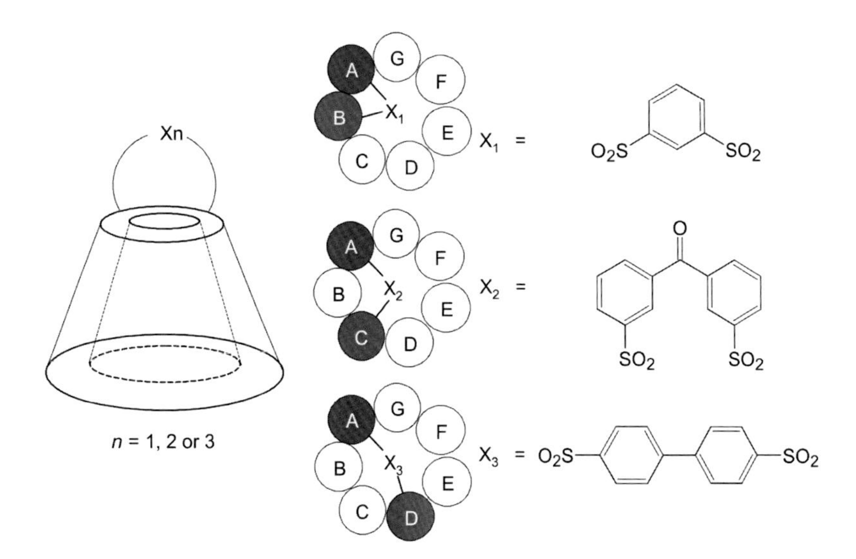

FIGURE 5.2 A schematic diagram showing the structures of the AB, AC, and AD isomers of 6-disulfonates of β-CD. *Reproduced from W.F. Lai, A. L. Rogach, W.T. Wong, Chemistry and engineering of cyclodextrins for molecular imaging, Chem. Soc. Rev. 46 (2017) 6379–6419 with permission from RSC, [3].*

Besides forming monofunctionalized CDs, disubstituted CD derivatives can be generated using appropriate disulfonyl chlorides, whose geometry can be exploited to control the regiochemistry and to generate AB, AC, or AD isomers (Fig. 5.2) [32].

Example protocols for experimental design

The method below is an example protocol for preparing Ts-CD. This protocol is based on the one previously reported by Tang and Ng [34].

1. Fit a three-necked, round-bottomed flask, into which a magnetic stir bar is added, with a Liebig condenser and a pressure equalizing addition funnel.
2. Fit the condenser with a rubber septum.
3. Degas the reaction setup by applying three cycles of pumping and nitrogen refilling.
4. Fill the flask with nitrogen.
5. Add 400 mL of freshly dried pyridine and 25 g of β-CD into the flask.
6. Turn on the magnetic stirrer.
7. Cool the flask to 0°C in an ice-water bath.
8. Fit a two-necked, round-bottomed flask, into which a magnetic stir bar is added, with a rubber septum.
9. Degas the flask by applying two cycles of pumping and nitrogen refilling.
10. Fill the flask with nitrogen.
11. Add 4 g of *p*-toluenesulfonyl chloride into the two-necked flask.
12. Add 30 mL of freshly dried pyridine into the flask using a glass syringe fitted with a hypodermic needle.
13. Dissolve *p*-toluenesulfonyl chloride in pyridine under magnetic stirring.

(Continued)

(Continued)

14. Add the obtained solution to the addition funnel (which has been fitted with the three-necked flask in Step 1) for one hour under reflux and magnetic stirring, during which the temperature of the reaction flask should be kept at 0°C.
15. Remove the water bath.
16. Keep the setup at ambient conditions for 24 hours.
17. Perform vacuum distillation under reduced pressure to remove most of the pyridine from the reaction mixture.
18. Add 600 mL of acetone into the flask with vigorous stirring for 30 minutes.
19. Filter the reaction mixture to obtain solid residues.
20. Wash the solid residues with acetone three times.
21. Use hot water to recrystalize the solid.
22. Dry the solid overnight in a vacuum oven at 60°C to obtain the product.

When functionalization of the larger rim is required, sulfonation or tosylation of the C-2 position is a vital step. To obtain 2-tosyl-β-CD, one possible approach is via a transesterification reaction using *m*-nitrophenyl tosylate in a DMF-water solution at the basic pH [38]. The reaction yield is, however, far from satisfactory. More recently, Teranishi has postulated a synthetic route to obtain disulfonated CDs at the C-2 position. Similar to the situation of 6-disulfonates as mentioned earlier, the geometry of the sulfonating agent has been exploited to control the regioselectivity of the sulfonation reaction [39,40]. Furthermore, strategies have been reported to generate 3-polyfunctionalized CDs [41], among which amino derivatives of CDs have received extensive research efforts. In particular, monoamine derivatives at the 3- or 6-position have been adopted to generate a series of CD derivatives with side chains via coupling reactions [42–46].

Roles in vector design

Native CDs have rarely been used directly as a nucleic acid carrier, particularly because of their failure to form stable complexes with nucleic acids for mediating transfection [47]. Derivatization of CDs is, therefore, usually performed prior to applications in nucleic acid transfer. One example of CD derivatives developed for nucleic acid delivery is the polycationic amphiphilic CD. This derivative is constructed by modifying the facial anisotropy of the truncated-cone CD torus through instillation of hydrophobic and cationic elements in the "jellyfish" or "skirt" configuration [48,49]. Its DNA complexation capacity and transfection efficiency can be tuned by modulating various parameters, including the nature of the functional groups, charge density, hydrophilic-hydrophobic balance, and spacer length [50–52]. More examples of CD derivatives are presented in Fig. 5.3. They have been prepared by incorporating β-CD with an alkylimidazole group, a methoxyethylamino group, a pyridylamino group, or a primary amine group at the 6-position of the glucose units [47]. Compared to native β-CD, these derivatives have been shown to be more effective in facilitating the uptake of the transgene [47]. In addition to functioning as nucleic acid carriers, CDs can be used as structural modifiers during vector development.

When CDs are applied as structural modifiers, they can be used in two ways. One is to be adopted as threading devices. This is exemplified by the case of supramolecular polyrotaxanes, in which around 12 α-CD rings can be found in each molecule of the poly((ethylene oxide)-ran-(propylene oxide)) (P(EO-r-PO)) random copolymer, with the rings being located selectively on EO segments [53]. The polyrotaxanes have displayed higher transfection efficiency than PEI 25 kDa in HEK293 cells [53], and may warrant further development as gene carriers for in vivo use. The second method of using CDs as structural modifiers is as pendants. A good example of nucleic acid carriers developed using this approach is the PAMAM dendrimer conjugates containing α-, β-, and γ-CDs. The conjugates can condense plasmids and protect the plasmids from DNase I-mediated degradation [54]. In vitro studies have shown that the conjugate with α-CD, namely α-CDE, is more effective in transfection

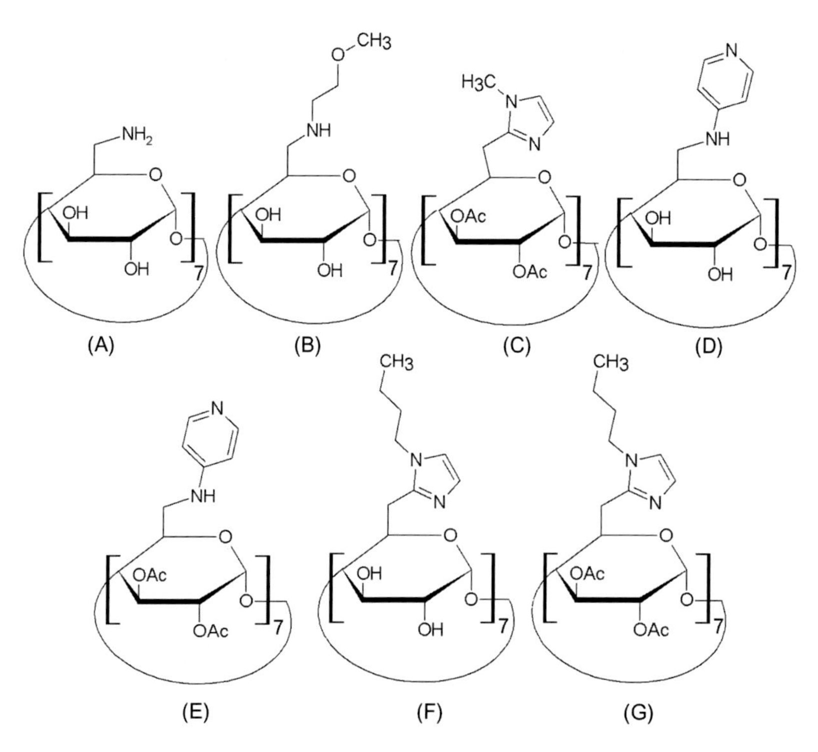

FIGURE 5.3 Structures of some CD derivatives. These derivatives include (A) heptakis(6-amino-6-deoxy)-β-CD, (B) heptakis(6-deoxy-6-methoxyethylamino)-β-CD, (C) heptakis[2,3-di-*O*-acetyl-6-deoxy-6-(1-methyl-1H-imidazol-2-yl)]-β-CD, (D) heptakis(6-deoxy-6-pyrid-4-ylamino)-β-CD, (E) heptakis(2,3-di-*O*-acetyl-6-deoxy-6-pyrid-4-ylamino)-β-CD, (F) heptakis[6-(1-*n*-butyl-1H-imidazol-2-yl)-6-deoxy]-β-CD, and (G) heptakis[2,3-di-*O*-acetyl-6-(1-*n*-butyl-1H-imidazol-2-yl)-6-deoxy]-β-CD. *Reproduced from W.F. Lai, Cyclodextrins in non-viral gene delivery, Biomaterials 35 (2014) 401–411 with permission from Elsevier B.V., [2].*

than those with β- and γ-CDs [54]. Its transfection efficiency in RAW264.7 and NIH3T3 cells has also been found to be much higher than that achieved by Lipofectin and the unmodified dendrimer [54]. To enhance the transfection efficiency of the conjugate, an earlier study has attempted to modulate the structural parameters by generating conjugates with different dendrimers. Results have revealed that α-CDE generated from the G3 dendrimer has exhibited higher transfection efficiency than those generated from G2 and G4 dendrimers [55]. In addition, conjugates having different degrees of substitution (DS) of α-CD have shown variations in cytotoxicity, transfection efficiency, and membrane-disruptive ability [56]. Compared with those having DS values of 1.1 and 5.4, the one having a DS value of 2.4 has been found to be more effective in transfection in NIH3T3 and HepG2 cells, and to show higher efficiency in delivering plasmids to liver, spleen, and kidney after i.v. administration [56]. These results have demonstrated the importance of structural optimization of a synthetic vector for nucleic acid transfer.

Apart from being used as structural modifiers, CDs can function as linking agents to covalently link other polymers together to form larger molecular constructs. For instance, linear CD-linked polymers have previously been generated from difunctionalized CDs and difunctionalized comonomers [57]. Among the polymers generated, the linear polymer containing 6 methylene units has displayed the highest transfection efficiency in BHK-21 cells. The efficiency of those containing 5, 4, 7, 8, 10 methylene units has been found to be only 6%, 22%, 50%, 64%, and 10% of the efficiency of the one containing 6 methylene units, respectively [29]. These results have shown that different levels of CD incorporation can influence the efficiency of the generated polymer in transfection. Another example of polymeric nucleic acid carriers developed by using CDs as linking agents is PEI-β-CD, which can be synthesized by first using tosyl chloride to generate amine-reactive tosyldeoxy-β-CD, followed by the reaction with PEI [24]. In vitro studies have demonstrated that PEI-β-CD exhibits higher transfection efficiency than unmodified PEI, and shows negligible toxicity in HEK293 cells at its working concentration for plasmid delivery [24]. Such high efficiency of PEI-β-CD in transfection can be further escalated upon conjugation of the polyplexes with human insulin that has been derivatized with a hydrophobic palmitate group [24]. Here it is worth mentioning that proper optimization of the grafting ratio of CDs is a prerequisite to the development of an effective PEI-β-CD-based carrier. This has been demonstrated by the fact that modification of 5%, 10%, and 16% of the amine groups in PEI with CDs may lead to a reduction in the luciferase activity by 1, 2, and 4 orders of magnitude, respectively [25]. This reduction has been attributed to the changes in the pKa values of the PEI amines, leading to a decrease in the efficiency in endosomal escape of the polyplexes. This hypothesis has been supported by the observation that the pH buffering capacity of PEI-β-CD is much lower than that of PEI [25].

Example protocols for experimental design

The method below is an example protocol for preparing PEI-β-CD. This protocol is based on the one previously reported by Forrest and coworkers [24].

1. Disperse 10 g of β-CD in 25 mL of H_2O.
2. Adjust the pH to 13 using a NaOH solution.
3. Stir the solution vigorously at 0°C in an ice-water bath.
4. Recrystallize 2.23 g of tosyl chloride from chloroform with petroleum ether, and then dissolve it in 5 mL of acetonitrile.
5. Add the solution dropwise to the CD solution under magnetic stirring.
6. Readjust the pH of the CD solution to 13.
7. Allow the mixture to react for 5 minutes.
8. Adjust the pH of the mixture to 5.5 using HCl.
9. Keep the reaction mixture at ambient conditions for 1 hour under vigorous stirring.
10. Filter the mixture to obtain the white precipitate.
11. Recrystallize the precipitate from boiling water.
12. Lyophilize to obtain tosyldeoxy-β-CD.
13. Dissolve 0.2 g of PEI 25 kDa in 6 mL of DMSO.
14. Add triethylamine into the PEI solution to reach a concentration of 100 mM.
15. Dissolve 368 mg of tosyldeoxy-β-CD in 1 mL of DMSO.
16. Add the solution in Step 15 into the PEI solution.
17. Keep the reaction mixture at ambient conditions for 12 hours.
18. Remove the solvent in vacuo at elevated temperature to obtain the product.

Besides native CDs, derivatives of CDs can be adopted as linking agents. This has been demonstrated by Huang et al.'s study, which has cross-linked PEI by using (2-hydroxypropyl)-β-CD (2-hy-β-CD) and (2-hydroxypropyl)-γ-CD (2-hy-γ-CD) [58]. The generated polymers have been found to be less cytotoxic than PEI 25 kDa, with the transfection efficiency in SKOV-3 cells being around 2 and 20 times higher than that achieved by PEI 25 kDa and PEI 600 Da, respectively. Apart from hydroxyl-alkylation, β-CD has been converted into the carboxymethyl-β-CD sodium salt, which has subsequently been combined with quaternized CS to generate a DNA carrier [59]. The carrier has been found to adsorb plasmids at a polymer/DNA mass-to-mass ratio of 4:1, and has reached 40% of the efficiency achieved by liposomes in transfection [59]. All polymers discussed above, however, have not been compared with those generated with native CDs. The effects of hydroxy- and carboxy-alkylation of CDs on the performance of the polymeric carriers still have not yet been fully elucidated. More recently, methyl β-CD has also been exploited as a crosslinking agent to generate a nucleic acid carrier from lPEI [60]. The product has exhibited a low membrane-disruptive capacity ex vivo and high transfection efficiency both in the presence and absence of serum [60]. It may be further developed as a promising delivery system for genetic manipulation.

Example protocols for experimental design

The method below is an example protocol for crosslinking lPEI with methyl β-CD.

1. Dissolve 0.05 g of methyl β-CD in 1 mL of degassed DMSO.
2. Dissolve 0.05 g of CDI in 1 mL of degassed DMSO.
3. Inject the two solutions into a flask that has been filled with nitrogen.
4. Add 100 μL of triethylamine into the flask.
5. Keep the reaction mixture in darkness under magnetic stirring for 3 hours.
6. Dissolve 0.67 g of lPEI (2.5 kDa) in 1.5 mL of H_2O and 11.5 mL of DMSO.
7. Heat the PEI solution at 60°C–70°C under vigorous stirring until complete dissolution.
8. Add the PEI solution into the reaction mixture.
9. Keep the reaction mixture at 60°C for 1 day under magnetic stirring in darkness.
10. Dialyze the reaction mixture (molecular weight cut-off = 12 kDa) against doubly deionized water for 2 days.
11. Lyophilize to obtain the product.

Cyclodextrin incorporation for multifunctionality

In preceding sections, we have discussed how CDs can be used in the design and modification of polymeric carriers. As a matter of fact, CDs show great potential for use in concomitant delivery of multiple agents. The possibility of this has been demonstrated by Hu et al. [61], who have conjugated tegafur to PEI-β-CD to generate a prodrug of tegafur for drug/gene codelivery. They have observed that at an optimal polymer–DNA ratio, the conjugate has condensed plasmids into complexes at the nanoscale; however, as primary amine groups on PEI have participated in the conjugation reaction, the DNA condensation capacity, pH buffering capacity, and transfection efficiency of PEI-β-CD have been reduced after incorporation with tegafur [61]. The 5-fluoro-2′-deoxyuridine (FdUrd)-PEI-β-CD conjugate, which has been synthesized from PEI-β-CD and FdUrd, is another example of CD-based carriers for gene/drug codelivery [62]. Compared to FdUrd, this polymer has displayed higher cellular internalization efficiency, higher cytotoxic effects, and stronger anti-proliferative activities in glioma cells [62]. Despite this, before PEI-β-CD-tegafur and FdUrd-PEI-β-CD can be adopted as prodrugs, their pharmacokinetic and pharmacodynamic profiles should be studied to examine the effect of PEI-β-CD conjugation on the properties of tegafur and FdUrd. Recently, through radical copolymerization of 2-vinyl-4,6-diamino-1,3,5-triazine (VDT) with PEG methacrylated β-CD, a hydrogen bonding strengthened hydrogel has been developed [63]. The hydrogel has been successfully loaded with ibuprofen. Furthermore, its surface has been found to anchor plasmids through hydrogen bonding between the DNA base pairs and diaminotriazine, thereby enabling reverse gene transfection in COS-7 cells cultured on the gel surface. This multifunctional feature enables the hydrogel to be exploited as a tissue engineering scaffold for drug/gene codelivery.

Apart from codelivering chemical drugs with nucleic acids, CDs can be incorporated with imaging agents to enable the fate of a carrier to be tracked. One example of these agents is cyanine dyes, which possess structures based on two aromatic or heterocyclic rings linked via a polymethine chain with conjugated carbon–carbon double bonds [64]. These compounds exhibit both colorimetric and fluorescent properties. Not only can they cover all wavelengths in the visible spectrum [65], but they also display high molar absorptivity and have narrow absorption bands [65]. They therefore have emerged as promising fluorescent probes for optical imaging [66,67]. In as early as the 1990s, synthesis of cyanine-β-CD derivatives has already been reported in an attempt to generate a fluorescent labeling reagent with enhanced photostability [68]. These compounds have later been successfully used as spectroscopic probes to recognize colorless guest molecules (e.g., 1-adamantanol, and vitamin B6) [69]. More recently, synthesis of two water-soluble cyanine dye/β-CD derivatives has been performed under simultaneous ultrasound/microwave irradiation via Cu(I)-catalyzed Huisgen 1,3-dipolar cycloaddition, in which monoazido CD derivatives have been linked together by using a 1,2,3-triazole moiety [70]. The stability constants for doxorubicin complexes with cyanine/β-CD derivatives have been found to be four orders of magnitude larger than the constants reported for those with native β-CD [70]. If the derivative is incorporated as part of a CD-containing polymeric carrier, it is plausible that the resulting polymer may enable the execution of imaging during gene/drug codelivery. We have reviewed the use of CDs in molecular imaging in detail elsewhere [3]. Readers may refer to that article for reference.

Manipulation of host–guest complexation

When CDs are incorporated into the design of a carrier, sometimes this is driven by the wish of taking advantage of the ability of CDs for host–guest complexation, either for improving the targeting specificity of the system [71] or for loading chemical drugs into the carrier for gene/drug codelivery [72]. To optimize the structure of these modified carriers, which need to make use of the complexation capacity of CDs, we need to understand properly the electrostatic potentials inside the CD cavity as well as the molecular interactions during inclusion complexation. As far as the complexation thermodynamics of CDs is concerned, it is thought to be contributed by three major mechanisms: (1) penetration of the hydrophobic moiety of a guest molecule into the CD cavity [73], (2) degradation of the guest molecule [73–75], and (3) conformational alternations or strain release experienced by the CD molecule upon complexation [76,77]. Apart from this, thermodynamic quantities of the inclusion complexation process may be influenced by the buffer used, the solvation of the chemical species, and the release of water molecules from the cavity to bulk water [74,78].

In fact, though α-, β-, and γ-CD have a similar height of torus (≈7.9 Å) [79], their cavity volumes are very different. As the number of α-D-glucopyranose units comprising the CD molecule increases from 6 for α-CD to 8 for γ-CD, the cavity volume increases from 174 to 427 $Å^3$ [79]. Taking the change in the size of the hydrophobic CD cavity into consideration, the general trend in the complexation thermodynamics of CDs can be partially explained by using the size-fit concept. The validity of this concept has been shown by Varghese and coworkers [80], who have studied changes in the photophysical properties of 3-naphthyl-1-phenyl-5-(4-carboxyphenyl)-2-pyrazoline (NPCP) when NPCP has drifted into the CD cavity from bulk water. Results have revealed that an increase in the concentration of β-CD (or γ-CD) in an aqueous solution has led to an increase in the intensity of fluorescence emission from NPCP, along with a shift to the lower wavelength side [80]. This phenomenon has been partly attributed to the deep inclusion of the pyrazoline fragment into the CD cavity (Fig. 5.4), thereby restricting the rotational and vibrational motions of NPCP and hence inhibiting the occurrence of nonradiative decay processes [81]. Under the stoichiometry of 1:1, the association constant K for NPCP/γ-CD complexes (1.3×10^4 mM^{-1}) has been reported to be much higher than that of NPCP/β-CD complexes (5.8×10^3 mM^{-1}). This has shown that the binding of NPCP to the cavity of γ-CD is stronger than that to the cavity of β-CD, and has been hypothesized to be related to the size fitting between the cavity and NPCP [80]. This hypothesis has been supported by semiempirical quantum mechanics calculations, which suggest that the binding energy of NPCP/γ-CD complexation for β-CD is more positive than that for γ-CD, though the process of inclusion complexation for both β-CD and γ-CD is exothermic [80].

In fact, by understanding the complexation thermodynamics of CDs, we can better design or manipulate the process of host–guest inclusion complexation. For instance, the diameter of the cavity of α-CD (≈4.7–5.3 Å) is much smaller than that of β-CD (≈6.0–6.5 Å) [79]. Along with the knowledge that the distance of separation is an important factor modulating the induction of van der Waals forces [82], we may expect that forces induced by a straight-chain guest molecule during complexation with β-CD will be smaller (less negative $\Delta H°$) than that with α-CD. We may also anticipate that the situation will go the other way around if we now change the straight-chain guest into an adamantyl guest. Here it is worth mentioning that if the guest molecule fails to be fully accommodated by

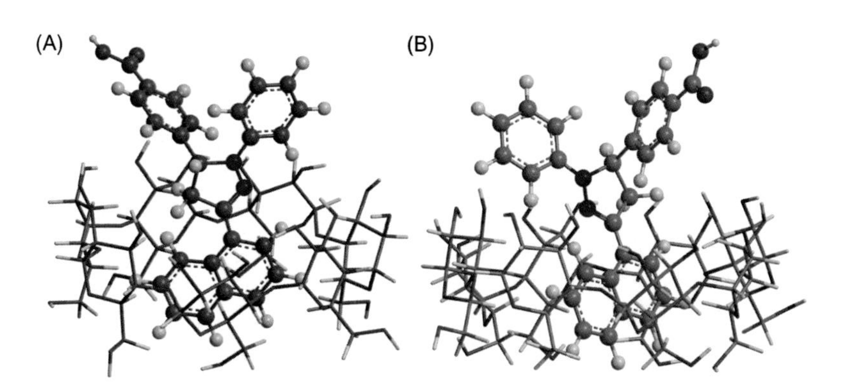

FIGURE 5.4 The most probable structure of the (A) NPCP/β-CD and (B) NPCP/γ-CD complex. The hydrogen, oxygen, carbon, and nitrogen atoms are colored in yellow, red, gray, and blue, respectively. *Reproduced from B. Varghese, S.N. Al-Busafi, F.O. Suliman, S.M.Z. Al-Kindy, Tuning the constrained photophysics of a pyrazoline dye 3-naphthyl-1-phenyl-5-(4-carboxyphenyl)-2-pyrazoline inside the cyclodextrin nanocavities: a detailed insight via experimental and theoretical approach, Spectrochim. Acta A, 173 (2017) 383–389 with permission from Elsevier B.V., [80].*

the CD cavity, the induction of van der Waals interactions can actually be affected by steric hindrance [82]. For this, it is a general trend that if the guest molecule is an acyclic compound, it may fit better into α-CD; whereas a cyclic aliphatic guest may fit better into β-CD. Yet, exceptions to this general trend exist. One example is imidazole, which possesses a five-membered ring and can fit better into the cavity of α-CD than that of β-CD [82]. In addition, if a guest molecule possesses a phenyl moiety, it usually exhibits a stronger affinity with β-CD than with α-CD, although the actual affinity of the guest molecule with CDs may be changed after the phenyl moiety has undergone substitution reactions [82].

The stability of an inclusion complex is affected not only by van der Waals forces but also by hydrogen bonding. The roles played by hydrogen bonding in host–guest inclusion complexation, however, are determined predominately by the type of functional groups present in the guest molecule. This has been suggested by a previous study, which has adopted various structurally related aromatic compounds (either with or without a phenolic hydroxyl group) as guest molecules. The study has revealed that, in general, charged and hydrophilic groups (except phenolic hydroxyl groups) present in the guest molecule stay in the bulk solution even after inclusion complexation [73]. In addition, the standard molar enthalpies and equilibrium constants for inclusion complexation between α-CD and either 1-*O*-hexyl-β-D-glucopyranoside or 1-hexanol have been observed to be basically the same, despite the difference in the types of hydrophilic groups present in those two guest molecules [83]. This observation has led to a hypothesis that only the hexyl group of these guest molecules can get into the CD cavity, from which both the glucopyranose moiety of hexyl glucopyranoside and the hydroxyl group of 1-hexanol are excluded owing to their high hydrophilicity.

Finally, the thermodynamics of host–guest chemistry is largely affected by the flexibility of a guest molecule. For example, when we see the presence of a double bond in the aliphatic chain residue of a guest molecule, we may expect that the conformational degree of freedom of that molecule will be lower than its saturated counterpart. The process of inclusion complexation with CDs will also become less favorable in the entropic term. This postulation has been supported experimentally by a previous study, which has found that the equilibrium constants of heptanoate and hexanoate for complexation with α-CD are substantially higher than those of 6-heptenoate and *trans*-3-hexenoate [75]. In addition to the flexibility of the molecular structure, the stereoisomerism of a guest molecule should be considered owing to the enantioselective nature of the inclusion complexation process [82]. All these have suggested that when CDs are incorporated into the design of a carrier in which the host–guest binding capacity is important, careful consideration of the complexation thermodynamics of CDs can help to facilitate carrier design and optimization.

Summary

Nucleic acid delivery is an expanding area of biotechnological research [84–86], partly because of its application potential to mediate genetic manipulation. Over the last several decades, while synthetic polymers have been extensively developed as carriers for nucleic acid transfer [87–92], the possible use of CDs in the design and optimization of these carriers has always been overlooked. This may be partly because native CDs fail to form stable complexes with plasmids [47], thereby being less effective in transfection as compared with other commonly used polymeric carriers, such as CS, PLGA, and PEI. In fact, CDs possess properties (e.g., forming inclusion complexes with chemical drugs for gene/drug codelivery, and modulating the cytotoxicity of other polymers) favorable for use in nucleic acid transfer. They cannot only modulate the performance of existing polymeric carriers, but can also serve as linking agents for the development of new polymers. CDs are, therefore, promising tools to enhance the efficiency of existing polymer-based nucleic acid delivery technologies for genetic manipulation in the future.

Directions for intervention development

CDs can modulate the properties of polymeric carriers, and can also function as linking agents during polymer fabrication. To exploit the possible benefits brought by CDs for genetic manipulation, the following steps may be taken during the design of a polymeric carrier.

1. Exploit the features of CDs and their relevance to the proposed intervention.
2. Consider how those features can facilitate the fabrication of the designed polymer or can improve the efficiency of the intervention.
3. Plan the procedure for the synthesis of the polymer by taking CDs as one of the components.
4. Characterize the structure of the product.
5. Evaluate the performance of the product.
6. Optimize the structure and properties of the product, if necessary, to maximize the efficiency of the proposed intervention.

References

[1] E.M.M. Del Valle, Cyclodextrins and their uses: a review, Process Biochem. 39 (2004) 1033–1046.

[2] W.F. Lai, Cyclodextrins in non-viral gene delivery, Biomaterials 35 (2014) 401–411.

[3] W.F. Lai, A.L. Rogach, W.T. Wong, Chemistry and engineering of cyclodextrins for molecular imaging, Chem. Soc. Rev. 46 (2017) 6379–6419.

[4] T. Loftsson, D. Duchene, Cyclodextrins and their pharmaceutical applications, Int. J. Pharmaceut. 329 (2007) 1–11.

[5] A. Villiers, Sur la fermentation de la fécule par l'action du ferment butyrique, Compte Rendu de l Acad'emie de Sciences 112 (1891) 536–538.

[6] F. Cramer, Einschlussverbindungen, Springer-Verlag, Berlin, 1954.

[7] H. He, S. Chen, J. Zhou, Y. Dou, L. Song, L. Che, et al., Cyclodextrin-derived pH-responsive nanoparticles for delivery of paclitaxel, Biomaterials 34 (2013) 5344–5358.

[8] S. Lepretre, F. Chai, J.C. Hornez, G. Vermet, C. Neut, M. Descamps, et al., Prolonged local antibiotics delivery from hydroxyapatite functionalised with cyclodextrin polymers, Biomaterials 30 (2009) 6086–6093.

[9] G. Fundueanu, M. Constantin, A. Dalpiaz, F. Bortolotti, R. Cortesi, P. Ascenzi, et al., Preparation and characterization of starch/cyclodextrin bioadhesive microspheres as platform for nasal administration of Gabexate Mesylate (Foy) in allergic rhinitis treatment, Biomaterials 25 (2004) 159–170.

[10] F. Quaglia, L. Ostacolo, A. Mazzaglia, V. Villari, D. Zaccaria, M. T. Sciortino, The intracellular effects of non-ionic amphiphilic cyclodextrin nanoparticles in the delivery of anticancer drugs, Biomaterials 30 (2009) 374–382.

[11] T.R. Thatiparti, A.J. Shoffstall, H.A. von Recum, Cyclodextrin-based device coatings for affinity-based release of antibiotics, Biomaterials 31 (2010) 2335–2347.

[12] Y.Y. Liu, X.D. Fan, Synthesis, properties and controlled release behaviors of hydrogel networks using cyclodextrin as pendant groups, Biomaterials 26 (2005) 6367–6374.

[13] I. Bechet, P. Paques, M. Fillet, P. Hubert, J. Crommen, Chiral separation of basic drugs by capillary zone electrophoresis with cyclodextrin additives, Electrophoresis 15 (1994) 818–823.

[14] J.W. Fredy, J. Scelle, G. Ramniceanu, B.T. Doan, C.S. Bonnet, E. Toth, et al., Mechanostereoselective one-pot synthesis of functionalized head-to-head cyclodextrin [3] rotaxanes and their application as magnetic resonance imaging contrast agents, Org. Lett. 19 (2017) 1136–1139.

[15] T. Loftsson, M.E. Brewster, M. Másson, Role of cyclodextrins in improving oral drug delivery, Am. J. Drug Deliv. 2 (2004) 261–275.

[16] J. Szejtli, Past, present and future of cyclodextrin research, Pure Appl. Chem. 76 (2004) 1825–1845.

[17] E. Redenti, C. Pietra, A. Gerloczy, L. Szente, Cyclodextrins in oligonucleotide delivery, Adv. Drug Deliv. Rev. 53 (2001) 235–244.

[18] C.R. Dass, Vehicles for oligonucleotide delivery to tumours, J. Pharm. Pharmacol. 54 (2002) 3–27.

[19] I. Habus, Q. Zhao, S. Agrawal, Synthesis, hybridization properties, nuclease stability, and cellular uptake of the oligonucleotide–amino-beta-cyclodextrins and adamantane conjugates, Bioconjugate Chem. 6 (1995) 327–331.

[20] Q. Zhao, J. Temsamani, S. Agrawal, Use of cyclodextrin and its derivatives as carriers for oligonucleotide delivery, Antisense Res. Dev. 5 (1995) 185–192.

[21] S. Abdou, J. Collomb, F. Sallas, A. Marsura, C. Finance, Beta-cyclodextrin derivatives as carriers to enhance the antiviral activity of an antisense oligonucleotide directed toward a coronavirus intergenic consensus sequence, Arch. Virol. 142 (1997) 1585–1602.

[22] M.E. Davis, The first targeted delivery of siRNA in humans via a self-assembling, cyclodextrin polymer-based nanoparticle: from concept to clinic, Mol. Pharmaceut. 6 (2009) 659–668.

[23] M.E. Davis, M.E. Brewster, Cyclodextrin-based pharmaceutics: past, present and future, Nat. Rev. Drug Discov. 3 (2004) 1023–1035.

[24] M.L. Forrest, N. Gabrielson, D.W. Pack, Cyclodextrin-polyethylenimine conjugates for targeted in vitro gene delivery, Biotechnol. Bioeng. 89 (2005) 416–423.

[25] S.H. Pun, N.C. Bellocq, A. Liu, G. Jensen, T. Machemer, E. Quijano, et al., Cyclodextrin-modified polyethylenimine polymers for gene delivery, Bioconjugate Chem. 15 (2004) 831–840.

[26] H. Gonzalez, S.J. Hwang, M.E. Davis, New class of polymers for the delivery of macromolecular therapeutics, Bioconjugate Chem. 10 (1999) 1068–1074.

[27] D.J. Freeman, R.W. Niven, The influence of sodium glycocholate and other additives on the in vivo transfection of plasmid DNA in the lungs, Pharm. Res. 13 (1996) 202–209.

[28] C. Formoso, The interaction of β-cyclodextrin with nucleic acid monomer units, Biochem. Biophys. Res. Commun. 50 (1973) 999–1005.

[29] S.J. Hwang, N.C. Bellocq, M.E. Davis, Effects of structure of beta-cyclodextrin-containing polymers on gene delivery, Bioconjugate Chem. 12 (2001) 280–290.

[30] M.A. Croyle, B.J. Roessler, C.P. Hsu, R. Sun, G.L. Amidon, Beta-cyclodextrins enhance adenoviral-mediated gene delivery to the intestine, Pharm. Res. 15 (1998) 1348–1355.

[31] M.P. Vaccher, J.P. Bonte, C. Vaccher, Capillary electrophoretic resolution of enantiomers of aromatic amino-acids with highly sulfated alpha-, beta- and gamma-cyclodextrins, Chromatographia 64 (2006) 51–55.

[32] F. Bellia, D. La Mendola, C. Pedone, E. Rizzarelli, M. Saviano, G. Vecchio, Selectively functionalized cyclodextrins and their metal complexes, Chem. Soc. Rev. 38 (2009) 2756–2781.

[33] A.R. Khan, P. Forgo, K.J. Stine, V.T. D'Souza, Methods for selective modifications of cyclodextrins, Chem. Rev. 98 (1998) 1977–1996.

[34] W. Tang, S.C. Ng, Facile synthesis of mono-6-amino-6-deoxy-alpha-, beta-, gamma-cyclodextrin hydrochlorides for molecular recognition, chiral separation and drug delivery, Nat. Protoc. 3 (2008) 691–697.

[35] K. Takahashi, K. Hattori, F. Toda, Monotosylated alpha-cyclodextrin and beta-cyclodextrin prepared in an alkaline aqueous-solution, Tetrahedron Lett. 25 (1984) 3331–3334.

[36] C.F. Potter, N.R. Russell, M. McNamara, Spectroscopic characterisation of metallo-cyclodextrins for potential chiral separation of amino acids and L/D-DOPA, J. Inclusion Phenom. Macrocycl. Chem. 56 (2006) 395–403.

[37] I. Baussanne, J.M. Benito, C.O. Mellet, J.M.G. Fernandez, H. Law, J. Defaye, Synthesis and comparative lectin-binding affinity of mannosyl-coated beta-cyclodextrin-dendrimer constructs, Chem. Commun. (2000) 1489–1490. Available from: https://doi.org/10.1039/b003765f.

[38] A. Ueno, R. Breslow, Selective sulfonation of a secondary hydroxyl group of beta-cyclodextrin, Tetrahedron Lett. 23 (1982) 3451–3454.

[39] K. Teranishi, Practical and convenient modifications of the A,C-secondary hydroxyl face of cyclodextrins, Tetrahedron 59 (2003) 2519–2538.

[40] K. Teranishi, Regioselective 2A-2D-disulfonylations of cyclodextrins for practical bifunctionalization on the secondary hydroxyl face, J. Inclusion Phenom. Macrocycl. Chem. 44 (2003) 313–316.
[41] A.R. Khan, L. Barton, V.T. Dsouza, Heptakis-2,3-epoxy-beta-cyclodextrin, a key intermediate in the synthesis of custom designed cyclodextrins, J. Chem. Soc. Chem. Commun. (1992) 1112–1114. Available from: https://doi.org/10.1039/c39920001112.
[42] H. Ikeda, Q. Li, A. Ueno, Chiral recognition by fluorescent chemosensors based on N-dansyl-amino acid-modified cyclodextrins, Bioorg. Med. Chem. Lett. 16 (2006) 5420–5423.
[43] C. Donze, E. Rizzarelli, G. Vecchio, Synthesis and intramolecular inclusion studies of tryptophan-modified-beta-cyclodextrins, J. Inclus. Phenom. Mol. 31 (1998) 27–41.
[44] C. Pean, C. Creminon, A. Wijkhuisen, J. Grassi, P. Guenot, P. Jehan, et al., Synthesis and characterization of peptidyl-cyclodextrins dedicated to drug targeting, J. Chem. Soc. Perk T 2 4 (2000) 853–863.
[45] G. Vecchio, D. La Mendola, E. Rizzarelli, The synthesis and conformation of β-cyclodextrins functionalized with enantiomers of Boc-carnosine, J. Supramol. Chem. 1 (2001) 87–95.
[46] H. Yamamura, S. Yamada, K. Kohno, N. Okuda, S. Araki, K. Kobayashi, et al., Preparation and guest binding of novel β-cyclodextrin dimers linked with various sulfur-containing linker moieties, J. Chem. Soc. Perk T 1 (1999) 2943–2948. Available from: https://doi.org/10.1039/a905067a.
[47] S.A. Cryan, A. Holohan, R. Donohue, R. Darcy, C.M. O'Driscoll, Cell transfection with polycationic cyclodextrin vectors, Eur. J. Pharm. Sci. 21 (2004) 625–633.
[48] F. Ortega-Caballero, C.O. Mellet, L. Le Gourrierec, N. Guilloteau, C. Di Giorgio, P. Vierling, et al., Tailoring beta-cyclodextrin for DNA complexation and delivery by homogeneous functionalization at the secondary face, Org. Lett. 10 (2008) 5143–5146.
[49] A. Diaz-Moscoso, P. Balbuena, M. Gomez-Garcia, C. Ortiz Mellet, J.M. Benito, L. Le Gourrierec, et al., Rational design of cationic cyclooligosaccharides as efficient gene delivery systems, Chem. Commun. (2008) 2001–2003. Available from: https://doi.org/10.1039/b718672j.
[50] A. Diaz-Moscoso, L. Le Gourrierec, M. Gomez-Garcia, J.M. Benito, P. Balbuena, F. Ortega-Caballero, et al., Polycationic amphiphilic cyclodextrins for gene delivery: synthesis and effect of structural modifications on plasmid DNA complex stability, cytotoxicity, and gene expression, Chemistry 15 (2009) 12871–12888.
[51] A. Mendez-Ardoy, M. Gomez-Garcia, C. Ortiz Mellet, N. Sevillano, M.D. Giron, R. Salto, et al., Preorganized macromolecular gene delivery systems: amphiphilic beta-cyclodextrin "click clusters", Org. Biomol. Chem. 7 (2009) 2681–2684.
[52] J.M. García Fernández, J.M. Benito, C.O. Mellet, Cyclodextrin-scaffolded glycotransporters for gene delivery, Pure Appl. Chem. 85 (2013) (in press).
[53] C. Yang, X. Wang, H. Li, S.H. Goh, J. Li, Synthesis and characterization of polyrotaxanes consisting of cationic alpha-cyclodextrins threaded on poly[(ethylene oxide)-ran-(propylene oxide)] as gene carriers, Biomacromolecules 8 (2007) 3365–3374.
[54] H. Arima, F. Kihara, F. Hirayama, K. Uekama, Enhancement of gene expression by polyamidoamine dendrimer conjugates with alpha-, beta-, and gamma-cyclodextrins, Bioconjugate Chem. 12 (2001) 476–484.
[55] F. Kihara, H. Arima, T. Tsutsumi, F. Hirayama, K. Uekama, Effects of structure of polyamidoamine dendrimer on gene transfer efficiency of the dendrimer conjugate with alpha-cyclodextrin, Bioconjugate Chem. 13 (2002) 1211–1219.
[56] F. Kihara, H. Arima, T. Tsutsumi, F. Hirayama, K. Uekama, In vitro and in vivo gene transfer by an optimized alpha-cyclodextrin conjugate with polyamidoamine dendrimer, Bioconjugate Chem. 14 (2003) 342–350.
[57] M.E. Davis, S.H. Pun, N.C. Bellocq, T.M. Reineke, S.R. Popielarski, S. Mishra, et al., Self-assembling nucleic acid delivery vehicles via linear, water-soluble, cyclodextrin-containing polymers, Curr. Med. Chem. 11 (2004) 179–197.
[58] H. Huang, G. Tang, Q. Wang, D. Li, F. Shen, J. Zhou, et al., Two novel non-viral gene delivery vectors: low molecular weight polyethylenimine cross-linked by (2-hydroxypropyl)-beta-cyclodextrin or (2-hydroxypropyl)-gamma-cyclodextrin, Chem. Commun. (2006) 2382–2384. Available from: https://doi.org/10.1039/b601130f.
[59] L.L. Ren, Y. Wu, D. Han, L.D. Zhao, Q.M. Sun, W.W. Guo, et al., Math1 gene transfer based on the delivery system of quaternized chitosan/Na-carboxymethyl-beta-cyclodextrin nanoparticles, J. Nanosci. Nanotechnol. 10 (2010) 7262–7265.
[60] W.F. Lai, D.W. Green, H.S. Jung, Linear poly(ethylenimine) cross-linked by methyl-beta-cyclodextrin for gene delivery, Curr. Gene. Ther. 14 (2014) 258–268.
[61] Q.D. Hu, H. Fan, W.J. Lou, Q.Q. Wang, G.P. Tang, Polyethylenimine-cyclodextrin-tegafur conjugate shows anticancer activity and a potential for gene delivery, J. Zhejiang Univ. Sci. B 12 (2011) 720–729.
[62] X. Lu, Y. Ping, F.J. Xu, Z.H. Li, Q.Q. Wang, J.H. Chen, et al., Bifunctional conjugates comprising beta-cyclodextrin, polyethylenimine, and 5-fluoro-2′- deoxyuridine for drug delivery and gene transfer, Bioconjugate Chem. 21 (2010) 1855–1863.
[63] X. Hu, N. Wang, L. Liu, W. Liu, Cyclodextrin-cross-linked diaminotriazine-based hydrogen bonding strengthened hydrogels for drug and reverse gene delivery, J. Biomater. Sci. Polym. Dd. (2013). Available from: https://doi.org/10.1080/09205063.2013.808150.
[64] M. Panigrahi, S. Dash, S. Patel, B.K. Mishra, Syntheses of cyanines: a review, Tetrahedron 68 (2012) 781–805.
[65] M. Henary, A. Levitz, Synthesis and applications of unsymmetrical carbocyanine dyes, Dyes Pigments 99 (2013) 1107–1116.
[66] T. Behnke, J.E. Mathejczyk, R. Brehm, C. Wurth, F.R. Gomes, C. Dullin, et al., Target-specific nanoparticles containing a broad band emissive NIR dye for the sensitive detection and characterization of tumor development, Biomaterials 34 (2013) 160–170.
[67] C.S. Huang, S. George, M. Lu, V. Chaudhery, R.M. Tan, R.C. Zangar, et al., Application of photonic crystal enhanced fluorescence to cancer biomarker microarrays, Anal. Chem. 83 (2011) 1425–1430.
[68] R. Guether, M.V. Reddington, Photostable cyanine dye beta-cyclodextrin conjugates, Tetrahedron Lett. 38 (1997) 6167–6170.
[69] J.L. Zhao, Y. Lv, H.J. Ren, W. Sun, Q. Liu, Y.L. Fu, et al., Synthesis, spectral properties of cyanine dyes-beta-cyclodextrin and their application as the supramolecular host with spectroscopic probe, Dyes Pigments 96 (2013) 180–188.
[70] T. Carmona, G. Marcelo, L. Rinaldi, K. Martina, G. Cravotto, F. Mendicuti, Soluble cyanine dye/beta-cyclodextrin derivatives: Potential carriers for drug delivery and optical imaging, Dyes Pigments 114 (2015) 204–214.
[71] C. Ma, T. Bian, S. Yang, C. Liu, T. Zhang, J. Yang, et al., Fabrication of versatile cyclodextrin-functionalized upconversion

luminescence nanoplatform for biomedical imaging, Anal. Chem. 86 (2014) 6508–6515.

[72] Y.H. Zhou, C.L. Wang, F. Wang, C.Z. Li, C. Dong, S.M. Shuang, Beta-cyclodextrin and its derivatives functionalized magnetic nanoparticles for targeting delivery of curcumin and cell imaging, Chin. J. Chem. 34 (2016) 599–608.

[73] P.D. Ross, M.V. Rekharsky, Thermodynamics of hydrogen bond and hydrophobic interactions in cyclodextrin complexes, Biophys. J. 71 (1996) 2144–2154.

[74] D. Hallen, A. Schon, I. Shehatta, I. Wadso, Microcalorimetric titration of alpha-cyclodextrin with some straight-chain alkan-1-Ols at 288.15-K, 298.15-K and 308.15-K, J. Chem. Soc. Faraday T 88 (1992) 2859–2863.

[75] M.V. Rekharsky, M.P. Mayhew, R.N. Goldberg, P.D. Ross, Y. Yamashoji, Y. Inoue, Thermodynamic and nuclear magnetic resonance study of the reactions of alpha- and beta-cyclodextrin with acids, aliphatic amines, and cyclic alcohols, J. Phys. Chem. B 101 (1997) 87–100.

[76] Y. Matsui, K. Mochida, Binding forces contributing to the association of cyclodextrin with alcohol in an aqueous-solution, B Chem. Soc. Jpn. 52 (1979) 2808–2814.

[77] W. Saenger, Cyclodextrin inclusion-compounds in research and industry, Angew. Chem. Int. Edit. 19 (1980) 344–362.

[78] G. Barone, G. Castronuovo, P. Delvecchio, V. Elia, M. Muscetta, Thermodynamics of formation of inclusion-compounds in water-alpha-cyclodextrin alcohol adducts at 298.15-K, J. Chem. Soc. Farad. T 1 82 (1986) 2089–2101.

[79] W.J. Shieh, A.R. Hedges, Properties and applications of cyclodextrins, J. Macromol. Sci. A A33 (1996) 673–683.

[80] B. Varghese, S.N. Al-Busafi, F.O. Suliman, S.M.Z. Al-Kindy, Tuning the constrained photophysics of a pyrazoline dye 3-naphthyl-1-phenyl-5-(4-carboxyphenyl)-2-pyrazoline inside the cyclodextrin nanocavities: a detailed insight via experimental and theoretical approach, Spectrochim. Acta A 173 (2017) 383–389.

[81] A. Samanta, N. Guchhait, S.C. Bhattacharya, Preferential molecular encapsulation of an ICT fluorescence probe in the suprannolecular cage of cucurbit[7]uril and beta-cyclodextrin: an experimental and theoretical approach, J. Phys. Chem. B 118 (2014) 13279–13289.

[82] M.V. Rekharsky, Y. Inoue, Complexation thermodynamics of cyclodextrins, Chem. Rev. 98 (1998) 1875–1918.

[83] M.V. Rekharsky, R.N. Goldberg, F.P. Schwarz, Y.B. Tewari, P.D. Ross, Y. Yamashoji, et al., Thermodynamic and nuclear-magnetic-resonance study of the interactions of alpha-cyclodextrin and beta-cyclodextrin with model substances - phenethylamine, ephedrines, and related substances, J. Am. Chem. Soc. 117 (1995) 8830–8840.

[84] W.F. Lai, Nucleic acid delivery: roles in biogerontological interventions, Ageing Res. Rev. 12 (2013) 310–315.

[85] W.F. Lai, Protein kinases as targets for interventive biogerontology: overview and perspectives, Exp. Gerontol. 47 (2012) 290–294.

[86] W.F. Lai, Nucleic acid therapy for lifespan prolongation: present and future, J. Biosci. 36 (2011) 725–729.

[87] E.A. Klausner, Z. Zhang, S.P. Wong, R.L. Chapman, M.V. Volin, R.P. Harbottle, Corneal gene delivery: chitosan oligomer as a carrier of CpG rich, CpG free or S/MAR plasmid DNA, J. Gene Med. 14 (2012) 100–108.

[88] W. Cheng, C. Yang, J.L. Hedrick, D.F. Williams, Y.Y. Yang, P. G. Ashton-Rickardt, Delivery of a granzyme B inhibitor gene using carbamate-mannose modified PEI protects against cytotoxic lymphocyte killing, Biomaterials 34 (2013) 3697–3705.

[89] B. Newland, M. Abu-Rub, M. Naughton, Y. Zheng, A.V. Pinoncely, E. Collin, et al., GDNF gene delivery via a 2-(dimethylamino)ethyl methacrylate based cyclized knot polymer for neuronal cell applications, ACS Chem. Neurosci. 4 (2013) 540–546.

[90] H.Y. Nam, K. Nam, M. Lee, S.W. Kim, D.A. Bull, Dendrimer type bio-reducible polymer for efficient gene delivery, J. Control Release 160 (2012) 592–600.

[91] W.F. Lai, In vivo nucleic acid delivery with PEI and its derivatives: current status and perspectives, Expert Rev. Med. Devic. 8 (2011) 173–185.

[92] W.F. Lai, M.C.M. Lin, Synthesis and properties of chitosan-PEI graft copolymers as vectors for nucleic acid delivery, J. Mater. Sci. Eng. 4 (2010) 34–41.

Chapter 6

Design of upconversion nanoparticles for intervention execution

Introduction

Rare earth elements comprise yttrium, scandium, and the 15 elements in the lanthanide series. Due to the presence of the $4f^n$ inner shell that enables internal 4f- or 4f–5d transitions [1,2], lanthanide ions (except erbium and ytterbium ions) exhibit unique luminescent properties. Owing to the long-life and real ladder-like energy levels of lanthanide ions embedded in the inorganic matrix host, the energy of the incident photons can be lower than the energy of the outcome photons. In reality, such an upconversion phenomenon, which is a nonlinear anti-Stokes process, can be achieved by three major mechanisms: (1) excited-state absorption (ESA), (2) energy transfer upconversion (ETU), and (3) photon avalanche (PA). PA is seldom observed in lanthanide materials at the nanoscale. Its importance to the development of nucleic acid carriers, therefore, is less significant. On the other hand, ETU is one of the most commonly used mechanisms to attain high upconversion efficiency in practice. It can be achieved in various ways, including energy transfer followed by ESA, successive energy transfers, cross-relaxation upconversion, cooperative sensitization, and cooperative luminescence [3]. Their general energy schemes are shown in Fig. 6.1.

With the foundation established by the preceding two chapters on polymer-mediated delivery, this chapter will touch upon the use of the designed polymers in engineering upconversion nanoparticles (UCNPs) to generate multifunctional nucleic acid carriers. Here it is worth noting that the objective of this chapter is to introduce the major principles of designing UCNPs for the execution of biogerontological interventions. The physics as well as the detailed mechanism of photoluminescence (PL) emission of UCNPs are beyond the scope of this chapter. These topics have been reviewed elsewhere [4–7] where readers can refer to those articles for details.

Major designing principles

Compared to delivering small-molecule compounds, which can still mediate the execution of the intervention even if the carrier fails to be internalized into the cell but simply releases the payload outside, delivery of nucleic acids is much more difficult because genetic manipulation can be performed only if the nucleic acids can properly undergo cellular internalization [8–10]. Because of this, the first two factors to be considered when UCNPs are designed for therapeutics delivery are the size and zeta potential of the nanoparticles. These factors determine the efficiency of cellular uptake of the carrier and may be altered after administration of the nanoparticles into the body. For instance, after nanoparticles have been administered into a body, the salt ions in the blood may complex with the exposed lanthanide ions on the UCNP surface, causing the nanoparticles to aggregate. The possible occurrence of this phenomenon should be evaluated when the nanoparticles are adopted for intervention development. The biodegradability, biocompatibility, and toxicity of the nanoparticles should also be examined before use [11–14].

To optimize the properties of UCNPs to favor cellular uptake and nucleic acid delivery, most of the strategies that have been introduced in Chapter 4 for structural modification of polymeric vectors can apply. For instance, an earlier study has documented that, in the SK-BR-3 cells that have overexpressed the Her2 receptor, UCNPs, whose surface has been modified with folic acid and anti-HER2 antibodies, have displayed higher transfection efficiency and gene silencing efficiency than the unmodified counterparts [15]. Furthermore, to increase the efficiency in subsequent loading of negatively charged agents (such as nucleic acids for genetic manipulation) and to facilitate the binding of UCNPs to anionic plasma membranes,

Delivery of Therapeutics for Biogerontological Interventions. DOI: https://doi.org/10.1016/B978-0-12-816485-3.00006-4

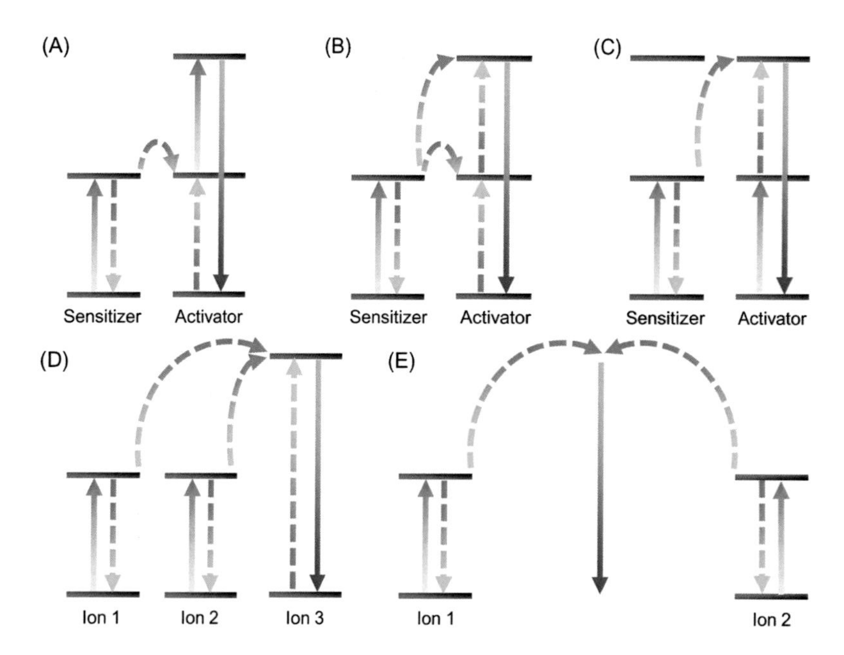

FIGURE 6.1 General energy schemes showing various ETU processes. *Green, orange,* and *violet arrows* represent the excitation light, energy transfer, and upconversion emission, respectively. (A) Energy transfer followed by excited-state absorption; (B) successive energy transfers; (C) cross-relaxation upconversion; (D) cooperative sensitization; and (E) cooperative luminescence. *Reproduced from Lai, W.F., Rogach, A.L., Wong, W.T., Molecular design of upconversion nanoparticles for gene delivery, Chem. Sci. 8 (2017) 7339–7358 with permission from the Royal Society of Chemistry [3].*

the zeta potential of UCNPs can be manipulated to be charged positively by incorporating polycations into the nanoparticle surface.

Finally, as cellular internalization is often mediated by endocytosis, the efficiency of UCNPs as carriers is partly affected by the capacity of the nanoparticles to elicit endo-lysosomal escape. This capacity can be estimated by using the pH buffering capacity assay. In addition, in order for the delivery process to have therapeutic effects, the released agent should be able to reach an appropriate site for action. In general, plasmids have to reach the nucleus; whereas RNA molecules have to be released in the cytoplasm. Failure to have the payload to be released at an appropriate intracellular location may lead to poor therapeutic outcomes [9]. To increase the spatial precision of the release process, careful manipulation of the UCNPs, especially the surface properties and the buffering capacity of the polymer coating, is particularly important.

Example protocols for experimental design

The method below is an example protocol for evaluating the pH buffering capacity of a carrier:

1. Dissolve the carrier in 30 mL of an NaCl solution (150 mM) to reach a concentration of 0.2 mg/mL.
2. Use an NaOH solution (0.1 M) to adjust the pH of the sample solution to 10.

(Continued)

(Continued)

3. Add 200 μL of an HCl solution (0.1 M) to the sample solution.
4. Measure the pH of the sample solution using a microprocessor pH meter.
5. Repeat Steps 3 and 4 until the change in pH is no longer significant even after the addition of the HCl solution.
6. Obtain a titration curve by plotting the pH value of the sample solution against the volume of the HCl solution totally added.
7. Compare the curve with that of a polymer (e.g., PEI 25 kDa) that is known to show a good pH buffering capacity.

Fabrication and morphological manipulation of upconversion nanoparticles

UCNPs can be fabricated by using various strategies, including the Ostwald-ripening method, thermal decomposition, and hydro(solvo)thermal synthesis. These strategies have been extensively reviewed elsewhere [7]. During fabrication, there are multiple ways we may use to control the morphology of the UCNPs generated. One of the easiest methods is to modulate the experimental conditions during nanoparticle fabrication. This is demonstrated by the case of $NaLuF_4$ microcrystals synthesized

FIGURE 6.2 Scanning electron micrographs of NaLuF4:Yb,Er microcrystals with different morphologies: (A) microspheres; (B) regular microplates; (C) mussel-like grains; (D) superstructures; (E) regular microtubes; and (F) long microtubes. The inset is the scanning electron micrograph with a higher magnification. *Reproduced from Li, W.B., Tan, C.B., Zhang, Y.T., Simultaneous phase and shape control of monodisperse NaLuF4:Yb, Er microcrystals and greatly enhanced upconversion luminescence from their superstructures, Opt. Commun. 295 (2013) 140–144 with permission from Elsevier B.V. [16].*

via the hydrothermal route [16]. A typical synthetic approach is first to add $Lu(NO_3)_3$, $Yb(NO_3)_3$, and Er $(NO_3)_3$ into an aqueous solution containing trisodium citrate [16]. After that, an aqueous solution containing NaF is added, followed by pH adjustment. Finally, the reaction proceeds in a hydrothermal autoclave reactor. The $NaLuF_4$:Yb,Er microcrystals fabricated are finally retrieved, washed, and dried [16]. Microcrystals with different morphologies (e.g., microspheres, microplates, mussel-like grains, and microtubes) have been reported to be obtained possibly by manipulating the citrate/Ln molar ratio, the pH value of the precursor solution, and the hydrothermal time (Fig. 6.2) [16]. A similar approach of modifying the morphology of UCNPs has been reported in the case of $BaYF_5$:Yb^{3+}/Er^{3+} UCNPs [17], which have been synthesized by first mixing $RE(NO_3)_3$ (RE = Er, Yb, and Y), $Ba(NO_3)_2$, NH_4HF_2, NaOH, ethanol, and oleic acid (OA) in deionized water, followed by a reaction in a hydrothermal autoclave reactor. Results have shown that the regularity of the rectangular shape of the nanoparticles has been improved by increasing the reaction temperature from 180°C to 220°C [17]. This can be explained by the occurrence of the Ostwald-ripening process during the formation of the UCNPs [17], in which irregular nanoparticles may be dissolved and absorbed by those that are larger. Apart from changing the reaction temperature, adjusting the amount of NaOH added can vary the morphology of the nanoparticles, changing the nanoparticles from irregular aggregated nanoparticles to quasi-spherical, rectangular, and nanosheet morphologies [17] (Fig. 6.3). This evolution of nanoparticle morphologies has been thought to be linked with the selective adhesion of OA on different parts of the nanoparticle surface, as well as the change in the concentration of monomers formed by ion exchange with sodium oleate in the growth medium [18].

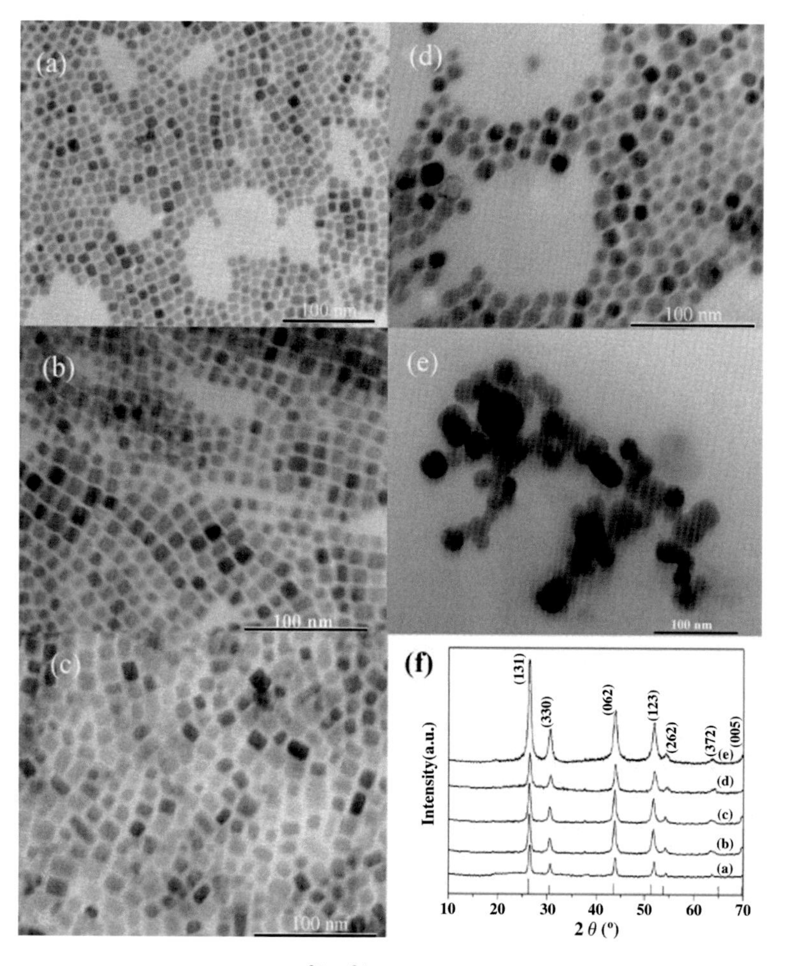

FIGURE 6.3 Transmission electron micrographs of $BaYF_5:Yb^{3+}/Er^{3+}$ UCNPs synthesized with different morphologies: (a) rectangular particles; (b) nanosheets; (c) elongated rectangular particles; (d) quasi-spheres, and (e) aggregated particles. The amounts of NaOH added during synthesis of nanoparticles (a)–(e) are 0.6, 0.9, 1.2, 0.3, and 0 g, respectively. (f) X-ray diffraction patterns of the UCNPs as shown in (a)–(e). *Reproduced from Sun, J.B., Xian, J.Y., Zhang, X.Y., Du, H.Y., Size and shape controllable synthesis of oil-dispersible $BaYF_5:Yb^{3+}/Er^{3+}$ upconversion fluorescent nanocrystals, J. Alloys Compd. 509 (2011) 2348–2354 with permission from Elsevier B.V. [17].*

Example protocols for experimental design

The method below is an example protocol for generating $NaYF_4:Yb^{3+},Er^{3+}$ UCNPs with a surface coating of PEI.

1. Dissolve ytterbium(III) chloride hexahydrate (0.216 mmol) and erbium trinitrate pentahydrate (0.024 mmol) in 80 mL of ethylene glycol.
2. Add yttrium nitrate hexahydrate (1.92 mmol), PEI 25 kDa (2.2 g), and $NaNO_3$ (0.408 g, 4.8 mmol) into the reaction mixture.
3. Heat the mixture to 80°C for 10 minutes under vigorous stirring.
4. Add an ethylene glycol solution of NH_4F (0.07 g/mL) to the mixture dropwise.
5. Incubate the mixture at 80°C for 10 minutes.
6. Transfer the reaction mixture to a Teflon-lined high-pressure stainless steel autoclave reactor.
7. Put the reactor into an oven at 180°C and incubate for 3 hours.

(Continued)

(Continued)

8. Let the reactor cool to room temperature.
9. Collect the products by centrifugation (10 minutes at 10,000 rpm).
10. Wash the pellet with distilled water.
11. Collect the products by centrifugation at 14,000 rpm for 20 minutes.
12. Wash the pellet with 95% ethanol.
13. Collect the products by centrifugation at 14,000 rpm for 20 minutes.
14. Repeat Steps 10–13 for two more times.
15. Dry the pellet in a vacuum oven at 80°C overnight to obtain the UCNPs.

Another strategy to change the morphology of UCNPs is to change the capping agent. This is documented by Zhou et al. [19], who have synthesized $NaLuF_4$:Yb,Er UCNPs in an OA-ionic liquids (OA-ILs) dual-phase

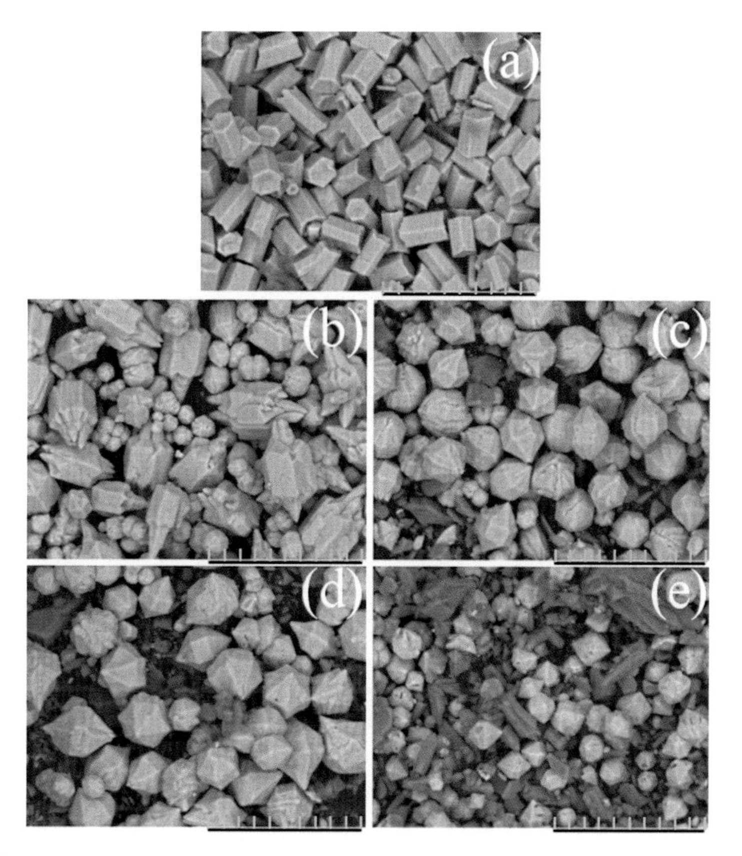

FIGURE 6.4 Scanning electron micrographs of the upconversion microcrystals codoped with Sc^{3+}. The nominal molar contents of Sc^{3+} ions in samples (a–e) are 0%, 18%, 38%, 58%, and 78%, respectively. Crystals having the monoclinic Na_3ScF_6 and/or hexagonal $NaScF_4$ crystal structures appear gray-black, whereas those having the hexagonal $NaYF_4$ crystal structure appear gray-white in the scanning electron micrographs. The scaling bar is 30 μm. *Reproduced from He, E.J., Chen, S.F., Zhang, M.L., Simultaneous morphology evolution and upconversion emission tuning of single Y-based fluoride microcrystal induced by Sc^{3+} co-doping, Mater. Res. Bull. 87 (2017) 61–71 with permission from Elsevier B.V. [22].*

reaction system and have improved the disparity of the nanoparticles via ligand exchange functionalization. In total, four surfactants have been tested, including sodium dodecyl sulfate (SDS), dodecyl dimethyl benzyl ammonium chloride (DDBAC), PEG, and sodium citrate. Those capped with sodium citrate are almost spherical, whereas capping the nanocrystals with SDS, DDBAC, and PEG can generate nanorods [19]. Except the PEG-capped UCNPs, which are at the microscale, all other nanoparticles are at the nanoscale [19]. Compared to those capped with SDS, DDBAC-capped particles have been found to be stockier [19]. The underlying principle of this capping agent-mediated modulation of particle morphology is that the viscosity of the interface of the dual-phase system, and hence the kinetics of particle formation, is affected by the capping agents adopted [19]. Furthermore, during the formation of $NaLuF_4$:Yb,Er UCNPs in the OA-ILs two-phase system, F^- anions have to compete with the surfactant chelators, and have to combine with Ln^{3+} ions for the formation of the $NaLuF_4$ lattice. The efficiency of this process, as well as the formation of UCNPs, is determined by the chelate constant (lg β), which varies among different capping agents [19].

Apart from the aforementioned, morphological manipulation of lanthanide-doped microcrystals can be achieved via ion codoping. The success of this has been suggested in previous studies, in which morphological evolution in β-$NaYF_4$:Yb^{3+},Er^{3+} microcrystals has been induced by K^+ and Cr^{3+} codoping for the enhancement of upconversion luminescence (UCL) [20,21]. More recently, this approach has been adopted to induce morphological evolution in upconversion microcrystals via Sc^{3+} codoping [22], which can vary the anisotropic growth of different crystal planes. For instance, without the use of Sc^{3+} codoping, the microcrystals fabricated are mostly in the form of short hexagonal prisms, with only a few microrods generated (Fig. 6.4A) [22]. When the degree of Sc^{3+} codoping is at 18%, the microcrystals are in the form of big hexagonal prisms with discrete tips, as well as small octadecahedrons with cracked ends (Fig. 6.4B) [22].

When the degree of codoping continues to increase to 38%–78%, the number of irregular octadecahedrons decreases while the number of amorphous gray-black crystals increases (Fig. 6.4C–E) [22]. X-ray crystallography has revealed that Sc^{3+}-free microcrystals consist of hexagonal $NaYF_4$; however, when Sc^{3+} codoping is adopted, more and more microcrystals taking the phase structure of monoclinic Na_3ScF_6 and/or hexagonal $NaScF_4$ appear [22]. This phenomenon can be explained by the change in the crystal stability of $NaYF_4$ upon the addition of different amounts of Sc^{3+} ions. Sc^{3+} has been reported to be able to accelerate the growth of $\{20\bar{2}1\}$ crystal planes, and to decelerate the growth of $\{10\bar{1}0\}$ and $\{0001\}$ crystal planes. Such anisotropic growth of crystal planes may attribute to morphological evolution of the microcrystals [22].

Generation of upconversion nanoparticle–based nucleic acid carriers

The surface of UCNPs per se in general does not have strong interactions with nucleic acids. When UCNPs are designed as nucleic acid carriers, the loading process, therefore, largely relies on electrostatic interactions between negatively charged nucleic acids and the cationic coating on the nanoparticle surface. This explains why, at the moment, UCNP-based nucleic acid carriers reported in literature have often been incorporated with a surface coating of positively charged functional groups or polycations. One example has been given by Jiang and coworkers [15], who have incorporated the surface of UCNPs with amine groups using *N*-[3-(trimethoxysilyl)propyl] ethylenediamine (AEAPTMS). The success of the nanoparticles in being loaded with RNA molecules, via electrostatic complexation between the positively charged amine groups and negatively charged RNA molecules, has been verified using the gel retardation assay.

As far as using polycations for surface coating is concerned, PEI has emerged as a favorable choice. This is partly because PEI exhibits a high proton buffering capacity [14,23]. In addition, it is one of the most extensively studied nonviral vectors in literature. Its efficiency in gene loading and delivery can, therefore, be more secured [24–28]. In an earlier study, He and colleagues have modified Gd^{3+}-doped UCNPs by first covalently conjugating PEG onto the nanoparticle surface, followed by the attachment of PEI using the layer-by-layer (LbL) assembly technique [29]. UCNPs surface-coated with two layers of PEI have been found to enable effective transfection even in a serum-containing environment. Besides PEI, there are few other polycations that can be used to coat the surface of UCNPs. For example, PLGA-poly(ethylene glycol) (PEG-PLGA), together with a multifunctional positively charged amphiphilic polymer generated by aminolyzing polysuccinimide (PSI) with *N*-(3-aminopropyl) imidazole (NAPI) and oleylamine (OAm), has been adopted to coat the surface of hydrophobic $NaYF_4$:Yb^{3+}/Er^{3+} UCNPs [30]. Because of the positive charge of the polymer coating, nucleic acids can effectively adsorb onto the UCNP surface [30], enabling the use of the nanoparticles for nucleic acid transfer.

Example protocols for experimental design

The method below is an example protocol for evaluating the cytotoxicity of a UCNP-based nucleic acid carrier:

1. Cultivate the cells in a 96-well culture plate at 37°C under a humidified atmosphere with 5% CO_2 until a confluence of 70%–80% is obtained.
2. Prepare a solution with a known concentration of the carrier.
3. Dilute the solution to obtain a series of solutions with different concentrations.
4. Add 50 μL of a solution with a known concentration to each well.
5. Incubate the plate at 37°C for 5 hours under a humidified atmosphere with 5% CO_2.
6. Add 20 μL of a filtered MTT reagent (0.5 mg/mL) to each well.
7. Remove the unreacted MTT reagent by aspiration after 5 hours.
8. Dissolve the violet crystals in each well in 100 μL of DMSO.
9. Determine the color intensity using an ELISA reader at a wavelength of 595 nm.
10. Calculate the cell viability (%) using the following formula, where $[A]_{Test}$ and $[A]_{Ctrl}$ represent the ELISA readings for the tested well and the control well, respectively.

$$\text{Cell viability (\%)} = \left(\frac{[A]_{Test}}{[A]_{Ctrl}}\right) \times 100 \quad (6.1)$$

Roles of upconversion nanoparticles in nucleic acid delivery

One advantage of incorporating UCNPs into polymeric vectors is to enhance the versatility of genetic manipulation during the execution of biogerontological interventions (e.g., by enabling imaging and tracking of the delivery process, achieving temporal-spatial confinement of genetic manipulation, and facilitating concomitant administration of multiple therapeutic agents). It is true that some of the aforementioned functions (e.g., applications in imaging and delivery process tracking) can also be accomplished by using conventional down-conversion fluorophores; however, UCNPs show distinct advantages (e.g., negligible photobleaching and photoblinking, and higher spatial resolution) which have rendered UCNPs promising to be adopted as a candidate to be integrated into the technologies of polymer-based nucleic acid delivery.

Light-mediated control of intervention execution

As discussed in Chapter 4, one of the prevailing strategies to offer target specificity to a polymeric carrier, at the moment, is ligand conjugation to the carrier surface. Yet, with the integration of UCNPs, which can convert near-infrared (NIR) or visible light to ultraviolet (UV) radiation, into carrier design to allow for controlled release of therapeutic nucleic acids, target specificity can be attained. This has been supported by the success in using UCNPs to mediate precise gene expression in tumors that have been transplanted to adult zebrafish [31]. Such controlled release of exogenous genes is beyond the reach of conventional polymeric carriers.

The temporal-spatial profile of RNA interference (RNAi) can be regulated using UCNPs as well. This has been shown by an earlier study, in which Yb^{3+}/Tm^{3+} codoped nanocrystals have first been encapsulated in a silica shell with surface amine groups, followed by surface functionalization with cationic photocaged linkers for subsequent small interfering RNA (siRNA) loading [32]. The loaded siRNA molecules can later be released when the photocaged linker on the nanoparticle surface is cleaved by the upconverted UV light [32]. More recently, a study has made siRNA molecules nonfunctional by caging the molecules using light-sensitive 4,5-dimethoxy-2-nitroacetophenone (DMNPE), followed by the delivery of the caged molecules using $NaYF_4$:Yb,Tm UCNPs [33]. Results have revealed that site-specific gene silencing has been achieved upon irradiation with NIR light, which has resulted in emission of upconverted UV light to uncage DMNPE. Based on the evidence presented above, it is not difficult to see the practical potential of integrating UCNPs into the design of polymeric carriers for precise manipulation of gene expression and hence spatial-temporal modulation of cellular processes.

In addition to spatial-temporal confinement of genetic manipulation, emission of upconverted light from the UCNPs can enable the execution of multimodal therapies to tackle age-associated diseases. For instance, a previous study has modified positively charged NaGdF4:Yb,Er UCNPs with multilayered polymer coatings via an LbL strategy for mediating gene therapy and photodynamic therapy (PDT) by carrying both chlorin e6 (Ce6) and siRNA molecules at the same time [34]. In the system, the siRNA molecules target the *Plk1* oncogene to induce apoptosis, whereas resonance energy transfer from UCNPs to Ce6 initiated by NIR excitation can generate singlet oxygen to kill cancer cells [34]. With the viability of mediating concomitant administration of multiple interventions using one system, the efficiency in tackling aging, which is known to be a multifactorial process, can be potentially enhanced.

Imaging and tracking of the delivery process

Besides enabling ligh-mediated control of intervention execution, another advantage of integrating UCNPs into the design of nucleic acid carriers is the possibility of tracking the delivery process via multimodal imaging. This has been evidenced by the Wang et al.'s UCNP-based gene carrier, which has been produced by coating the UCNP with PEI using an LbL approach (Fig. 6.5) [35]. Under the laser scanning UCL microscope, UCL signals have been found in the intracellular region of HeLa cells that have been incubated with the DNA/carrier complex. The signal-to-noise ratio has been estimated to be more than 300. This has suggested that the carrier can give low background fluorescence during live cell imaging. Apart from mediating a single imaging modality, PEI-coated UCNPs have been exploited to function as contrast agents for multiple imaging modalities, ranging from MRI and UCL imaging to computed tomography (CT), while delivering nucleic acids in vitro effectively [36]. Lately, $NaLuF_4$:Gd,Yb,Er UCNPs, in which carboxyl-containing glutarate has been employed as a surface ligand, have been further incorporated with cypate (a dicarboxylic acid–containing carbocyanine fluorophore with high photothermal conversion efficiency) through a hydrazide bond for the development of a multifunctional carrier (Fig. 6.6A) [37]. By taking advantage of their magnetic and optical properties, the nanoparticles have been reported as a dual-modality contrast agent for UCL imaging and MRI (Fig. 6.6B), and have successfully delivered siRNA to manipulate cell functions [37]. Together with the photothermal ablation possibly led by cypate (Fig. 6.6C), the nanoparticles have shown potential for genetic manipulation and for the treatment of diseases (e.g., cancers) that require elimination of malfunctional cells [37].

To enhance the tracking performance, one may tune the optical properties of the UCNPs by (1) making use of Förster resonance energy transfer (FRET) and luminescence resonance energy transfer (LRET) between UCNPs and the coupled emitters (e.g., dyes and quantum dots (QDs)), (2) incorporating UCNPs with the core/shell architecture, (3) changing the dopant concentration, (4) using appropriate energy transferor migration pathways, (5) manipulating the host/activator combination, (6) designing the size- and shape-induced surface effects, or (7) controlling the relaxation processes induced by phonons of the surrounding ligands. Here it is worth

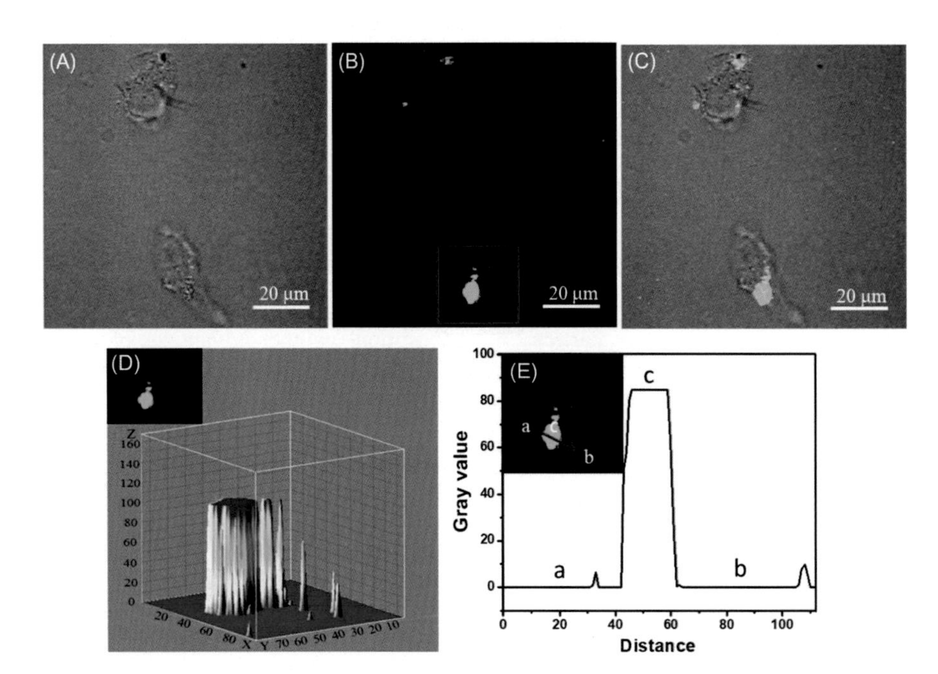

FIGURE 6.5 The (A) bright-field image, (B) fluorescence image, and (C) overlay-field image of the cells transfected with the complexes formed between DNA and the UCNP-based carrier. (D) The 3D surface plot of the UCL intensity distribution for the UCL image of the HeLa cell in the top left corner. (E) The luminescence gray value distribution from the region a to the region b. *Reproduced from Wang, Y., Cao, P., Li, S., Zhang, X., Hu, J., Yang, M., et al., 2018. Layer-by-layer assembled PEI-based vector with the upconversion luminescence marker for gene delivery, Biochem. Biophys. Res. Commun. 503, 2504–2509 with permission from Elsevier B.V. [35].*

noting that deleterious cross-relaxation events among lanthanide ions, and hence a loss of the luminescent efficiency, may result when the emission of UCNPs is tuned [7]. Nevertheless, by combining multiple strategies as mentioned above to tackle different aspects of the color tuning process and by suppressing surface deactivations, it is possible to minimize the effects, or even occurrence, of cross relaxation.

Optimization of upconversion nanoparticle–based nucleic acid carriers

Due to the possibility of using UCNPs to control and track the process of genetic manipulation, integrating UCNPs into the development of polymeric vectors emerges as one of the noteworthy approaches to vector design. In fact, the practical potential of this has been supported by the case of silica-coated surface-functionalized $NaYF_4$:Yb/Er UCNPs [38]. The nanoparticles have been reported to condense the plasmid, pcDNA3.1/VP1-GFP, and have elicited humoral and cellular immune responses, upon intramuscular administration of the UCNPs to guinea pigs, to protect the guinea pigs from attack by the foot-and-mouth disease virus [38]. Despite this promising potential, effective use of UCNPs in intervention execution has predominately been mediated by PL emission from the nanoparticles. Due to the low quantum yield and the low extinction coefficient, at the moment the excitation density provided by the reported UCNPs can reach only around $10^{-2} - 10^{-1}$ W/cm^2. This has jeopardized the efficiency of the nanoparticles in mediating imaging and light-controlled intervention execution. Moreover, owing to the nano-size of the UCNPs and hence the high surface-to-volume ratio, the emission efficiency of the generated carrier is highly sensitive to surface-related deactivations. These deactivations can be caused by various mechanisms. For example, they can be resulted from the quenching centers neighboring to the photoexcited dopants located on or around the UCNP surface, or can occur when the energy possessed by the photoexcited dopants located in the center of nanophosphors migrates to the dopant on or around the UCNP surface or directly to the surface quenching sites.

To tackle these problems, there are two possible strategies. One is to compensate for the low extinction coefficient resulting from 4f to 4f optical transitions in lanthanide ions by using the antenna effect from other species (e.g., QDs, plasmons, and dyes) that give strong absorption. Another approach is to incorporate the UCNPs with a core/shell structure, in which the host material of the shell should exhibit a low lattice mismatch with the core material, to suppress the occurrence of surface-related quenching mechanisms. These core–shell UCNPs can be generated by using a two-stage process.

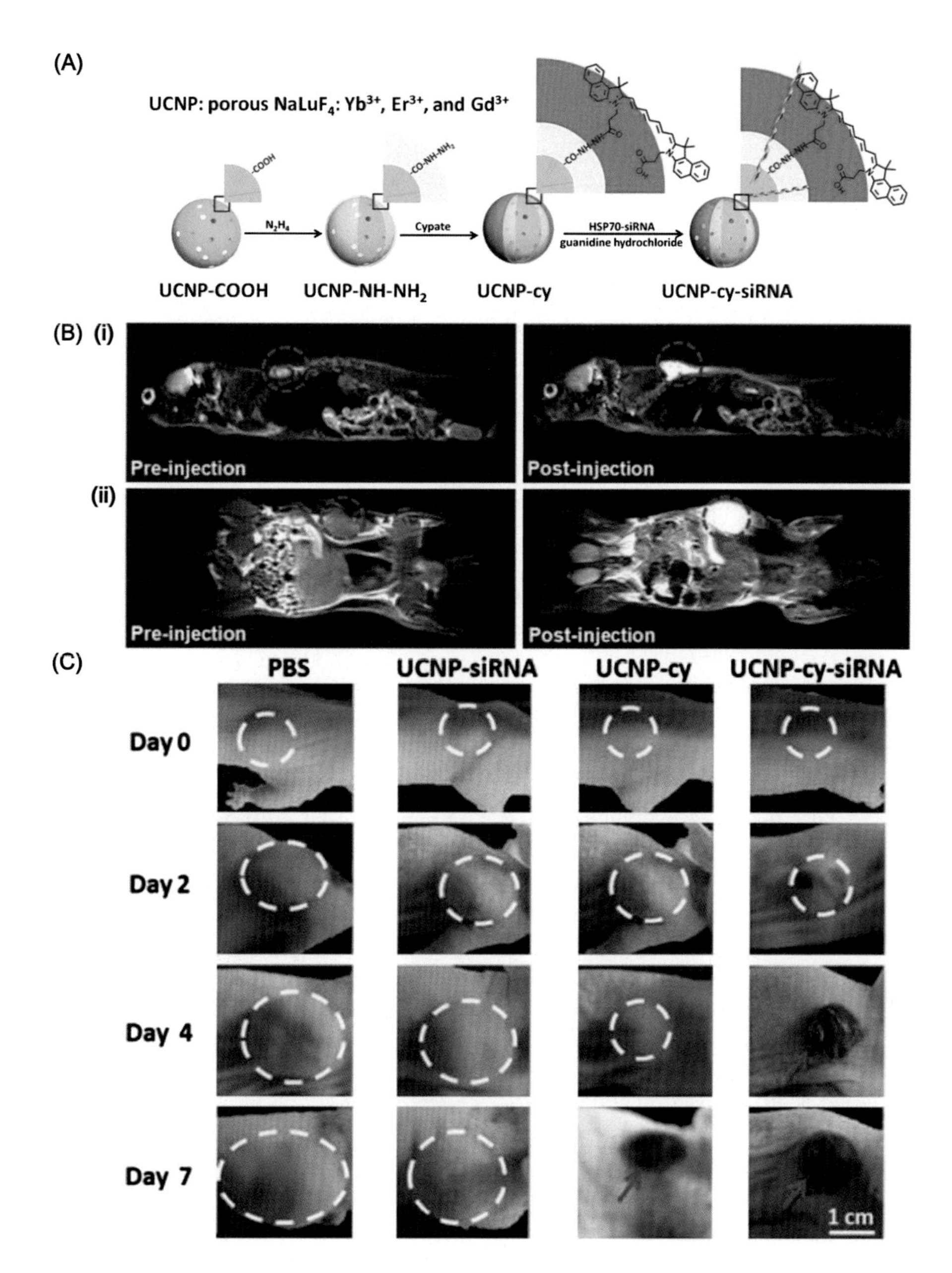

FIGURE 6.6 (A) A schematic diagram showing how cypate and siRNA molecules are loaded into UCNPs. (B) (i) Sagittal and (ii) coronal plane MRI of the mouse before and after i.v. administration of the cypate-conjugated UCNPs (10 mg/kg). (C) Images depicting tumor development in mice treated with phosphate buffered saline (PBS), siRNA-loaded UCNPs, cypate-conjugated UCNPs, and siRNA-loaded cypate-conjugated UCNPs. *Reproduced from Wang, L., Gao, C., Liu, K., Liu, Y., Ma, L., Liu, L., et al., Cypate-conjugated porous upconversion nanocomposites for programmed delivery of heat shock protein 70 small interfering RNA for gene silencing and photothermal ablation, Adv. Funct. Mater. 26 (2016) 3480–3489 with permission from John Wiley and Sons [37].*

The first stage involves the generation of UCNPs as usual. The nanoparticles generated can then function as core nanocrystals in the second stage, during which shell layers are deposited onto the core nanocrystals via epitaxy, which can be done by using chemical reactions similar to those used to fabricate the core nanocrystals.

One of the common approaches to attain epitaxial shell coating is the heat-up strategy, with which UCNPs with a multishelled structure can be fabricated by either performing the same synthetic approach repeatedly, or by combining dissimilar synthetic protocols for depositing shells with different properties onto the core crystals [39]. The effectiveness of using this method to generate UCNPs with the core–shell architecture has been demonstrated by a previous study [40] in which the hexagonal $NaGdF_4$ shell has been deposited onto the surface of the $NaYF_4$:Yb/Er nanocrystal by heating the nanocrystal in an OA/1-octadecence (ODE) solution containing different precursors. Apart from the heat-up strategy, epitaxial shell coating can be achieved by using the hot-injection strategy, in which shell precursors are injected into a reaction system, which generates the core nanocrystals, for one-pot synthesis of multishelled nanostructures [41]. The feasibility of this approach has been partially shown by the successful generation of $NaYF_4$:Yb/Er@$NaYF_4$ core–shell UCNPs [41]. During synthesis, NaYF4:Yb/Er core nanoparticles are first grown by heating related rare earth trifluoroacetates in OAm. After that, epitaxial deposition of the undoped $NaYF_4$ shell layer onto the nanoparticles is achieved by injecting an OAm solution of shell precursors into the reaction mixture. Other examples of core–shell UCNPs successfully generated by using the hot-injection strategy include $NaGdF_4$@$NaGdF_4$ and $LiLuF_4$@$LiLuF_4$ [42–44].

In addition to epitaxy, shell layers can be deposited onto the surface of pre-synthesized UCNPs in a nonepitaxial manner via either surface polymerization or chemical bonding [45,46]. Core–shell UCNPs that have been reported to be generated by using the nonepitaxial approach include $NaYF_4$:Yb/Er@SiO_2 core–shell nanoparticles [46], and $NaYF_4$:Yb/Tm UCNPs with tunable surface coverage of gold nanoparticles [47]. The enhancement in the PL efficiency of UCNPs upon incorporation of the multishelled architecture has been evidenced in various studies in literature. For example, $NaYF_4$:Yb,Er and $NaYF_4$:Yb,Tm UCNPs have exhibited a 7.4- and 29.6-fold increase, respectively, in their PL efficiency after being incorporated with an undoped $NaYF_4$ shell [41]. $NaYF_4$:Yb/Er UCNPs have also shown a significant increase in the overall emission intensity and nanoparticle lifetime upon incorporation with a hexagonal $NaGdF_4$ shell [40], possibly because surface defects of $NaYF_4$:Yb/Er nanocrystals have been passivated during the shell deposition process [40]. All these have demonstrated the technical feasibility of enhancing the emission intensity of UCNPs by incorporating the nanoparticles with the core–shell architecture.

Apart from the PL efficiency, another factor to be considered when UCNPs are incorporated into the design of polymeric carriers is the biodegradability of the nanoparticles. In fact, UCNPs generally can hardly be degraded into biologically benign molecules. To ensure effective clearance from a body, the hydrodynamic size of the UCNPs may need to be less than 10 nm [7]; however, this hydrodynamic size may not be the most optimal size for polymer-mediated nucleic acid delivery. The optimization of the size of the UCNP to maximize both the delivery efficiency as well as the biodegradability is a challenge to be tackled when UCNP-assisted intervention execution is adopted. Apart from this, at the moment excitation of UCNPs is often performed using light at 980 nm. This wavelength matches well with the absorption of Yb^{3+}, which is a commonly used sensitizer. Unfortunately, light at this wavelength can also be absorbed by water, resulting in the generation of heat to cause damage to biological tissues. To tackle this problem, one may tune the nanoparticle composition so that the UCNPs can be excited using light at more tissue transparent wavelengths. The technical viability of this has been supported by the success in blue-shifting the excitation wavelength by sensitizing Yb^{3+} using the second sensitizer, Nd^{3+} [48,49]. Future efforts to tune the excitation wavelength to the one that is transparent to tissues are highly desirable because this can secure the safety use of the nanoparticles when they are applied in practice.

Improvement in upconversion nanoparticle fabrication

Apart from increasing the emission intensity, developing simple methods to enable the generation of uniform, biocompatible, and hydrophilic UCNPs will greatly facilitate the use of UCNPs in interventions [50]. Few studies have addressed this need by using CDs [51–53]. For example, one study has tried to increase the hydrophilicity of UCNPs by first capping the nanoparticles with adamantaneacetic acid, followed by complexation with β-CD [52]. In a chloroform/water biphasic system under continuous-wave (CW) excitation at 980 nm, the uncomplexed adamantaneacetic acid–capped UCNPs in the organic phase have displayed three UCL signals [52]. Two of the signals have been found to be at 521 and 540 nm. They have been assigned to $^4H_{11/2}$–$^4I_{15/2}$ and $^4S_{3/2}$–$^4I_{15/2}$ transitions, respectively [52]. One signal has been detected at 654 nm, and has been attributed to the $^4F_{9/2}$–$^4I_{15/2}$ transition [52]. After the addition of β-CD, phase transfer of the nanoparticles has been observed, with the UCL signals detected in the aqueous layer. Transmission electron microscopy (TEM) has shown that host–guest inclusion complexation with β-CD does not affect the morphology of the UCNPs [52].

More recently, derivatives of β-CD have been adopted for surface modification of OA-capped $NaYF_4$:Yb,Er UCNPs for enhancing the dispensability of the nanoparticles in water [50]. During the modification process, OA-capped UCNPs are first subjected to ligand exchange [50] (in which 2-azidoethylphosphonic acid ligands replace the OA ligands on the nanoparticle surface), followed by conjugation of 6-propargylamino-6-deoxy-β-CD to the nanoparticles [50]. Apart from enhancing the hydrophilicity and biocompatibility of the UCNPs, the CD moieties on UCNPs facilitate subsequent incorporation of functional elements. For instance, with the presence of CD moieties, the cell targeting capacity of the UCNPs can be enhanced by incorporating cyclic

RGD-conjugated adamantane (Ad-RGD) onto the UCNP surface via host–guest inclusion complexation [50]. This has facilitated the optimization and modification of UCNPs during the design and development of an intervention.

Summary

The advancement of delivery technologies for genetic manipulation can bring hopes to modulate the aging process, and hence plays an important role in the development of biogerontological interventions. In this chapter, we have described feasible strategies to integrate UCNPs into the design of polymeric carriers to extend the possibilities in nucleic acid delivery. In the next chapter, we will switch our focus from nucleic acid transfer to delivery of nongenetic materials. As mentioned in Chapter 2, besides genetic materials, nongenetic materials such as small-molecule drugs, proteins, and peptides can be used as therapeutic agents. By attaining the technical knowledge of developing delivery technologies for both genetic and nongenetic materials, the flexibility of intervention design to manipulate the aging process can be enhanced. Different aspects of the multifactorial aging process can also be tackled in an easier and more comprehensive manner.

Directions for intervention development

Incorporation of UCNPs into vector design can potentially enhance the versatility of polymeric carriers. To take advantage of the unique properties of these nanoparticles, the following steps can be taken:

1. Develop an effective polymeric vector by following the steps depicted in preceding chapters.
2. Determine the role potentially played by UCNPs in the proposed intervention.
3. Select the appropriate composition and structure of the UCNPs for attaining the desired purpose.
4. Choose an appropriate method to fabricate the UCNPs and to integrate the nanoparticles into the polymeric vector.
5. Evaluate the performance and properties of the resulting vector.
6. Optimize the performance of the vector for intervention execution.

References

[1] F. Wang, X. Liu, Recent advances in the chemistry of lanthanide-doped upconversion nanocrystals, Chem. Soc. Rev. 38 (2009) 976–989.

[2] S.V. Eliseeva, J.C. Bunzli, Lanthanide luminescence for functional materials and bio-sciences, Chem. Soc. Rev. 39 (2010) 189–227.

[3] W.F. Lai, A.L. Rogach, W.T. Wong, Molecular design of upconversion nanoparticles for gene delivery, Chem. Sci. 8 (2017) 7339–7358.

[4] P.D. Nguyen, S.J. Son, J. Min, Upconversion nanoparticles in bioassays, optical imaging and therapy, J. Nanosci. Nanotechnol. 14 (2014) 157–174.

[5] M. Gonzalez-Bejar, L. Frances-Soriano, J. Perez-Prieto, Upconversion nanoparticles for bioimaging and regenerative medicine, Front. Bioeng. Biotechnol. 4 (2016) 47.

[6] X. Chen, D. Peng, Q. Ju, F. Wang, Photon upconversion in core-shell nanoparticles, Chem. Soc. Rev. 44 (2015) 1318–1330.

[7] G. Chen, H. Qiu, P.N. Prasad, X. Chen, Upconversion nanoparticles: design, nanochemistry, and applications in theranostics, Chem. Rev. 114 (2014) 5161–5214.

[8] W.F. Lai, Z.D. He, Design and fabrication of hydrogel-based nanoparticulate systems for in vivo drug delivery, J. Control Release 243 (2016) 269–282.

[9] W.F. Lai, Nucleic acid delivery: roles in biogerontological interventions, Ageing Res. Rev. 12 (2013) 310–315.

[10] W.F. Lai, Nucleic acid therapy for lifespan prolongation: present and future, J. Biosci. 36 (2011) 725–729.

[11] W.F. Lai, M.C. Lin, Nucleic acid delivery with chitosan and its derivatives, J. Control Release 134 (2009) 158–168.

[12] W.F. Lai, Microfluidic methods for non-viral gene delivery, Curr. Gene Ther. 15 (2015) 55–63.

[13] W.F. Lai, Cyclodextrins in non-viral gene delivery, Biomaterials. 35 (2014) 401–411.

[14] W.F. Lai, In vivo nucleic acid delivery with PEI and its derivatives: current status and perspectives, Expert Rev. Med. Devices 8 (2011) 173–185.

[15] S. Jiang, Y. Zhang, K.M. Lim, E.K. Sim, L. Ye, NIR-to-visible upconversion nanoparticles for fluorescent labeling and targeted delivery of siRNA, Nanotechnology. 20 (2009) 155101.

[16] W.B. Li, C.B. Tan, Y.T. Zhang, Simultaneous phase and shape control of monodisperse NaLuF4:Yb, Er microcrystals and greatly enhanced upconversion luminescence from their superstructures, Opt. Commun. 295 (2013) 140–144.

[17] J.Y. Sun, J.B. Xian, X.Y. Zhang, H.Y. Du, Size and shape controllable synthesis of oil-dispersible $BaYF_5:Yb^{3+}/Er^{3+}$upconversion fluorescent nanocrystals, J. Alloys Compd. 509 (2011) 2348–2354.

[18] X. Peng, Mechanisms for the shape-control and shape-evolution of colloidal semiconductor nanocrystals, Adv. Mater. 15 (2003) 459–463.

[19] N. Zhou, P. Qiu, K. Wang, H. Fu, G. Gao, R. He, et al., Shape-controllable synthesis of hydrophilic NaLuF4:Yb,Er nanocrystals by a surfactant-assistant two-phase system, Nanoscale Res. Lett. 8 (2013) 518.

[20] M. Ding, D. Chen, S. Yin, Z. Ji, J. Zhong, Y. Ni, et al., Simultaneous morphology manipulation and upconversion luminescence enhancement of beta-$NaYF4:Yb^{3+}/Er^{3+}$ microcrystals by simply tuning the KF dosage, Sci. Rep. 5 (2015) 12745.

[21] C.Y. Wang, X.H. Cheng, Influence of Cr^{3+} ions doping on growth and upconversion luminescence properties of β-$NaYF4:Yb^{3+}/Er^{3+}$ microcrystals, J. Alloys Compd. 649 (2015) 196–203.

[22] E.J. He, S.F. Chen, M.L. Zhang, Simultaneous morphology evolution and upconversion emission tuning of single Y-based fluoride microcrystal induced by Sc^{3+} co-doping, Mater. Res. Bull. 87 (2017) 61–71.

[23] M. Neu, D. Fischer, T. Kissel, Recent advances in rational gene transfer vector design based on poly(ethylene imine) and its derivatives, J. Gene Med. 7 (2005) 992–1009.

[24] I. Chemin, D. Moradpour, S. Wieland, W.B. Offensperger, E. Walter, J.P. Behr, et al., Liver-directed gene transfer: a linear polyethlenimine derivative mediates highly efficient DNA delivery to primary hepatocytes in vitro and in vivo, J. Viral Hepat. 5 (1998) 369–375.
[25] M.E. Davis, Non-viral gene delivery systems, Curr. Opin. Biotechnol. 13 (2002) 128–131.
[26] O. Boussif, F. Lezoualc'h, M.A. Zanta, M.D. Mergny, D. Scherman, B. Demeneix, et al., A versatile vector for gene and oligonucleotide transfer into cells in culture and in vivo: polyethylenimine, Proc. Natl. Acad. Sci. U.S.A. 92 (1995) 7297–7301.
[27] D. Goula, J.S. Remy, P. Erbacher, M. Wasowicz, G. Levi, B. Abdallah, et al., Size, diffusibility and transfection performance of linear PEI/DNA complexes in the mouse central nervous system, Gene Ther. 5 (1998) 712–717.
[28] S.M. Zou, P. Erbacher, J.S. Remy, J.P. Behr, Systemic linear polyethylenimine (L-PEI)-mediated gene delivery in the mouse, J. Gene Med. 2 (2000) 128–134.
[29] L. He, L. Feng, L. Cheng, Y. Liu, Z. Li, R. Peng, et al., Multilayer dual-polymer-coated upconversion nanoparticles for multimodal imaging and serum-enhanced gene delivery, ACS Appl. Mater. Interfaces 5 (2013) 10381–10388.
[30] X. Bai, S. Xu, J. Liu, L. Wang, Upconversion luminescence tracking of gene delivery via multifunctional nanocapsules, Talanta 150 (2016) 118–124.
[31] M.K. Jayakumar, A. Bansal, B.N. Li, Y. Zhang, Mesoporous silica-coated upconversion nanocrystals for near infrared light-triggered control of gene expression in zebrafish, Nanomedicine. 10 (2015) 1051–1061.
[32] Y. Yang, F. Liu, X. Liu, B. Xing, NIR light controlled photorelease of siRNA and its targeted intracellular delivery based on upconversion nanoparticles, Nanoscale 5 (2013) 231–238.
[33] H. Guo, D. Yan, Y. Wei, S. Han, H. Qian, Y. Yang, et al., Inhibition of murine bladder cancer cell growth in vitro by photocontrollable siRNA based on upconversion fluorescent nanoparticles, PLoS One 9 (2014) e112713.
[34] X. Wang, K. Liu, G. Yang, L. Cheng, L. He, Y. Liu, et al., Near-infrared light triggered photodynamic therapy in combination with gene therapy using upconversion nanoparticles for effective cancer cell killing, Nanoscale 6 (2014) 9198–9205.
[35] Y. Wang, P. Cao, S. Li, X. Zhang, J. Hu, M. Yang, et al., Layer-by-layer assembled PEI-based vector with the upconversion luminescence marker for gene delivery, Biochem. Biophys. Res. Commun. 503 (2018) 2504–2509.
[36] L. Wang, J. Liu, Y. Dai, Q. Yang, Y. Zhang, P. Yang, et al., Efficient gene delivery and multimodal imaging by lanthanide-based upconversion nanoparticles, Langmuir 30 (2014) 13042–13051.
[37] L. Wang, C. Gao, K. Liu, Y. Liu, L. Ma, L. Liu, et al., Cypate-conjugated porous upconversion nanocomposites for programmed delivery of heat shock protein 70 small interfering RNA for gene silencing and photothermal ablation, Adv. Funct. Mater. 26 (2016) 3480–3489.
[38] H. Guo, R. Hao, H. Qian, S. Sun, D. Sun, H. Yin, et al., Upconversion nanoparticles modified with aminosilanes as carriers of DNA vaccine for foot-and-mouth disease, Appl. Microbiol. Biotechnol. 95 (2012) 1253–1263.
[39] F. Wang, R. Deng, J. Wang, Q. Wang, Y. Han, H. Zhu, et al., Tuning upconversion through energy migration in core-shell nanoparticles, Nat. Mater. 10 (2011) 968–973.
[40] F. Zhang, R. Che, X. Li, C. Yao, J. Yang, D. Shen, et al., Direct imaging the upconversion nanocrystal core/shell structure at the subnanometer level: shell thickness dependence in upconverting optical properties, Nano. Lett. 12 (2012) 2852–2858.
[41] G.S. Yi, G.M. Chow, Water-soluble $NaYF_4$:Yb,Er(Tm)/$NaYF_4$/polymer core/shell/shell nanoparticles with significant enhancement of upconversion fluorescence, Chem. Mater. 19 (2007) 341–343.
[42] F. Vetrone, R. Naccache, V. Mahalingam, C.G. Morgan, J.A. Capobianco, The active-core/active-shell approach: a strategy to enhance the upconversion luminescence in lanthanide-doped nanoparticles, Adv. Funct. Mater. 19 (2009) 2924–2929.
[43] G.Y. Chen, T.Y. Ohulchanskyy, S. Liu, W.C. Law, F. Wu, M.T. Swihart, et al., Core/Shell NaGdF4:Nd^{3+}/NaGdF4 nanocrystals with efficient near-infrared to near-infrared downconversion photoluminescence for bioimaging applications, ACS Nano 6 (2012) 2969–2977.
[44] P. Huang, W. Zheng, S.Y. Zhou, D.T. Tu, Z. Chen, H.M. Zhu, et al., Lanthanide-doped LiLuF4 upconversion nanoprobes for the detection of disease biomarkers, Angew. Chem. Int. Ed. Engl. 53 (2014) 1252–1257.
[45] W. Zou, C. Visser, J.A. Maduro, M.S. Pshenichnikov, J.C. Hummelen, Broadband dye-sensitized upconversion of near-infrared light, Nat. Photonics 6 (2012) 560–564.
[46] Z.Q. Li, Y. Zhang, S. Jiang, Multicolor core/shell-structured upconversion fluorescent nanoparticles, Adv. Mater. 20 (2008) 4765–4769.
[47] H. Zhang, Y.J. Li, I.A. Ivanov, Y.Q. Qu, Y. Huang, X.F. Duan, Plasmonic modulation of the upconversion fluorescence in NaYF4:Yb/Tm hexaplate nanocrystals using gold nanoparticles or nanoshells, Angew. Chem. Int. Ed. 49 (2010) 2865–2868.
[48] J. Shen, G. Chen, A.M. Vu, W. Fan, O.S. Bilsel, C.C. Chang, et al., Engineering the upconversion nanoparticle excitation wavelength: cascade sensitization of tri-doped upconversion colloidal nanoparticles at 800 nm, Adv. Opt. Mater. 1 (2013) 644–650.
[49] X. Xie, N. Gao, R. Deng, Q. Sun, Q.H. Xu, X. Liu, Mechanistic investigation of photon upconversion in Nd^{3+}-sensitized core-shell nanoparticles, J. Am. Chem. Soc. 135 (2013) 12608–12611.
[50] C. Ma, T. Bian, S. Yang, C. Liu, T. Zhang, J. Yang, et al., Fabrication of versatile cyclodextrin-functionalized upconversion luminescence nanoplatform for biomedical imaging, Anal. Chem. 86 (2014) 6508–6515.
[51] G. Tian, W.L. Ren, L. Yan, S. Jian, Z.J. Gu, L.J. Zhou, et al., Red-emitting upconverting nanoparticles for photodynamic therapy in cancer cells under near-infrared excitation, Small 9 (2013) 1929–1938.
[52] Q.A. Liu, C.Y. Li, T.S. Yang, T. Yi, F.Y. Li, "Drawing" upconversion nanophosphors into water through host-guest interaction, Chem. Commun. 46 (2010) 5551–5553.
[53] Q. Liu, M. Chen, Y. Sun, G.Y. Chen, T.S. Yang, Y. Gao, et al., Multifunctional rare-earth self-assembled nanosystem for tri-modal upconversion luminescence/fluorescence/positron emission tomography imaging, Biomaterials 32 (2011) 8243–8253.

Chapter 7

Design of hydrogel-based nanoparticles for intervention execution

Introduction

In carrier development, the size of the carrier designed should be decided based on actual needs. Carriers at the nanoscale are often more favorable because they can undergo cellular internalization, and can deliver agents to modulate biological processes at the metabolic and genetic levels. Generally speaking, nanoparticles are defined as submicronic colloidal systems with the size less than 1 μm [1]. Over the years, diverse types of materials (including metal oxides [2], metals [3], polymeric materials [4], and lipids [5]) have been described in literature to generate nanoparticles. Nanoparticles generated from hydrogels have demonstrated unique values in drug delivery applications.

Hydrogels consist of three-dimensional hydrophilic networks of polymers. This enables hydrogels to absorb a substantial amount of fluids [6]. In addition, due to the presence of cross-links among polymer chains, hydrogels are resistant to dissolution [6]. Some hydrogels have displayed high drug encapsulation efficiency and release sustainability [7,8], and have shown great potential for applications in the pharmaceutical formulation. The aim of this chapter is to present an overview of the latest progress in the development of hydrogel-based nanoparticulate systems as carriers, and to discuss the opportunities and challenges of applying these systems to biogerontological interventions.

Development of hydrogels for drug delivery

The term "hydrogel" emerged in 1894 [9]. At that time the term was used to describe a colloidal gel of inorganic salts, rather than a three-dimensional, water-swollen, cross-linked network of polymers [9]. In the 1950s, a list of material properties was proposed. Examples of these properties include the absence of extractable impurities, good permeability for water-soluble substances, and good resistance to enzymatic degradation. These properties are thought to be needed by a synthetic material when that material is to be applied in direct contact with living tissues [10]. Keeping these properties in mind, the first "real" hydrogel, which has been fabricated by free radical polymerization of 2-hydroxyethyl methacrylate (HEMA) in an aqueous solution using ethylene glycol dimethacrylate (EGDMA) as a cross-linker, has later emerged [11]. Until now, hydrogels have become an important class of materials involving in wide biomedical applications, ranging from tissue engineering to controlled delivery of bioactive agents [12–16].

Over the years, different systems (e.g., liposomes [17–19] and emulsions [20]) have been applied as drug carriers; however, hydrogels show unique advantages that make them favorable for drug delivery. Hydrogels can be generated in an all-aqueous environment. This avoids the use of volatile organic solvents. In addition, compared to systems such as emulsions, whose applications may be hampered by surfactant toxicity and droplet coalescence, hydrogels are relatively stable and less toxic. Because of these, a number of hydrogel-based carriers have proceeded to preclinical studies. One example is the polyacrylamide hydrogel nanoparticles, which have a size of approximately 20–30 nm [21]. To enhance tumor targeting, the surface of the nanoparticles has been modified with the F3 peptide, which is a subcomponent of the HMGN2 protein [22–24]. The nanoparticles have been given to an ovarian tumor mice model via i.v. infusion. Results have shown that 24 hours after administration at a dose of 100 mg/kg, effective binding of the nanoparticles to tumor vessels was observed [21]. Moreover, i.v. infusion of the nanoparticles, which have been loaded with cisplatin, has been found to lead to a significant decline in tumor volume [21]. Histological analysis of the tumor has revealed that the nanoparticles have not only caused the inside of the tumor to undergo hemorrhage and necrosis, but have also reduced the size of tumor islets [21]. In addition to delivering therapeutic agents to cancer cells, hydrogels

Delivery of Therapeutics for Biogerontological Interventions. DOI: https://doi.org/10.1016/B978-0-12-816485-3.00007-6

have been used to carry therapeutics to diverse organs [25–33]. This has evidenced the application potential of hydrogels in treatment development.

Design of hydrogels for intervention development

When hydrogels are used as carriers, they can be given to a body systematically or locally based on actual needs. While bulk hydrogels can be adapted to mediate sustained drug release via topical or local administration [34,35], if we need the hydrogels to enter cells or the blood circulation, they have to be at the nanoscale. Because of their submicron size, hydrogel nanoparticles have a high surface-to-volume ratio. This renders hydrogel nanoparticles effective in adsorbing poorly soluble drugs (e.g., cyclosporine, amphotericin B, and paclitaxel) [36]. Owing to the presence of a large surface area, surface functionalization can be easily accomplished as well, rendering nanoparticles target-specific during drug delivery. Along with the intrinsically high biodegradability and biocompatibility of hydrogels, hydrogel nanoparticles have attracted extensive research interests as candidates for carrier development.

As far as nanoparticle development is concerned, proper design of the fabrication procedure is pivotal because it affects the physical properties of the nanoparticles obtained and ultimately influences the performance of the nanoparticles at each stage of the drug delivery process (Fig. 7.1) [37]. Although the fenestration size and vasculature may be changed by some pathological conditions, such as cancers [38], which may change the biodistribution pattern of the hydrogel nanoparticles upon administration, it is generally expected that particles having a size less than 20–30 nm can be removed from the body easily via renal excretion [39,40]. Particles having a size of 30–150 nm are expected to get accumulated in the stomach [41], kidney [39,40], heart [42], and bone marrow [43]. For those having a size of 150–300 nm, they may be present largely in the liver [44] and spleen [45]. In fact, particle size cannot only affect the biodistribution profile, but can also change the surface-to-volume ratio, which may influence the drug loading efficiency as well as the ease of particle aggregation. These problems shall be addressed so as to maximize the performance of the nanoparticles as drug carriers. Possible methods of optimizing different parts of the nanoparticle development process will be detailed in the following sections.

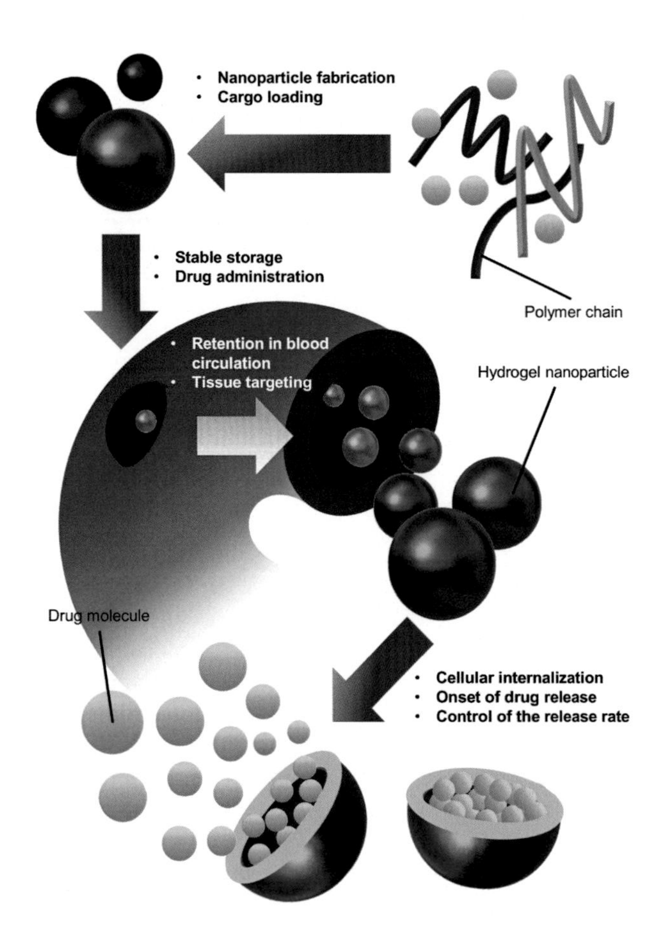

FIGURE 7.1 A schematic diagram showing important stages of the drug delivery process mediated by hydrogel nanoparticles. *Reproduced from W.F. Lai, Z.D. He, Design and fabrication of hydrogel-based nanoparticulate systems for in vivo drug delivery, J. Control Release 243 (2016) 269–282 with permission from Elsevier B.V. [37].*

Generation of hydrogel nanoparticles

Electrospray, which is a technique for generating aerosols via electrostatic dispersion of liquids [46], is one of the possible methods to generate nanoparticulate systems. Examples of nanoparticulate systems that have been reported to be generated by using electrospray include solid collagen nanoparticles [47], solid CS nanoparticles [48], drug-loaded poly(L-lactic acid) nanoparticles [49], and silk fibroin nanoparticles [50]. In addition to generating solid polymeric nanoparticles, electrospray can be used to generate hydrogel-based systems. By spraying CMC, CS, or Alg into a solution in which the counterions are present, hydrogel particles can be produced. This technique has been documented in a previous study for the generation of doxorubicin (DOX)-loaded CS nanoparticles [51], with tripolyphosphate used as a stabilizer. The nanoparticles have successfully prolonged drug release over at least seven hours, with the encapsulation efficiency estimated to be around 60%–70% at 1%–0.25% DOX loading. This has corroborated the possible use of the electrospray technique to generate hydrogel nanoparticles in drug delivery research. Despite this, owing to the swelling property of hydrogels, the size of the droplets generated by electrospray has to be small enough if hydrogel nanoparticles are to be fabricated. Yet, various parameters (particularly the electric field strength and the diameter of the nozzle) have limited the achievable small droplet size. Moreover, the viscosity of a polymer solution may influence the droplet formation process and hence the diameter of the hydrogel particles obtained. The impact of the viscosity of a solution to droplet formation can be estimated by using a dimensionless number π_μ [52]:

$$\pi_\mu = \frac{\sqrt{\gamma^2 \rho \left(\frac{\varepsilon \varepsilon_0}{K}\right)}}{\mu} \tag{7.1}$$

where ε, μ, γ, ρ, and K are dielectric constant, viscosity, surface tension, solution density, and conductivity, respectively. As shown in the equation, the influence of viscosity on the size of the droplets obtained is negligible when the π_μ value of a solution is much larger than 1 [48]. Many gel-forming solutions, however, have a π_μ value smaller than 1 [48]. This implies that the size of the hydrogel particles formed can be increased when the viscosity of a polymer solution increases. Because of the technical difficulty of obtaining droplets at the nanoscale, electrospray-mediated production of hydrogel nanoparticles is still technically challenging right now.

Example protocols for experimental design

The method below is an example protocol for generating Alg-based hydrogel particles using electrospray:

1. Use a micropipette puller to taper a cylindrical capillary.
2. Polish the tip of the capillary to a desired diameter using a piece of sand paper.
3. Use the capillary to make a device (Fig. 7.2A and B).
4. Make the setup for electrospray as shown in Fig. 7.2C.

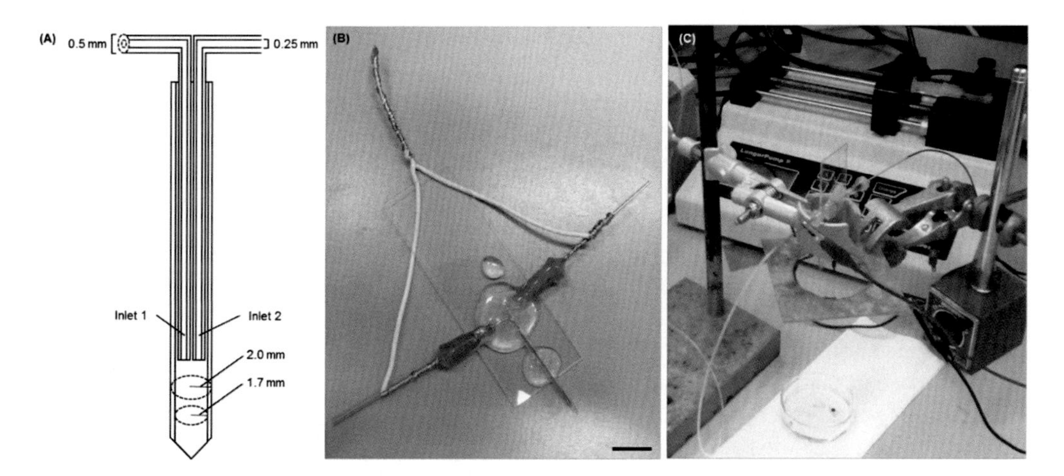

FIGURE 7.2 (A) Dimensions of the capillary device used for microfluidic electrospray. (B) A photo of the device. The scale bar is 1 cm. (C) The setup for microfluidic electrospray.

5. Dissolve Alg in distilled water to obtain a 4% w/v solution.
6. Drive the Alg solution into the setup using a syringe pump.
7. Form a high-strength electric field between the nozzle and a ground circular electrode (which is connected to a high voltage power supply).
8. Adjust the electric field strength until a tapered tip driven by the electrostatic force is formed.
9. Continue to adjust the electric field strength so that the jet with the tapered tip breaks up into micro-droplets.
10. Collect the droplets in a collection bath that contains a $CaCl_2$ solution (3% w/v).
11. Wait for 20 minutes so that the droplets fully undergo ionic gelation.
12. Retrieve the generated hydrogel particles by centrifugation.

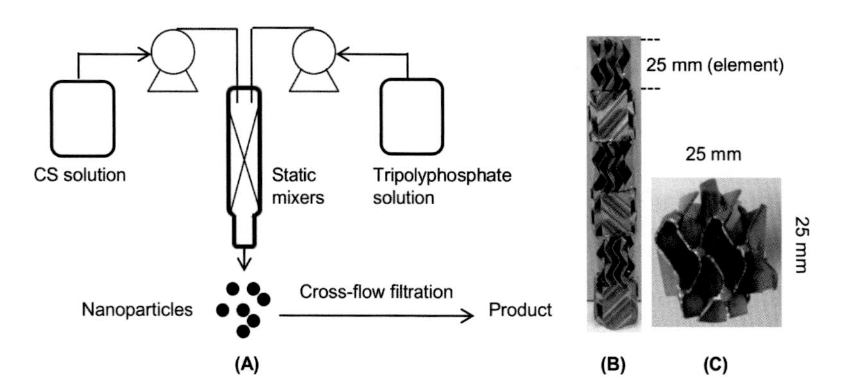

FIGURE 7.3 (A) Preparation of CS-based hydrogel nanoparticles by using a static mixer. The (B) side view and (C) top view of one segment, in which six mixing elements have been welded together. *Reproduced from Y.C. Dong, W.K. Ng, S.C. Shen, S. Kim, R.B.H. Tan, Scalable ionic gelation synthesis of chitosan nanoparticles for drug delivery in static mixers, Carbohydr. Polym. 94 (2013) 940–945 with permission from Elsevier B.V. [53].*

Another approach to generate hydrogel nanoparticles is to mix a polymer solution rapidly with a solution of an ionic cross-linker. This approach has been documented in a previous study, in which hydrogel nanoparticles have been produced by using a static mixer [53], which has enabled liquid flows to be mixed rapidly. In the mixer, six mixing elements have been welded together to form one segment, and 1–3 segments have been used during the mixing process (Fig. 7.3). The size of the particles generated has been found to be at the nanoscale, and has been affected by the CS concentration, the CS/triphenylphosphonium (TPP) volume ratio, and the flow rate. After the drug loading process, the nanoparticles have shown a triphasic release profile, with 90% of the loaded drug released within the first 8 hours [53]. Although further efforts to improve the drug release sustainability of the nanoparticles are needed before the nanoparticles can be effectively used in interventions, this reported method of nanoparticle fabrication has offered a simple procedure for continuous and possibly large scale production of ionically cross-linked hydrogel nanoparticles in the future. In addition to using an ionic cross-linker, one can generate hydrogel nanoparticles via complex coacervation. One example is first to mix Alg with Ca^{2+} to form pregel nuclei [54], which can then undergo electrostatic interactions with polycations, such as PEI, to generate sponge-like hydrogel nanoparticles for sustained release of bioactive agents [55].

Example protocols for experimental design

The method below is an example protocol for generating Alg/PEI nanoparticles:

1. Dissolve 6 mg of $CaCl_2$ in 6 mL of distilled water.
2. Prepare 45 mL of an Alg solution with a concentration of 0.6 mg/mL.
3. Dissolve an appropriate amount of a bioactive agent, which is to be loaded into the nanoparticles, in the Alg solution.

(Continued)

(Continued)

4. Add the $CaCl_2$ solution dropwise into the Alg solution.
5. Wait for 30 minutes at ambient conditions.
6. Dissolve PEI in distilled water to reach a concentration of 0.68 mg/mL.
7. Add 8 mL of the PEI solution into the solution mixture in Step 5 under constant stirring at 600 rpm.
8. Put the resulting mixture at ambient conditions for 30 minutes.
9. Put the mixture at 4°C for 16 hours.
10. Retrieve the nanoparticles by centrifugation for 40 minutes at 10,000 × g at 4°C.

No matter electrospray or mechanical mixing is used for nanoparticle fabrication; the process of nanoparticle generation is driven largely by physical interactions (e.g., Van der Waals forces, hydrogen bonds, electrostatic interactions, and chain entanglements) between polymer chains. Because the process of physical self-assembly can be easily influenced by changes in environmental parameters (e.g., pH and temperature) [56] or polymer concentrations [57–59], the experimental conditions have to be properly regulated in order to control the size of the generated nanoparticles. Apart from relying on physical self-assembly, hydrogel nanoparticles can be produced by cross-linking polymer chains chemically. Compared to those produced by physical self-assembly, nanoparticles fabricated by chemical cross-linking are, in general, more stable, though postsynthetic treatments to eliminate unreacted chemical residues are needed to ensure that the nanoparticles can be safely used in a biological body. Some of the commonly used chemical methods of nanoparticle fabrication include ring-opening polymerization (ROP), aqueous dispersion polymerization, and emulsion polymerization.

Ring-opening polymerization

ROP is a widely adopted polymerization technique that involves ring opening of the cyclic monomers for polymer generation. ROP reactions are largely driven by the ring strain and the associated steric interactions [60]. Until now, ROP has been used to produce various polymers, including polysiloxane, polyphosphazene, polynorbornene, PEI, and olycyclooctene [60]. In a recent study, ROP has also been exploited to generate PEG-armed hydrogel nanoparticles [61]. During nanoparticle production, a core cross-linked star polymer [in which PEG monomethyl ether has been adopted as the arm to a difunctional phosphate monomer, namely 3,6-dioxaoctan-1,8-diyl *bis*(ethylene phosphate)] has been produced [61]. After its swelling in an aqueous solution, PEG-armed hydrogel nanoparticles have been fabricated [61]. As the core material has consisted of polyphosphoester, which is known to be biodegradable [62,63], the generated nanoparticles have displayed a high biological safety profile for biomedical applications [61].

Aqueous dispersion polymerization

Aqueous dispersion polymerization, as a variant of radical cross-linking, enables the production of water-insoluble polymers from water-soluble monomers [64]. During the polymerization process, the aqueous phase is the major site in which initial and primary chain growth happens [65]. Due to the insolubility of the polymer obtained, aggregation of the aqueous chains occurs. When the threshold molecular weight is reached, nucleation results [65]. The dissolved radicals adsorb onto the surface of these nuclei, causing the particle surface to be swollen with monomers. Polymerization then proceeds in the swollen layer. The sorption process becomes irreversible when the chain end grows into the particle [65]. This polymerization mechanism has been exploited to produce temperature-responsive block copolymer hydrogel nanoparticles [66]. The nanoparticles consist of poly(*N*,*N*-diethylacrylamide) (PDEAAm), and are generated by copolymerization of *N*,*N*-diethylacrylamide (DEAAm) with *N*,*N'*-methylene bisacrylamide (MBA) [66]. The reaction is carried out in the presence of poly(ethylene oxide) (PEO)-*b*-poly(*N*,*N*-dimethylacrylamide) (PEO-*b*-PDMAAm) macromolecular reversible addition-fragmentation chain transfer agents (macroRAFT agents) end-capped by a trithiocarbonate reactive group, with the reaction temperature set to be at 70°C, which is above the lower critical solution temperature (LCST) of PDEAAm [66]. Because DEAAm can be dissolved in the aqueous phase, the reaction medium is homogeneous in the beginning of the polymerization process. When polymerization proceeds, PDEAAm forms a separate phase [66]. Finally, with self-stabilization by PEO-*b*-PDMAAm chains, core cross-linked PDEAAm nanoparticles are produced. Because the nanoparticles can be generated in one-pot, aqueous dispersion polymerization gives a reproducible, direct, and efficient method of the fabrication of hydrogel nanoparticles in a surfactant-free aqueous environment.

Emulsion polymerization

As a variant of free radical polymerization, emulsion polymerization occurs in an emulsion containing monomers and surfactants [67]. Due to the presence of surfactants, tedious postsynthetic treatments for the removal of surfactants may be required; however, emulsion polymerization enables the generation of a polymer with a high molecular weight at a reasonably high polymerization rate [67]. In addition to producing polymers, over the last several decades emulsion polymerization has been exploited for nanoparticle fabrication. A good example is the poly(oligo(ethylene oxide) monomethyl ether methacrylate) (POEOMA) hydrogel nanoparticles, which have been produced in a cyclohexane inverse mini-emulsion via ATRP, in the presence of dithiopropionyl PEG dimethacrylate as a cross-linker [68]. Because of the presence of disulfide linkages in the nanoparticles produced, the nanoparticles have been found to be degradable in the presence of glutathione [68], which is a peptide that can be found basically in all cell compartments ranging from the endoplasmic reticulum to the cytosol [69,70]. This has lowered the cytotoxicity of the nanoparticles for applications in drug delivery.

Another example is the hydrogel nanoparticles produced by inverse microemulsion polymerization using acrylate monomers, including 2-hydroxyethylacrylate (HEA), PEG-diacrylate (PEGDA), and 2-acryloxyethyltrimethylammonium chloride (AETMAC). During the synthetic process, AETMAC has been used to enhance the electrostatic interactions between the hydrogel system generated and the phosphate groups in gene drugs [71]. HEA and PEGDA have helped to lower the quaternary ammonium ion content of the system and hence the cytotoxicity [71]. Characterization of the nanoparticles has revealed that the size distribution of the nanoparticles is narrow [71], with negligible aggregation observed even after 1-month storage at ambient conditions or after 6-month storage at 4°C in an aqueous medium [71]. Because of the presence of quaternary ammonium ion side chains, hydrogel nanoparticles formed with AETMAC have been found to possess a positive surface charge. This has made cellular internalization of the nanoparticles more effective [71].

Methods for drug loading

Practically, different methods (including covalent bonding and physical entrapment) can be used to load drug molecules into nanoparticles. These methods can either be physical or chemical in nature. They possess different advantages and limitations imposed on subsequent in vivo applications of the nanoparticles generated.

Strategies for cargo loading

To perform drug loading chemically, drug molecules can be bonded to a hydrogel network. This approach has previously been used to load CS-based hydrogel nanoparticles with 5-fluorouracil (5-FU) derivatives (aminopentyl-carbamoyl-5-FU or aminopentyl-ester-methylene-5-FU) [72]. During the loading process, CS has been cross-linked with glutaraldehyde, which has also enabled the binding of the 5-FU derivatives to the nanoparticles via Schiff's base formation. Although this loading method enables stable immobilization of drug molecules to the hydrogel nanoparticles; side reactions with functional groups in the loaded drug may occur, causing a reduction in the therapeutic action of the loaded agent. Therefore when drug molecules are involved, careful planning of the loading process and thorough understanding of the drug action are necessary to make the chemical loading process successful.

Drug loading can also be carried out physically by either attaching drug molecules to the nanoparticle surface or by using the hydrogel matrix to entrap drug molecules [73–75]. In comparison with the chemical loading method, physical entrapment of drug molecules does not need to involve chemical or photochemical triggering during the loading process, thereby being able to minimize structural changes possibly introduced to some fragile drug molecules [76,77]. The ability of physical entrapment to maintain the drug activities has been evidenced by a previous study [78], in which insulin-loaded CS-based hydrogel nanoparticles have been generated by first dissolving insulin in an NaOH solution that has later been mixed with a TPP solution, followed by ionotropic gelation of CS with TPP anions. In vivo studies have revealed the maintenance of the activity of the insulin after the loading process, and have shown the capacity of the CS-based nanoparticles to enhance nasal insulin absorption [78]. Recently, a similar drug loading procedure has been adopted by Sarei et al. [79] to obtain diphtheria toxoid-loaded nanoparticles, which have been fabricated by simply adding the Alg solution with the toxoid-containing calcium chloride solution under homogenization. After the loading process, the antigenicity and stability of toxoid have been found to be maintained [79], and a higher humoral immune response in guinea pigs has resulted after immunization with the toxoid that has been loaded into the nanoparticles [79].

In addition to carrying therapeutic agents, hydrogel nanoparticles can carry contrast agents to enhance tracking of the fate of the nanoparticles inside a body or to mediate complementary imaging. This has been evidenced by an earlier study, in which a brain tumor model has been established from an 8-week-old Sprague Dawley rat [80]. When the tumor radius has reached 1–2 mm, polyacrylamide hydrogel nanoparticles, into which Coomassie Blue has been covalently incorporated, have been given to the rat via i.v. administration [80]. By taking advantage of the tumor-specific visible color staining offered by those nanoparticles, color-guided tumor resection has been performed in real time [80]. In addition, owing to the presence of PEG and F3 peptides that have been conjugated to the nanoparticle surface, the blood circulation time and tumor targeting capacity of the nanoparticles have been enhanced [80]. This study has revealed that hydrogel nanoparticles can be used not only in drug delivery but also in diagnosis.

Factors influencing the loading efficiency

The process of drug loading is one of the factors to be considered during the development of nanoparticles because it may affect the in vivo performance of the nanoparticles by causing a change in the physical properties (and hence the pharmacodynamics) of the nanoparticles. This has been suggested by the variation in biodistribution between QD- and fluorescein isothiocyanate (FITC)-incorporated hydrogel nanoparticles, with the former exhibiting a higher level of accumulation at the tumor site and a lower rate of renal excretion [81]. The hydrophilicity of the hydrogel matrix is another factor to be taken into account during nanoparticle development. As a matter of fact, if the affinity of the drug molecules with the hydrophilic hydrogel matrix is low, the time in which the hydrogel can retain the loaded drug will be limited. Under this situation, the loaded drug may diffuse out of the hydrogel matrix before the process of drug loading is complete. This limits the loading yield obtained. To tackle this problem, one of the possible methods is to structurally modify the polymer constituents of the hydrogel so as to increase the affinity of the drug molecules with the hydrogel matrix. The viability of this approach has been shown by a previous study [82], which has studied the changes in the efficiency of loading acrylic acid hydrogels with triamcinolone acetonide (TA) after acrylamidomethyl-γ-CD (γ-CD-NMA) incorporation. Due to the greater solubility and affinity of γ-CD-NMA than sodium acrylate in 40% ethanol, in this solution the acrylic

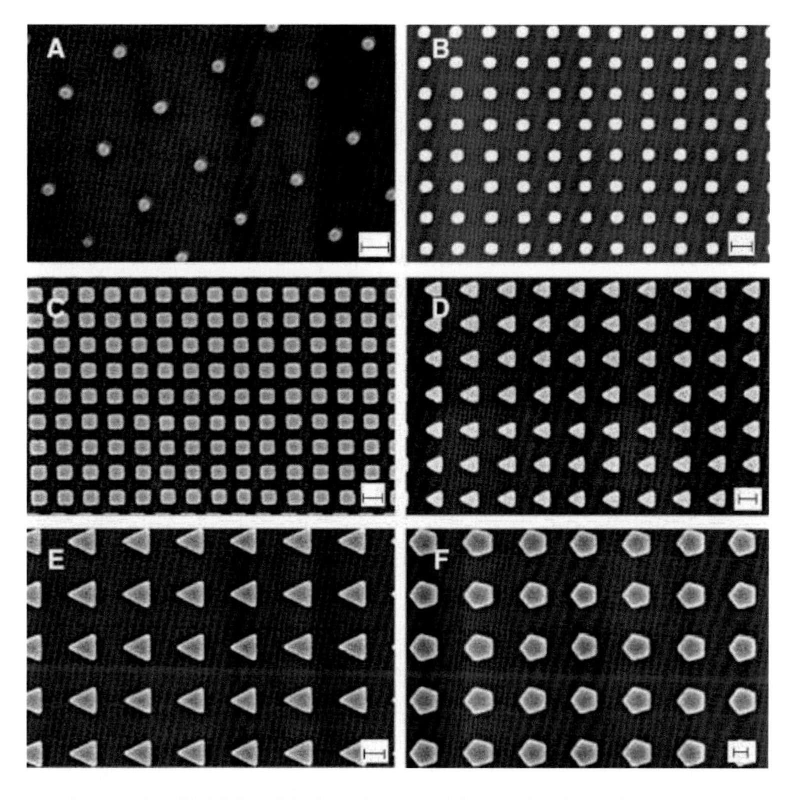

FIGURE 7.4 Scanning electron micrographs of PEG-based hydrogel nanoparticles produced by using S-FIL: square nanoparticles with the size of (A) 50 nm, (B) 100 nm, and (C) 200 nm; triangular nanoparticles with the size of (D) 200 nm and (E) 400 nm; pentagonal nanoparticles with the size of (F) 400 nm. The scale bar in (A) is 100 nm; the bars in (B), (D), and (F) are 200 nm, and the bars in (C) and € are 300 nm. *Reproduced from L.C. Glangchai, M. Caldorera-Moore, L. Shi, K. Roy, Nanoimprint lithography based fabrication of shape-specific, enzymatically-triggered smart nanoparticles, J. Control Release 125 (2008) 263–272 with permission from Elsevier B.V. [88].*

acid hydrogels prepared with γ-CD-NMA have been found to swell more remarkably than those prepared without γ-CD-NMA [82]. Furthermore, the loading efficiency of hydrogels possessing 9.4% γ-CD-NMA has been found to be over 50-fold higher than that of conventional acrylic acid hydrogels [82]. This method of modulating the affinity of the hydrogel matrix with drug molecules has also been adopted by Byrne and colleagues [83], who have generated hydrogels from multiple functional monomers and template molecules to increase the loading efficiency. All these studies have supported the feasibility of facilitating the loading efficiency at the hydrogel design level.

Modulation of drug release

In addition to high loading efficiency, high drug release sustainability is needed when hydrogel nanoparticles are applied as drug carriers. This can help to reduce the total dose of drug administration [84–87]. Drug release from hydrogel nanoparticles can be achieved via both physical and chemical mechanisms. In the former, one of the major mechanisms of drug release is diffusion. To modulate the kinetics of drug release in this scenario, one can manipulate the diffusion coefficient by engineering the polymer constituents of hydrogel nanoparticles. Alternatively, one may consider manipulating the geometry of the hydrogel nanoparticles to change the diffusion profile. Nonspherical nanoparticles can be generated by using step and flash imprint lithography (S-FIL) (Fig. 7.4) [88]. During nanoparticle fabrication, a quartz template with relief images is used to first mold photopolymerizable macromers into patterns on a substrate, followed by the removal of the template to reveal the nanoparticles [88]. Upon the etching of the thin residual layer by using oxygen plasma, nonspherical nanoparticles can be obtained [88]. In addition to S-FIL, other methods that can be used to generate nonspherical particles include the mini-emulsion technique [89], template-assisted assembly [90], particle replication in nonwetting templates [91,92], and film stretching [93]. Finally, the rate of diffusion can be altered by changing the mesh size within the hydrogel matrix [94]. As changing the mesh size may lead to alternations in other physical properties (including the mechanical strength, diffusivity, and degradability) of the hydrogel systems [95,96], this may, however, make the optimization process a bit complicated.

Apart from the physical means, drug release from hydrogel nanoparticles can be mediated via chemical reactions upon incorporation of stimuli-responsive elements into the nanoparticles. For example, by incorporating the polymer constituents of the hydrogel nanoparticles with either acidic or basic moieties that can undergo ionization at specific pH values [97,98], pH-sensitive nanoparticles can be obtained. By manipulating the temperature-dependent balance of different interactions (including hydrogen bonding, physical entanglements, and hydrophobic interactions) that lead to the cross-linking of polymer chains [99], one can engineer the thermo-sensitive gelation behavior of the hydrogel nanoparticles to fabricate a delivery system that is sensitive to temperature. Besides temperature and pH, hydrogel nanoparticles can be engineered to show sensitivity to the surrounding biomolecules or even ionic strength. These systems undergo conformational changes in response to the concentrations of surrounding biomolecules or ions [100]. One example is the previously reported hydrogel nanoparticles prepared by the interactions between pregel nuclei (which are formed by interactions between Alg and calcium ions) and PEI [55]. Because the formation of these nanoparticles is a result of complex coacervation, which is driven by attractive electrostatic forces and entropically favorable molecular rearrangements [101,102], the entropic driving force for complex formation (as well as the stability of the nanoparticles) can be easily changed when the ionic strength of the surrounding medium changes [102–104]. Such sensitivity can be exploited to enable the onset of drug release to be tuned to occur only when the nanoparticles reach the blood circulation. Another example is glucose-sensitive nanoparticles. These nanoparticles can possibly be generated by incorporating both glucose oxidase and pH-sensitive moieties into the hydrogel. After that, glucose molecules diffusing into the hydrogel can be converted to gluconic acid. This leads to a decrease in pH, followed by protonation of the amine functionalities present in the system. The final result is the swelling of the hydrogel nanoparticles and the release of the loaded bioactive agents [105,106].

Example protocols for experimental design

The method below is an example protocol for evaluating the drug release sustainability of the hydrogel nanoparticles:

1. Mix the drug-loaded nanoparticles with a known volume of a releasing medium in a container.
2. Incubate the nanoparticles at 37°C and 5% CO_2 with saturated humidity.
3. Remove a known volume of the releasing medium from the container at intervals of 10 min.

(Continued)

(Continued)

4. Each time after sampling, add the fresh releasing medium, in a volume equal to the sampling volume adopted in Step 3, back to the container.
5. Determine the amount of the drug released from the nanoparticles using an appropriate method of quantification.
6. Calculate the cumulative drug release using the following equation, where m_t is the mass of the drug released from the nanoparticles at time t, and m_∞ is the mass of the drug loaded into the nanoparticles.

$$\text{Cumulative protein release } (\%) = \frac{\sum_{t=0}^{t} m_t}{m_\infty} \times 100\% \qquad (7.2)$$

Summary

In this chapter, we have presented an overview of the recent development of hydrogel-based nanoparticulate systems for drug delivery, and have discussed different factors that should be considered when the properties of hydrogel nanoparticles are modulated. With the continuous advances in materials engineering, techniques to manipulate materials properties are expected to be more sophisticated in the forthcoming decades. The functional versatility of hydrogel materials may also be extended remarkably by emerging microfabrication technologies. Taken this into consideration, to design carrier systems for the execution of biogerontological interventions we may no longer want to limit the possibility and efforts only to the manipulation of chemical structures. Physical parameters such as the size and system geometry may also be optimized so that a more ideal system can be obtained for intervention development.

Directions for intervention development

Based on the practical needs (e.g., the properties of the agent to be delivered, and the site of action of the agent) of a proposed intervention, we may get closer to put the intervention into practice by following the steps below:

1. List the practical needs of the proposed intervention.
2. Select a hydrogel material that can meet those needs.
3. Choose a production method suitable for the fabrication of the nanoparticulate system.
4. Generate the nanoparticulate system from the selected hydrogel material.
5. Fine-tune and characterize the system generated.
6. Examine the performance of the system in vitro and in vivo.
7. Optimize the performance and properties of the system to improve the efficiency of the intervention.

References

[1] G. Lambert, E. Fattal, P. Couvreur, Nanoparticulate systems for the delivery of antisense oligonucleotides, Adv. Drug Deliv. Rev. 47 (2001) 99–112.

[2] M. Rizwan, S. Ali, M.F. Qayyum, Y.S. Ok, M. Adrees, M. Ibrahim, et al., Effect of metal and metal oxide nanoparticles on growth and physiology of globally important food crops: a critical review, J. Hazard. Mater. (2016). Available from: https://doi.org/10.1016/j.jhazmat.2016.05.061.

[3] M.F. Hornos Carneiro, F. Barbosa Jr., Gold nanoparticles: a critical review of therapeutic applications and toxicological aspects, J. Toxicol. Environ. Health B Crit. Rev. 19 (2016) 129–148.

[4] R.C. Beck, A.F. Ourique, S.S. Guterres, A.R. Pohlmann, Spray-dried polymeric nanoparticles for pharmaceutics: a review of patents, Recent Pat. Drug Deliv. Formul. 6 (2012) 195–208.

[5] S. Weber, A. Zimmer, J. Pardeike, Solid lipid nanoparticles (SLN) and nanostructured lipid carriers (NLC) for pulmonary application: a review of the state of the art, Eur. J. Pharm. Biopharm. 86 (2014) 7–22.

[6] E.M. Ahmed, Hydrogel: preparation, characterization, and applications: a review, J. Adv. Res. 6 (2015) 105–121.

[7] W.F. Lai, H.C. Shum, Hypromellose-graft-chitosan and Its polyelectrolyte complex as novel systems for sustained drug delivery, ACS Appl. Mater. Interfaces 7 (2015) 10501–10510.

[8] Z. Lin, W. Gao, H. Hu, K. Ma, B. He, W. Dai, et al., Novel thermo-sensitive hydrogel system with paclitaxel nanocrystals: high drug-loading, sustained drug release and extended local retention guaranteeing better efficacy and lower toxicity, J. Control Release 174 (2014) 161–170.

[9] J.M. van Bemmelenm, Das hydrogel und das krystallinische hydrat des kupferoxyds, Z. Anorg. Chem. 5 (1894) 466–483.

[10] O. Wichterle, Hydrogels, in: H.F. Mark, N.G. Gaylord (Eds.), Encyclopedia of Polymer Science and Technology, Interscience, New York, 1971, p. 273.

[11] O. Wichterle, D. Lím, Hydrophilic gels for biological use, Nature 185 (1960) 117–118.

[12] Z. Li, W. Ning, J. Wang, A. Choi, P.Y. Lee, P. Tyagi, et al., Controlled gene delivery system based on thermosensitive biodegradable hydrogel, Pharm. Res. 20 (2003) 884–888.

[13] N.J. Meilander-Lin, P.J. Cheung, D.L. Wilson, R.V. Bellamkonda, Sustained in vivo gene delivery from agarose hydrogel prolongs nonviral gene expression in skin, Tissue Eng. 11 (2005) 546–555.

[14] C.A. Holden, P. Tyagi, A. Thakur, R. Kadam, G. Jadhav, U.B. Kompella, et al., Polyamidoamine dendrimer hydrogel for enhanced delivery of antiglaucoma drugs, Nanomedicine 8 (2012) 776–783.

[15] J. Kim, A. Conway, A. Chauhan, Extended delivery of ophthalmic drugs by silicone hydrogel contact lenses, Biomaterials 29 (2008) 2259–2269.

[16] A.K. Banga, Y.W. Chien, Hydrogel-based iontotherapeutic delivery devices for transdermal delivery of peptide/protein drugs, Pharm. Res. 10 (1993) 697–702.

[17] S.P. Vyas, K. Khatri, Liposome-based drug delivery to alveolar macrophages, Expert Opin. Drug Deliv. 4 (2007) 95–99.

[18] G.K. Khuller, M. Kapur, S. Sharma, Liposome technology for drug delivery against mycobacterial infections, Curr. Pharm. Des. 10 (2004) 3263–3274.

[19] D.S. Mahrhauser, G. Reznicek, S. Gehrig, A. Geyer, M. Ogris, R. Kieweler, et al., Simultaneous determination of active component and vehicle penetration from F-DPPC liposomes into porcine skin layers, Eur. J. Pharm. Biopharm 97 (2015) 90–95.

[20] C.X. Zhao, Multiphase flow microfluidics for the production of single or multiple emulsions for drug delivery, Adv. Drug Deliv. Rev. 65 (2013) 1420–1446.

[21] I. Winer, S. Wang, Y.E. Lee, W. Fan, Y. Gong, D. Burgos-Ojeda, et al., F3-targeted cisplatin-hydrogel nanoparticles as an effective therapeutic that targets both murine and human ovarian tumor endothelial cells in vivo, Cancer Res. 70 (2010) 8674–8683.

[22] S. Christian, J. Pilch, M.E. Akerman, K. Porkka, P. Laakkonen, E. Ruoslahti, Nucleolin expressed at the cell surface is a marker of endothelial cells in angiogenic blood vessels, J. Cell Biol. 163 (2003) 871–878.

[23] G.R. Reddy, M.S. Bhojani, P. McConville, J. Moody, B.A. Moffat, D.E. Hall, et al., Vascular targeted nanoparticles for imaging and treatment of brain tumors, Clin. Cancer Res. 12 (2006) 6677–6686.

[24] K. Porkka, P. Laakkonen, J.A. Hoffman, M. Bernasconi, E. Ruoslahti, A fragment of the HMGN2 protein homes to the nuclei of tumor cells and tumor endothelial cells in vivo, P. Natl. Acad. Sci. U. S. A. 99 (2002) 7444–7449.

[25] R. Tripathi, B. Mishra, Development and evaluation of sodium alginate-polyacrylamide graft-co-polymer-based stomach targeted hydrogels of famotidine, AAPS PharmSciTech 13 (2012) 1091–1102.

[26] W.K. Bae, M.S. Park, J.H. Lee, J.E. Hwang, H.J. Shim, S.H. Cho, et al., Docetaxel-loaded thermoresponsive conjugated linoleic acid-incorporated poloxamer hydrogel for the suppression of peritoneal metastasis of gastric cancer, Biomaterials 34 (2013) 1433–1441.

[27] J. Li, C. Gong, X. Feng, X. Zhou, X. Xu, L. Xie, et al., Biodegradable thermosensitive hydrogel for SAHA and DDP delivery: therapeutic effects on oral squamous cell carcinoma xenografts, PLoS One 7 (2012) e33860.

[28] R. Pignatello, A.H. Stancampiano, C.A. Ventura, G. Puglisi, Dexamethasone sodium phosphate-loaded chitosan based delivery systems for buccal application, J. Drug Target. 15 (2007) 603–610.

[29] S. Heilmann, S. Kuchler, C. Wischke, A. Lendlein, C. Stein, M. Schafer-Korting, A thermosensitive morphine-containing hydrogel for the treatment of large-scale skin wounds, Int. J. Pharm. 444 (2013) 96–102.

[30] B.K. Abdul Rasool, E.F. Abu-Gharbieh, S.A. Fahmy, H.S. Saad, S.A. Khan, Development and evaluation of ibuprofen transdermal gel formulations, Trop. J. Pharm. Res. 9 (2010) 355–363.

[31] V. Sabale, S. Vora, Formulation and evaluation of microemulsion-based hydrogel for topical delivery, Int. J. Pharm. Investig. 2 (2012) 140–149.

[32] S. Das, U. Subuddhi, Cyclodextrin mediated controlled release of naproxen from pH-sensitive chitosan/poly(vinyl alcohol) hydrogels for colon targeted delivery, Ind. Eng. Chem. Res. 52 (2013) 14192–14200.

[33] Y.L. Lo, C.Y. Hsu, H.R. Lin, pH- and thermo-sensitive pluronic/poly(acrylic acid) in situ hydrogels for sustained release of an anticancer drug, J. Drug Target. 21 (2013) 54–66.

[34] J.L. Georgii, T.P. Amadeu, A.B. Seabra, M.G. de Oliveira, A. Monte-Alto-Costa, Topical S-nitrosoglutathione-releasing hydrogel improves healing of rat ischaemic wounds, J. Tissue Eng. Regen. Med. 5 (2011) 612–619.

[35] K. Reimer, P.M. Vogt, B. Broegmann, J. Hauser, O. Rossbach, A. Kramer, et al., An innovative topical drug formulation for wound healing and infection treatment: in vitro and in vivo investigations of a povidone-iodine liposome hydrogel, Dermatology 201 (2000) 235–241.
[36] L. Jia, H. Wong, C. Cerna, S.D. Weitman, Effect of nanonization on absorption of 301029: ex vivo and in vivo pharmacokinetic correlations determined by liquid chromatography/mass spectrometry, Pharm. Res. 19 (2002) 1091–1096.
[37] W.F. Lai, Z.D. He, Design and fabrication of hydrogel-based nanoparticulate systems for in vivo drug delivery, J. Control Release 243 (2016) 269–282.
[38] S.K. Hobbs, W.L. Monsky, F. Yuan, W.G. Roberts, L. Griffith, V. P. Torchilin, et al., Regulation of transport pathways in tumor vessels: role of tumor type and microenvironment, Proc. Natl. Acad. Sci. U. S. A. 95 (1998) 4607–4612.
[39] R. Nakaoka, Y. Tabata, T. Yamaoka, Y. Ikada, Prolongation of the serum half-life period of superoxide dismutase by poly (ethylene glycol) modification, J. Control Release 46 (1997) 253–261.
[40] S.M. Moghimi, A.C. Hunter, J.C. Murray, Long-circulating and target-specific nanoparticles: theory to practice, Pharmacol. Rev. 53 (2001) 283–318.
[41] T. Banerjee, S. Mitra, A. Kumar Singh, R. Kumar Sharma, A. Maitra, Preparation, characterization and biodistribution of ultrafine chitosan nanoparticles, Int. J. Pharm. 243 (2002) 93–105.
[42] M. Gaumet, A. Vargas, R. Gurny, F. Delie, Nanoparticles for drug delivery: the need for precision in reporting particle size parameters, Eur. J. Pharm. Biopharm. 69 (2008) 1–9.
[43] S.M. Moghimi, Exploiting bone marrow microvascular structure for drug delivery and future therapies, Adv. Drug Deliv. Rev. 17 (1995) 61–73.
[44] Y. Takakura, R.I. Mahato, M. Hashida, Extravasation of macromolecules, Adv. Drug Deliv. Rev. 34 (1998) 93–108.
[45] S.M. Moghimi, Mechanisms of splenic clearance of blood cells and particles: towards development of new splenotropic agents, Adv. Drug Deliv. Rev. 17 (1995) 103–115.
[46] A.K. Stark, M. Schilling, D. Janasek, J. Franzke, Characterization of dielectric barrier electrospray ionization for mass spectrometric detection, Anal. Bioanal. Chem. 397 (2010) 1767–1772.
[47] U. Nagarajan, K. Kawakami, S. Zhang, B. Chandrasekaran, B. Unni Nair, Fabrication of solid collagen nanoparticles using electrospray deposition, Chem. Pharm. Bull. 62 (2014) 422–428.
[48] S. Zhang, K. Kawakami, One-step preparation of chitosan solid nanoparticles by electrospray deposition, Int. J. Pharm. 397 (2010) 211–217.
[49] H. Valo, L. Peltonen, S. Vehvilainen, M. Karjalainen, R. Kostiainen, T. Laaksonen, et al., Electrospray encapsulation of hydrophilic and hydrophobic drugs in poly(L-lactic acid) nanoparticles, Small 5 (2009) 1791–1798.
[50] J. Qu, Y. Liu, Y. Yu, J. Li, J. Luo, M. Li, Silk fibroin nanoparticles prepared by electrospray as controlled release carriers of cisplatin, Mater. Sci. Eng. C Mater. Biol. Appl. 44 (2014) 166–174.
[51] K. Songsurang, N. Praphairaksit, K. Siraleartmukul, N. Muangsin, Electrospray fabrication of doxorubicin-chitosan-tripolyphosphate nanoparticles for delivery of doxorubicin, Arch. Pharm. Res. 34 (2011) 583–592.
[52] J. Rosell-Llompart, J. Fernández la Mora, Generation of monodisperse droplets 0.3 to 4 μm in diameter from electrified cone-jets of highly conducting and viscous liquids, J. Aerosol Sci. 25 (1994) 1093–1119.
[53] Y.C. Dong, W.K. Ng, S.C. Shen, S. Kim, R.B.H. Tan, Scalable ionic gelation synthesis of chitosan nanoparticles for drug delivery in static mixers, Carbohydr. Polym. 94 (2013) 940–945.
[54] M. Rajaonarivony, C. Vauthier, G. Couarraze, F. Puisieux, P. Couvreur, Development of a new drug carrier made from alginate, J. Pharm. Sci. 82 (1993) 912–917.
[55] W.F. Lai, H.C. Shum, A stimuli-responsive nanoparticulate system using poly(ethylenimine)-graft-polysorbate for controlled protein release, Nanoscale 8 (2016) 517–528.
[56] S.Y. Yu, J.H. Hu, X.Y. Pan, P. Yao, M. Jiang, Stable and pH-sensitive nanogels prepared by self-assembly of chitosan and ovalbumin, Langmuir 22 (2006) 2754–2759.
[57] K. Kuroda, K. Fujimoto, J. Sunamoto, K. Akiyoshi, Hierarchical self-assembly of hydrophobically modified pullulan in water: gelation by networks of nanoparticles, Langmuir 18 (2002) 3780–3786.
[58] E. Akiyama, N. Morimoto, P. Kujawa, Y. Ozawa, F.M. Winnik, K. Akiyoshit, Self-assembled nanogels of cholesteryl-modified polysaccharides: effect of the polysaccharide structure on their association characteristics in the dilute and semidilute regimes, Biomacromolecules 8 (2007) 2366–2373.
[59] K. Nagahama, Y. Mori, Y. Ohya, T. Ouchi, Biodegradable nanogel formation of polylactide-grafted dextran copolymer in dilute aqueous solution and enhancement of its stability by stereocomplexation, Biomacromolecules 8 (2007) 2135–2141.
[60] O. Nuyken, S.D. Pask, Ring-opening polymerization-an introductory review, Polymers (Basel) 5 (2013) 361–403.
[61] M.H. Xiong, J. Wu, Y.C. Wang, L.S. Li, X.B. Liu, G.Z. Zhang, et al., Synthesis of PEG-armed and polyphosphoester core-cross-linked nanogel by one-step ring-opening polymerization, Macromolecules 42 (2009) 893–896.
[62] Y.C. Wang, L.Y. Tang, T.M. Sun, C.H. Li, M.H. Xiong, J. Wang, Self-assembled micelles of biodegradable triblock copolymers based on poly(ethyl ethylene phosphate) and poly(epsilon-caprolactone) as drug carriers, Biomacromolecules 9 (2008) 388–395.
[63] X.Z. Yang, Y.C. Wang, L.Y. Tang, H. Xia, J. Wang, Synthesis and characterization of amphiphilic block copolymer of polyphosphoester and poly(L-lactic acid), J. Polym. Sci. Pol. Chem. 46 (2008) 6425–6434.
[64] S. Sugihara, A. Blanazs, S.P. Armes, A.J. Ryan, A.L. Lewis, Aqueous dispersion polymerization: a new paradigm for in situ block copolymer self-assembly in concentrated solution, J. Am. Chem. Soc. 133 (2011) 15707–15713.
[65] H.F. Mark, Encyclopedia of Polymer Science and Technology, Wiley-Interscience, Hoboken, 2007.
[66] J. Rieger, C. Grazon, B. Charleux, D. Alaimo, C. Jerome, Pegylated thermally responsive block copolymer micelles and nanogels via in situ RAFT aqueous dispersion polymerization, J. Polym. Sci. Pol. Chem. 47 (2009) 2373–2390.
[67] I. Capek, L. Fialova, D. Berek, On the kinetics of inverse emulsion polymerization of acrylamide, Des. Monomers Polym. 11 (2008) 123–137.
[68] J.K. Oh, D.J. Siegwart, H.I. Lee, G. Sherwood, L. Peteanu, J.O. Hollinger, et al., Biodegradable nanogels prepared by atom transfer radical polymerization as potential drug delivery carriers: synthesis, biodegradation, in vitro release, and bioconjugation, J. Am. Chem. Soc. 129 (2007) 5939–5945.

[69] H.D. Humes, D.A. Buffington, S.M. MacKay, A.J. Funke, W.F. Weitzel, Replacement of renal function in uremic animals with a tissue-engineered kidney, Nat. Biotechnol. 17 (1999) 451–455.
[70] C.H. Foyer, G. Noctor, Redox sensing and signalling associated with reactive oxygen in chloroplasts, peroxisomes and mitochondria, Physiol. Plantarum 119 (2003) 355–364.
[71] K. McAllister, P. Sazani, M. Adam, M.J. Cho, M. Rubinstein, R. J. Samulski, et al., Polymeric nanogels produced via inverse microemulsion polymerization as potential gene and antisense delivery agents, J. Am. Chem. Soc. 124 (2002) 15198–15207.
[72] Y. Ohya, M. Shiratani, H. Kobayashi, T. Ouchi, Release behavior of 5-fluorouracil from chitosan-gel nanospheres immobilizing 5-fluorouracil coated with polysaccharides and their cell-specific cytotoxicity, J. Macromol Sci. A A31 (1994) 629–642.
[73] A. Alsughayer, A.Z.A. Elassar, F. Al Sagheer, S. Mustafa, Synthesis and characterization of polysulfanilamide and its copolymers: bioactivity and drug release, Pharm. Chem. J. 46 (2012) 418–428.
[74] X. Lu, Y. Ping, F.J. Xu, Z.H. Li, Q.Q. Wang, J.H. Chen, et al., Bifunctional conjugates comprising beta-cyclodextrin, polyethylenimine, and 5-fluoro-2'- deoxyuridine for drug delivery and gene transfer, Bioconj. Chem. 21 (2010) 1855–1863.
[75] W.F. Lai, Cyclodextrins in non-viral gene delivery, Biomaterials 35 (2014) 401–411.
[76] X.L. Zhang, J. Huang, P.R. Chang, J.L. Li, Y.M. Chen, D.X. Wang, et al., Structure and properties of polysaccharide nanocrystal-doped supramolecular hydrogels based on cyclodextrin inclusion, Polymer 51 (2010) 4398–4407.
[77] D. Sarkar, Fabrication of an optimized fluorescer encapsulated polymer coated gelatin nanoparticle and study of its retarded release properties, J. Photochem. Photobiol. A 252 (2013) 194–202.
[78] R. Fernandez-Urrusuno, P. Calvo, C. Remunan-Lopez, J.L. Vila-Jato, M.J. Alonso, Enhancement of nasal absorption of insulin using chitosan nanoparticles, Pharm. Res. 16 (1999) 1576–1581.
[79] F. Sarei, N.M. Dounighi, H. Zolfagharian, P. Khaki, S.M. Bidhendi, Alginate nanoparticles as a promising adjuvant and vaccine delivery system, Indian J. Pharm. Sci. 75 (2013) 442–449.
[80] G. Nie, H.J. Hah, G. Kim, Y.E. Lee, M. Qin, T.S. Ratani, et al., Hydrogel nanoparticles with covalently linked coomassie blue for brain tumor delineation visible to the surgeon, Small 8 (2012) 884–891.
[81] R. Bakalova, B. Nikolova, S. Murayama, S. Atanasova, Z. Zhelev, I. Aoki, et al., Passive and electro-assisted delivery of hydrogel nanoparticles in solid tumors, visualized by optical and magnetic resonance imaging in vivo, Anal. Bioanal. Chem. 408 (2016) 905–914.
[82] U. Siemoneit, C. Schmitt, C. Alvarez-Lorenzo, A. Luzardo, F. Otero-Espinar, A. Concheiro, et al., Acrylic/cyclodextrin hydrogels with enhanced drug loading and sustained release capability, Int. J. Pharm. 312 (2006) 66–74.
[83] S. Venkatesh, S.P. Sizemore, M.E. Byrne, Biomimetic hydrogels for enhanced loading and extended release of ocular therapeutics, Biomaterials 28 (2007) 717–724.
[84] J.E. Oh, Y.S. Nam, K.H. Lee, T.G. Park, Conjugation of drug to poly(D,L-lactic-co-glycolic acid) for controlled release from biodegradable microspheres, J. Control Release 57 (1999) 269–280.
[85] Y.S. Nam, J.Y. Park, S.H. Han, I.S. Chang, Intracellular drug delivery using poly(D,L-lactide-co-glycolide) nanoparticles derivatized with a peptide from a transcriptional activator protein of HIV-1, Biotechnol. Lett. 24 (2002) 2093–2098.
[86] R.C. Luo, Y. Cao, P. Shi, C.H. Chen, Near-infrared light responsive multi-compartmental hydrogel particles synthesized through droplets assembly induced by superhydrophobic surface, Small 10 (2014) 4886–4894.
[87] S.H. Kim, J.W. Kim, D.H. Kim, S.H. Han, D.A. Weitz, Polymersomes containing a hydrogel network for high stability and controlled release, Small 9 (2013) 124–131.
[88] L.C. Glangchai, M. Caldorera-Moore, L. Shi, K. Roy, Nanoimprint lithography based fabrication of shape-specific, enzymatically-triggered smart nanoparticles, J. Control Release 125 (2008) 263–272.
[89] Z. Yang, W.T. Huck, S.M. Clarke, A.R. Tajbakhsh, E.M. Terentjev, Shape-memory nanoparticles from inherently non-spherical polymer colloids, Nat. Mater. 4 (2005) 486–490.
[90] D. Kohler, M. Schneider, M. Kruger, C.M. Lehr, H. Mohwald, D. Wang, Template-assisted polyelectrolyte encapsulation of nanoparticles into dispersible, hierarchically nanostructured microfibers, Adv. Mater. 23 (2011) 1376–1379.
[91] J. Xu, D.H. Wong, J.D. Byrne, K. Chen, C. Bowerman, J.M. DeSimone, Future of the particle replication in nonwetting templates (PRINT) technology, Angew. Chem. Int. Ed. Engl. 52 (2013) 6580–6589.
[92] J.Y. Kelly, J.M. DeSimone, Shape-specific, monodisperse nanomolding of protein particles, J. Am. Chem. Soc. 130 (2008) 5438–5439.
[93] C.C. Ho, A. Keller, J.A. Odell, R.H. Ottewill, Preparation of monodisperse ellipsoidal polystyrene particles, Colloid Polym. Sci. 271 (1993) 469–479.
[94] T. Canal, N.A. Peppas, Correlation between mesh size and equilibrium degree of swelling of polymeric networks, J. Biomed. Mater. Res. 23 (1989) 1183–1193.
[95] B. Amsden, Solute diffusion within hydrogels. mechanisms and models, Macromolecules 31 (1998) 8382–8395.
[96] M.N. Mason, A.T. Metters, C.N. Bowman, K.S. Anseth, Predicting controlled-release behavior of degradable PLA-b-PEG-b-PLA hydrogels, Macromolecules 34 (2001) 4630–4635.
[97] Y. Qiu, K. Park, Environment-sensitive hydrogels for drug delivery, Adv. Drug Deliv. Rev. 53 (2001) 321–339.
[98] P. Gupta, K. Vermani, S. Garg, Hydrogels: from controlled release to pH-responsive drug delivery, Drug Discov. Today 7 (2002) 569–579.
[99] G.B.W.L. Ligthart, O.A. Scherman, R.P. Sijbesma, E.W. Meijer, Supramolecular polymer engineering, in: K. Matyjaszewski, Y. Gnanou, L. Leibler (Eds.), Macromolecular Engineering: Precise Synthesis, Materials Properties, Applications, Wiley-VCH Verlag GmbH & Co. KGaA, Germany, 2007, pp. 351–399.
[100] T. Miyata, T. Uragami, K. Nakamae, Biomolecule-sensitive hydrogels, Adv. Drug Deliv. Rev. 54 (2002) 79–98.
[101] J.T. Overbeek, M.J. Voorn, Phase separation in polyelectrolyte solutions: theory of complex coacervation, J. Cell. Physiol. Suppl. 49 (1957) 7–22. discussion, 22–26.
[102] S.L. Perry, Y. Li, D. Priftis, L. Leon, M. Tirrell, The effect of salt on the complex coacervation of vinyl polyelectrolytes, Polymers (Basel) 6 (2014) 1756–1772.

[103] J. van der Gucht, E. Spruijt, M. Lemmers, M.A.C. Stuart, Polyelectrolyte complexes: bulk phases and colloidal systems, J. Colloid Interfaces Sci. 361 (2011) 407–422.
[104] P.M. Biesheuvel, M.A.C. Stuart, Electrostatic free energy of weakly charged macromolecules in solution and intermacromolecular complexes consisting of oppositely charged polymers, Langmuir 20 (2004) 2785–2791.
[105] K. Ishihara, M. Kobayashi, N. Ishimaru, I. Shinohara, Glucose-induced permeation control of insulin through a complex membrane consisting of immobilized glucose-oxidase and a poly (amine), Polym. J. 16 (1984) 625–631.
[106] C.M. Hassan, F.J. Doyle, N.A. Peppas, Dynamic behavior of glucose-responsive poly(methacrylic acid-g-ethylene glycol) hydrogels, Macromolecules 30 (1997) 6166–6173.

Part III

From Technologies to Interventions

Chapter 8

Use of delivery technologies to mediate RNA degradation

Introduction

In Part II of this book, we have discussed different strategies for designing, developing, and engineering viral and nonviral carriers. Undoubtedly, attainment of an effective carrier is essential to intervention development, but it is only the first step of the whole developmental process. How to use the carrier properly to bring the proposed intervention to practice is another challenge that we have to overcome. From this chapter onward, we will highlight a number of approaches, which, when combined with the use of an effective delivery system, may show the potential to be turned into an antiaging intervention. This chapter will first focus on the modulation of gene expression via RNA degradation.

The feasibility of manipulating gene expression for intervening in the aging network has been demonstrated by the case of HuR. This protein belongs to a family consisting of neuron-specific proteins, HuD, HuC, Hel-N1, and others [1,2]. It contains three characteristic RNP2/RNP1-type RNA binding motifs. One RNA binding domain allows the protein to bind to the long chain poly (A) tail of an mRNA transcript [3]; whereas the other two domains render the protein capable of binding to Au-rich elements (AREs) present in the 3′-untranslated region (UTR) of an mRNA molecule to enhance the stability and/or translation of the transcript [3]. Examples of transcripts possessing AREs are those encoding for VEGF, glucose transporter 1 (GLUT1), p21, c-fos, cyclin A, and cyclin B1 [3–6]. Physiologically, HuR and AMP-activated kinase (AMPK) work together to modulate the proliferative status of a cell [7]. In brief, AMPK can be activated in response to cellular stress (e.g., hypoglycemia and serum/growth factor depletion) that increases the AMP/ATP ratio [8–10]. Such activation results in a decline in the cytoplasmic level of HuR, causing a reduction in the levels of cyclin A, cyclin B1, and p21. This leads to a reduction in the proliferative status of a cell. The cytoplasmic level of HuR also affects the process of senescence [11]. This has been illustrated by the fact that suppression of HuR expression in IDH4 human fibroblasts can lead to signs of cellular senescence, including a lower level of basal cyclin-dependent protein kinase (CDK) activity, and a higher level of senescence-associated β-galactosidase (SA-β-gal) activity [12]. These changes can be reversed upon overexpression of HuR, which can cause senescent cells to display a younger phenotype. This observation has revealed that the onset of age-associated phenotypes is related to the HuR level, and has demonstrated the viability of rejuvenating the aging phenotype via manipulation at the genetic level.

Manipulation of the aging network by RNAi

To regulate the cytoplasmic level of mRNA, precise control of the rates of RNA synthesis and degradation is required [11]. In higher eukaryotic cells, deadenylation-dependent degradation is a major mechanism of RNA degradation during which removal of the poly(A) tail from the 3′ end of the RNA molecule occurs, leading to the formation of a deadenylated or oligo-adenylated product. Finally, the 5′ end of the product is removed, followed by the occurrence of 5′–3′ and/or 3′–5′ exoribonuclease-mediated degradation [11]. Apart from deadenylation-dependent degradation, nonsense-mediated decay and nonstop decay pathways exist for RNA degradation. These pathways are especially important to the degradation of mRNA molecules that contain nonsense codons or lack stop codons, respectively [11]. In addition to the aforementioned mechanisms, RNA degradation can be achieved using RNAi, which can be mediated by using RNA duplexes such as siRNA. With the action of the RNA-induced silencing complex (RISC) in the cytoplasm, the two strands of the duplex separate, with the antisense strand bound to the RISC. The antisense strand subsequently anneals to an mRNA transcript that possesses homologous nucleotide sequences, causing

Delivery of Therapeutics for Biogerontological Interventions. DOI: https://doi.org/10.1016/B978-0-12-816485-3.00008-8

endoribonucleolytic cleavage of the transcript [13]. The action of siRNA is highly specific and efficient [14], and has been exploited in diverse biomedical applications (including therapeutic target validation [15,16] and cancer treatment [17]).

The advent of RNAi technologies has facilitated aging research by making genetic manipulation more technically practicable, thereby furthering our understanding of the mechanisms of aging and age-related diseases. This has been exemplified by studies on SOD enzymes, which play frontline defensive roles against oxidative stress by converting superoxide into oxygen and hydrogen peroxide. Three forms of SOD (viz., SOD1, SOD2, and SOD3) exist in humans; however, the role played by SOD2 has been poorly elucidated until recently when Kirby and coworkers have silenced the expression of *Sod2* in *D. melanogaster* using RNAi [18]. After gene silencing, the fruit flies have been found to experience a loss of essential enzymatic components of the mitochondrial respiratory chain and the Krebs cycle, with an increase in the sensitivity to oxidative stress and an increase in early-onset mortality in young adults. Similar hypersensitivity to oxidative stress has later been reported in a *Drosophila* model with a *Sod2* null mutation [19]. This has demonstrated the practical potential of RNAi in basic research to elucidate the roles of players involved in physiological processes. Such potential has also been demonstrated by studies on sirtuins. The role of sirtuins in modulating longevity has been suggested by the observation that either specific mutations on *Sir4* or the elevation of the *Sir2* activity can increase the replicative life span of *S. cerevisiae* [20,21]. Later, RNAi has been adopted to study the effect of the loss of function of three *Sir2* and *Sir2-like* genes on the life span of fruit flies [22]. Results have indicated that suppression of *Sir2* in neurons, or ubiquitous silencing of the *Sir2-like* genes, has led to life span shortening. In fact, there are many examples of genes whose functions have been deciphered by using RNAi [23–25]. It is anticipated that biogerontological endeavors will continue to benefit from the use of RNAi technologies in the future.

Enhancement of the siRNA performance

To enhance the performance of siRNA duplexes, one strategy is to manipulate the structure of RNA directly. After chemical modification, the cytotoxicity, hemolytic effect, stability, binding affinity, and gene silencing efficiency can be compared with the unmodified counterpart for evaluating the performance of the modified structure. Over the years, diverse chemical modification strategies have been reported. Many of those strategies have focused on the modification of the sugar moiety, particularly the 2′ position. Upon 2′-*O*-methylation, the binding affinity and nuclease stability of the duplex have been reported to be increased [26–28], although few studies have suggested that a large number of 2′-*O*-methylation modifications may decrease the activity of the duplex [29–32]. The 2′-position of the siRNA duplex can also be substituted with fluorine to enhance the serum stability and binding affinity [33]. Apart from the sugar moiety, phosphate linkages in the siRNA duplex can be replaced by phosphodiester linkages. Incorporation of phosphodiester linkages, however, into the duplex center may lead to a reduction in the potency of the siRNA molecule [34]. In addition, improper control of the level of phosphodiester linkage incorporation may cause an increase in the cytotoxicity of the duplex [35]. Instead of incorporating phosphodiester linkages, the siRNA duplex can be incorporated with boranophosphate linkages. The modified duplex not only displays good nuclease stability over the native one, but also shows higher potency as compared to the phosphodiester-modified counterpart [36].

In addition to modifying the backbone, modifications on other parts of the siRNA molecule (e.g., base, overhangs, and termini) have been adopted. This has been exemplified by the case of the 2,4-difluorotoluyl ribonucleoside (rF)-incorporated siRNA duplex [37]. The gene silencing capacity of the duplex has been found not to be affected upon 5′-modification of the guide strand with rF [37]. Internal uridine-to-rF substitutions have also been shown to be well-tolerated [37]. Upon rF modification, the nuclease resistance of the duplex in serum has been enhanced, while the capacity of gene silencing has been maintained [37]. Apart from chemical modification, manipulation of the structure of the duplex has been reported. A good example is the small internally segmented interfering RNA, which is made from an intact antisense strand complemented with two shorter sense strands [38]. This siRNA construct enables less genome-wide off-target effects to be induced, and shows the potential to improve the safety of RNAi [38]. Finally, to enhance the potency of an siRNA duplex, the duplex length can be increased so as to facilitate cellular Dicer-mediated cleavage of the duplex [39]. The success of this has been reported by an earlier study, in which the reported siRNA duplex with 25–30 nucleotides in length has been found to be 100-fold more potent than the corresponding conventional 21-mer siRNA duplex [39]. Although duplexes that are too long may trigger the interferon response [40], the possibility of enhancing the potency simply by length manipulation has provided a route to easily improve the effectiveness of RNAi-mediated target cleavage in practice.

Example protocols for experimental design

The method below is an example protocol for examining the hemolytic effect of an RNA analog.

1. Collect blood from a subject using a heparin-containing blood collection tube.
2. Centrifuge the tube for 10 minutes at 2000 × *g* at 4°C.
3. Discard the supernatant.
4. Wash the collected erythrocytes using PBS (pH = 7.4).
5. Centrifuge the tube for 10 minutes at 2000 × *g* at 4°C.
6. Discard the supernatant.
7. Repeat Steps 4–6 until the supernatant becomes colorless.
8. Add the RNA analog into the collected erythrocytes to reach an appropriate concentration.
9. Incubate the mixture at 37°C for 1 hour.
10. Centrifuge the mixture at 2000 × *g* for 15 minutes.
11. Record the absorbance of the supernatant at 414 nm.
12. Determine the hemolytic effect of the RNA analog by assuming that hemolysis led by PBS and 0.1% Triton X-100 is 0% and 100%, respectively.

Delivery of siRNA therapeutics

The wide application of siRNA therapeutics has been limited, partly, by the occurrence of innate immune responses elicited by RNA. For instance, RNA molecules with certain GU-rich sequence motifs (e.g., 5′-GUCCUUCAA-3′) may induce inflammatory cytokines to be secreted in a cell type-specific and sequence-specific manner. Big double-stranded RNA molecules (e.g., those having a sequence length larger than 30 nucleotides) may induce interferon responses as well [41]. Even though siRNA therapeutics can get to the proximity of the target site, some of their unfavorable properties (e.g., large size and negative charge) will lower the efficiency in cellular internalization [42]. This, together with the off-target effects and the extra vulnerability of RNA to enzymatic degradation (which may cause the half-life of RNA molecules in human plasma to last for no more than an hour) [33,43], has imposed barriers to systemic administration of siRNA therapeutics during intervention execution.

To address these problems, the most straightforward approach is to use a carrier to facilitate the delivery process. Due to the similar electrostatic properties between DNA and RNA, a carrier that can be used to deliver DNA often shows the capacity to deliver RNA [43]. This has been evidenced by the case of the Starburst dendrimer conjugate with α-CD (namely α-CDE). After the incorporation of folate into α-CDE by using PEG as a linker (Fig. 8.1), the generated product (namely Fol-PαC) has been adopted as a carrier for siRNA delivery [44]. In C6 tumor-bearing mice, both FITC-labeled siRNA and tetramethylrhodamine isothiocyanate (TRITC)-labeled Fol-PαC have been found to accumulate in tumor tissues after i.v. administration of the polyplexes (Fig. 8.2) [45]. More detailed discussions of carriers for RNA transfer will be provided in the next chapter. To further enhance the therapeutic performance, siRNA can be conjugated to diverse types of molecules, ranging from proteins and peptides to lipids [46]. For instance, an earlier study has conjugated siRNA to an amphipathic poly(vinyl ether) (which consists of butyl and amino vinyl ethers, and is reversibly attached to PEG and *N*-acetylgalactosamine via a bifunctional maleamate linkage) through a reversible disulfide linkage [47]. This siRNA conjugate has been shown in vivo to be nontoxic, and has led to effective knockdown of two endogenous genes in the liver: *apoB* and *ppara* [47]. Knockdown of *apoB* has reduced the level of serum cholesterol, and has increased fat accumulation in the liver [47]; whereas knockdown of *ppara* has led to phenotypic changes consistent with its known function [47]. Similar success in using siRNA conjugates to knock down *apoB* has been reported by Soutschek et al. [48], who have covalently conjugated cholesterol to the 3′-terminus of the sense strand of an siRNA duplex *via* a pyrrolidone linkage. Upon conjugation, the stability of the siRNA molecule has been found to be increased. Compared to naked siRNA that has failed to elicit the siRNA activity in tissues at 24 hour after i.v. administration, injection of the conjugate to mice via the same administration route has resulted in silencing of *apoB* in the liver and jejunum, and has caused a reduction in the total cholesterol level [48]. Apart from the aforementioned examples, other conjugates showing improved delivery efficiency has been reported. One example is the lipophilic siRNA duplex that has been conjugated with derivatives of lithocholic acid, cholesterol, or lauric acid [49]. The lipid moieties have been linked to the 5′-end of the duplex using phosphoramidite chemistry. The generated duplex has effectively inhibited the expression of the reporter gene in vitro [49]. Another example is the thiol-containing siRNA duplex that has been conjugated with penetratin or transportan via a disulfide bond [50]. The modified duplex has been found to reduce the expression of the reporter gene in several mammalian cell types [50]. All these have evidenced the possibility of conjugating appropriate molecules to the siRNA duplex for delivery purposes.

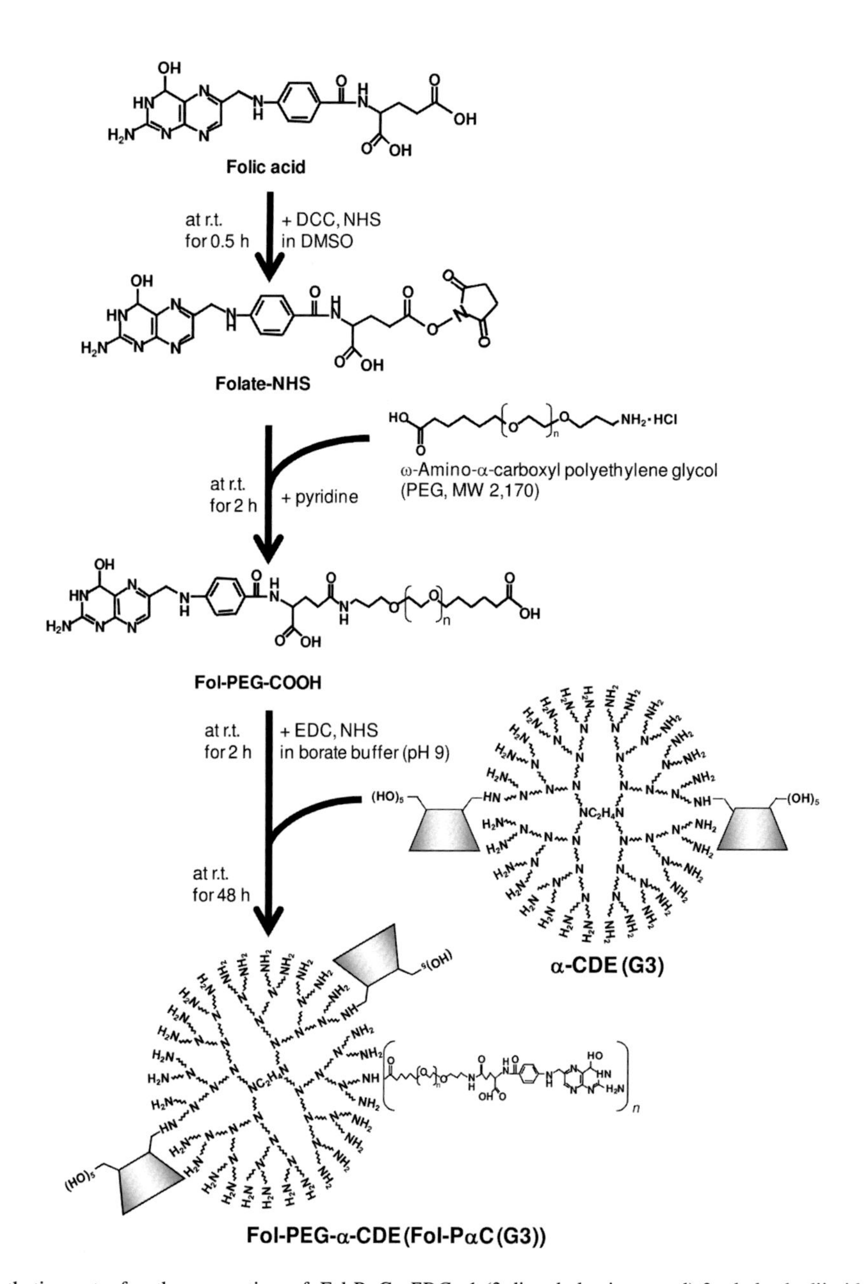

FIGURE 8.1 The synthetic route for the generation of Fol-PαC. *EDC*, 1-(3-dimethylaminopropyl)-3-ethylcarbodiimide hydrochloride; *NHS*, *N*-hydroxysuccinimide; r.t., room temperature; *DMSO*, dimethyl sulfoxide; *DCC*, *N,N*-dicyclohexylcarbodiimide. *Reproduced from H. Arima, A. Yoshimatsu, H. Ikeda, A. Ohyama, K. Motoyama, T. Higashi, et al., Folate-PEG-appended dendrimer conjugate with alpha-cyclodextrin as a novel cancer cell-selective siRNA delivery carrier, Mol. Pharmaceut. 9 (2012) 2591–2604 with permission from the American Chemical Society [45].*

Delivery of siRNA can also be enhanced upon conjugation with an aptamer, which is an oligonucleic acid or peptide that binds to target molecules. By using this approach, an earlier study has designed an aptamer-siRNA chimera for tackling cancers [51]. In the chimera, the aptamer portion can specifically bind to PSMA (which is a cell surface receptor overexpressed in prostate cancer cells and tumor vascular endothelia); whereas the siRNA portion can silence target genes. Results have shown that the chimera has targeted PSMA-positive cells, leading to depletion of the target proteins [51]. In addition, it has inhibited the growth of tumors in the xenograft model of prostate cancer [51]. More recently, upon the incorporation of structural modifications (e.g., appending a PEG moiety to the chimera to prolong the circulating half-life, adding 2-nucleotide 3'-overhangs, and truncating the aptamer portion to facilitate large-scale chemical synthesis), the performance of the chimera has been enhanced, resulting in remarkable regression of PSMA-expressing tumors in athymic mice after systemic administration of the chimera [52].

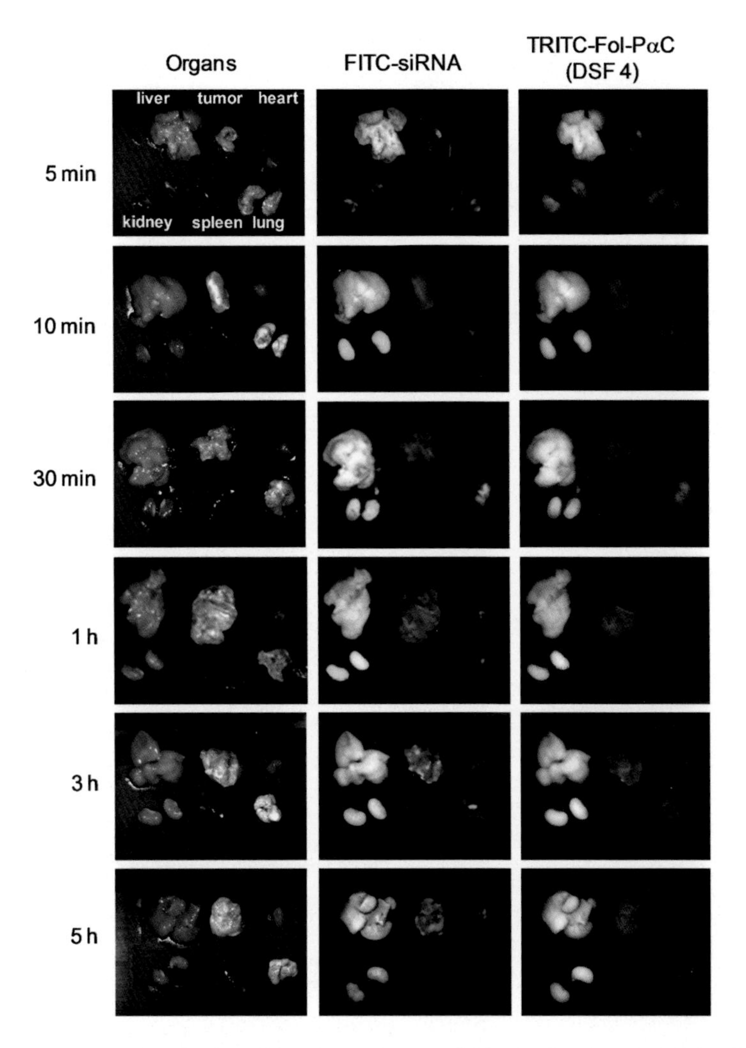

FIGURE 8.2 In vivo imaging after i.v. injection of Fol-PαC/siRNA complexes to C-26 tumor-bearing mice. Fol-PαC has been labeled with TRITC; whereas siRNA has been labeled with FITC. *Reproduced from H. Arima, A. Yoshimatsu, H. Ikeda, A. Ohyama, K. Motoyama, T. Higashi, et al., Folate-PEG-appended dendrimer conjugate with alpha-cyclodextrin as a novel cancer cell-selective siRNA delivery carrier, Mol. Pharmaceut. 9 (2012) 2591–2604 with permission from the American Chemical Society [45].*

Apart from the strategies presented above, by using antibodies to complex with siRNA, target specificity of RNAi can be elevated, minimizing the occurrence of off-target effects. This has been demonstrated by the case of the fusion protein, namely F105-P, which has been generated by fusion of protamine to the C terminus of the heavy chain Fab fragment of an antibody that can act against the HIV-1 envelope. The fusion protein has complexed with siRNA, with the complex generated found to induce gene silencing only in cells expressing the HIV-1 envelope [53]. By acting against the HIV-1 capsid gene *gag*, the complex has successfully inhibited HIV replication in HIV-infected primary T cells which have been difficult to be transfected [53]. This work has evidenced the feasibility of employing the properties of antibodies for achieving cell-type-specific siRNA delivery. An overview of the advantages and drawbacks of using siRNA conjugates for delivery, along with the extracellular and intracellular barriers that impede the delivery efficiency of the conjugates, is presented in Fig. 8.3.

Example protocols for experimental design

The method below is an example protocol for examining the effect of gene silencing mediated by the siRNA duplex.

1. Seed C166-GFP cells in a 24-well plate.
2. Incubate the plate at 37°C under a humidified atmosphere with 5% CO_2 until a confluence of 70%–80% is obtained.
3. Replace the medium in each well with 300 mL of the Opti-MEM I reduced serum medium in which the siRNA duplex (or its complexes with a carrier) is added at a concentration of 25 nM.

(Continued)

(A)

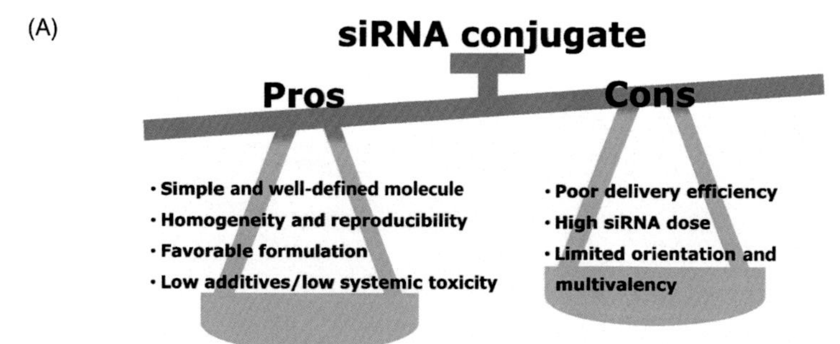

(B)

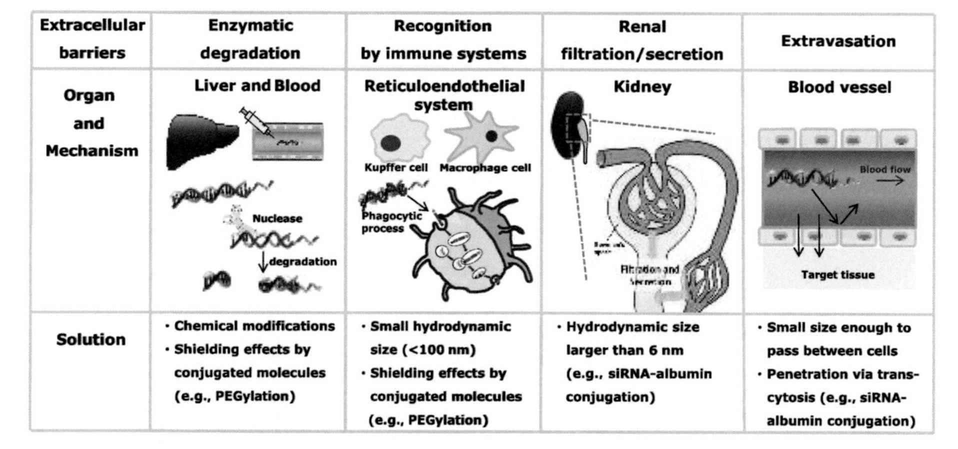

(C)

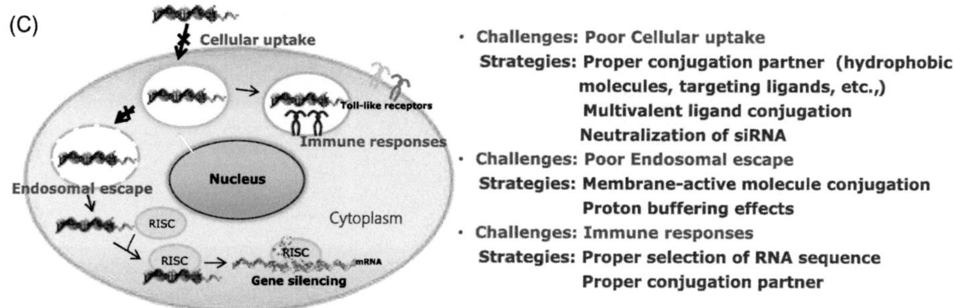

FIGURE 8.3 (A) The advantages and drawbacks of using siRNA conjugates for delivery, along with the (B) extracellular and (C) intracellular barriers that impede the delivery efficiency. *Reproduced from S.H. Lee, Y.Y. Kang, H.E. Jang, H. Mok, Current preclinical small interfering RNA (siRNA)-based conjugate systems for RNA therapeutics, Adv. Drug Deliv. Rev. 104 (2016) 78–92 with permission from Elsevier B.V. [46].*

(Continued)

4. Incubate the plate at 37°C under a humidified atmosphere with 5% CO_2.
5. Changes in the expression of EGFP are monitored regularly by using a fluorescence microscope.
6. After 3 days of incubation with the siRNA duplex (or its complexes with a carrier), trypsinize and resuspend the cells in PBS.
7. Use a flow cytometer to perform fluorescence-activated cell sorting (FACS) for quantification of EGFP-positive cells. Nontreated cells are adopted to set the background.

Clinical use of siRNA therapeutics

With the accumulation of research efforts, some siRNA therapeutics have reached clinical trials. For example, one clinical trial has been performed on bevasiranib, which is a siRNA duplex that targets VEGF-A. This siRNA has previously been exploited to mediate maintenance therapy (after the use of anti-VEGF therapy mediated by Lucentis); however, the trial has been terminated later because preliminary data has suggested that the possibility of reaching the primary endpoint is very low. Other examples of trials on siRNA therapeutics are those performed on VEGFA165b siRNA, which targets the

VEGF A165 isoform, and on AGN211745 (also known as Sirna-027), which is a chemically modified siRNA duplex that targets VEGF receptor-1. The mechanism of using anti-VEGF siRNA therapeutics for treating age-related macular degeneration (AMD) has recently been challenged. This is because intravitreal administration of an siRNA duplex having 21 nucleotides or longer has been found to suppress neovascularization in mice, but such suppression has been achieved irrespective of the nucleotide sequence or the intended target of the duplex injected. This phenomenon has been attributed partly to the induction of IFN-γ and IL-12, and also to nonspecific stimulation of the cell surface Toll-like receptor TLR3 pathway [54].

Apart from AMD, the use of siRNA therapeutics for cancer treatment has been evaluated clinically. A good example is CALAA-01, which is a polymer-based nanoparticle containing siRNA that targets the M2 subunit of ribonucleotide reductase. The nanoparticles have been surface-functionalized with transferrin for targeting tumor cells [55]. Results have shown that upon i.v. administration, the nanoparticles have successfully silenced the target gene in tumor tissues [56]. Another example is Atu027, which is an siRNA formulation that targets protein kinase N3 in patients with advanced solid tumors [57]. Repeated bolus injections or infusions of the formulation via the i.v. route have led to effective silencing of the target gene in various animal models. In orthotopic mouse models of pancreatic and prostate cancers, the formulation has successfully inhibited the formation of lymph node metastasis, and has reduced tumor growth significantly [58]. The performance of Atu027 has recently been evaluated in the clinical context [57]. Results have suggested that the formulation has displayed a high safety profile, with the concentration of the siRNA antisense strand in patients' plasma exhibiting a dose-dependent manner. All these works have established a foundation for translating the RNAi technology from research to practical applications.

Implications for the execution of biogerontological interventions

The possible use of siRNA in tackling aging can be illustrated by the case of neurodegeneration, which is partly resulting from age-associated alternations in gene expression and protein synthesis. Over the years, diverse age-associated neuroanatomical changes (e.g., a decline in the motor nerve conductance velocity [59], and the onset of neuron loss in the cerebral and cerebellar cortex [60,61]) have been identified. Some of these changes are thought to be caused by chromosomal alternations [62], a lack of trophic factors [63], gene silencing [64], and a decline in the RNA polymerase level [65]. Among different neurodegenerative diseases, one common disease associated with the advanced age is Alzheimer's disease (AD), which leads to progressive memory loss and cognitive decline. The brain of an AD patient is usually characterized by the presence of neurofibrillary tangles, and myloid plaques. As far as AD is concerned, BACE1, an aspartic acid protease, is an enzyme that involves in the generation of amyloid-β from the amyloid precursor protein (APP) through a series of proteolytic cleavage events. Owing to the association between the aggregation of amyloid-β peptides and AD, various methods have been proposed to silence the expression of *BACE1* or *APP* for AD treatment [66–68].

The success in restoring cognitive function in the AD model has been reported by Alvarez-Erviti and coworkers, who have employed exosomes as carriers for administration of siRNA that targets *BACE1* [69]. The reported exosomes have enabled the delivered siRNA molecules to be internalized by neurons, microglia, and oligodendrocytes in the brain. Upon i.v. injection of siRNA-loaded exosomes, silencing of the expression of *BACE1* has been achieved. This success has evidenced the feasibility of intervening in the pathogenesis of AD by using siRNA as the mediator. The technology of siRNA-mediated RNAi has also benefited treatment of other age-associated neurodegenerative diseases such as Parkinson's disease (PD), which is characterized by the death of dopamine (DA)-producing neurons in the substantia nigra. One feature of PD is protein accumulation and aberrant protein clearance, leading to the formation of intracytoplasmic inclusions, namely Lewy bodies (LB). As α-synuclein is one of the major components of LB, an earlier study has delivered α-synuclein siRNA to rat pheochromocytoma PC12 cells by using PEG-PEI as a carrier [70]. The siRNA duplex has been found to reduce the level of mRNA encoding α-synuclein, and has reduced the occurrence of apoptosis in the *N*-methyl-4-phenylpyridinium (MPP^+)-induced cellular model of PD.

Example protocols for experimental design

The method below is an example protocol for producing siRNA-loaded exosomes.

1. Prepare an electroporation buffer by mixing potassium chloride (25 mM) and potassium phosphate (1.15 mM, pH 7.2) with the 21% (v/v) OptiPrep density gradient medium.
2. Store the buffer at 4°C for subsequent use.
3. Mix the exosomes (which can be produced by using methods reported in literature [69,71–73]) and siRNA in a 1:1 mass-to-mass ratio in the electroporation buffer so that the final concentration of exosomes reaches 0.5 μg/μL.

(Continued)

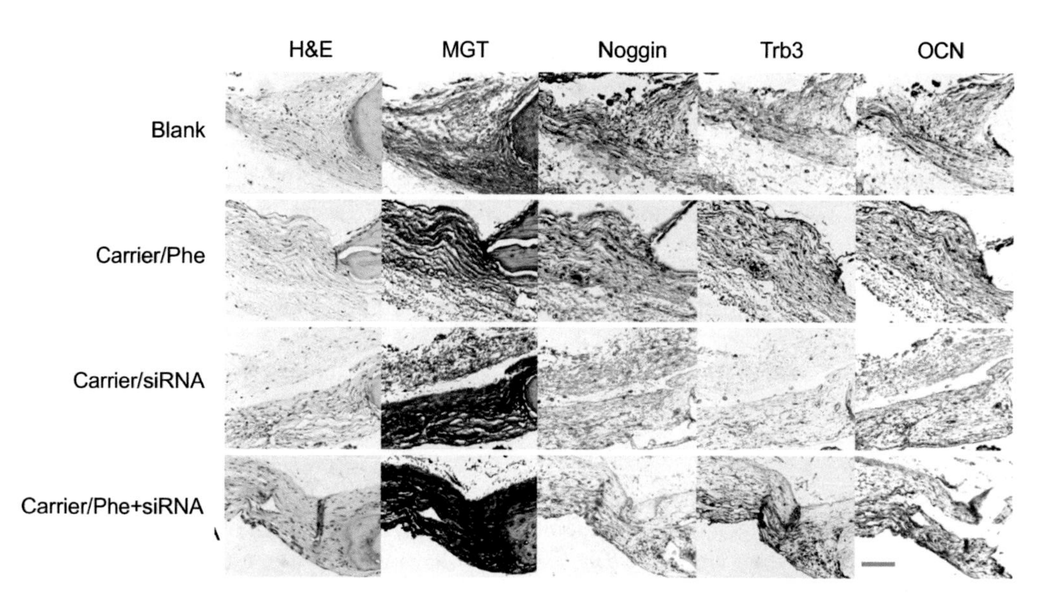

FIGURE 8.4 Images of hematoxylin and eosin (H&E) staining, Masson-Goldner trichrome staining, and immunohistochemical staining of noggin, Trb3, and OCN. The liposomal system formulated with stearylamine and cholesterol has been adopted as the carrier. Scale bar = 50 μm. *Reproduced from Z.K. Cui, J.A. Sun, J.J. Baljon, J. Fan, S. Kim, B.M. Wu, et al., Simultaneous delivery of hydrophobic small molecules and siRNA using Sterosomes to direct mesenchymal stem cell differentiation for bone repair, Acta Biomater. 58 (2017) 214–224 with permission from Elsevier B.V. [76].*

(Continued)

4. Electroporate the mixture at 400 mV and 125 μF capacitance, with the pulse time set to 10–15 ms.
5. Resuspend the exosomes in 20 mL of an appropriate resuspension buffer.
6. Ultracentrifuge at 100,000 × *g* for 90 minutes at 4°C to collect the exosome pellet.
7. Repeat Steps 5 and 6 twice.
8. Use the siRNA-loaded exosomes for intended applications.

In addition to tackling neurodegeneration, RNAi can be adopted to enhance tissue regeneration. For instance, by implanting mesenchymal stem cells (MSCs) to a mouse model of acute liver failure, along with systemic administration of siRNA to silence the expression of the gene for IL-1β, the tissue regeneration capacity of the implanted cells has been found to be enhanced [74]. Furthermore, upon treatment of cochleae with siRNA to silence *Hes1*, the level of *Hes1* mRNA has been successfully reduced [75]. This, along with the up-regulation of the expression of *Atoh1*, has been shown to induce the regeneration of mammalian cochlear and vestibular hair cells [75]. More recently, by using a liposomal system formulated with stearylamine and cholesterol for codelivery of phenamil and siRNA to MSCs, the osteogenic differentiation of the cells has been enhanced [76]. In the in vivo context, codelivery of Phe and siRNA (which silences the expression of the gene for noggin) has been reported to facilitate bone repair in the mouse model of calvarial defects [76]. Compared to the blank group in which only fibrous-like tissues with minimal bone formation have been detected, formation of the osteoid matrix has occurred in the edges of defects in the group treated with phenamil (Phe)/siRNA codelivery [76]. Results of immunohistochemical staining have suggested that the siRNA duplex has silenced the expression of the gene for noggin [76]. On the other hand, Phe has promoted the BMP activity by inducing the expression of the gene for Trb3, as demonstrated by Trb3 staining [76]. Finally, intense staining of the late osteogenic marker OCN has been found in the group treated with the Phe/siRNA formulation (Fig. 8.4) [76]. These findings have evidenced the possibility of stimulating tissue repair and regeneration via RNAi-mediated genetic modulation to combat age-associated damage in the body.

Summary

Aging is a complex process controlled largely by genetic materials. Recent advances in RNAi technologies have made the intervention of the aging network possible. Despite this, RNA molecules are susceptible to degradation, along with other technical barriers (e.g., poor cellular uptake of RNA, and stimulation of the innate immune response), the efficiency in using siRNA directly for intervention execution is limited. In this chapter, we have highlighted some important roles played by RNAi in aging research, and have discussed various strategies to modify RNA molecules structurally to improve the properties and performance. With the advances in chemical technologies, it is expected that more stable and effective RNA analogs for intervention development will emerge continuously in the forthcoming decades. In fact, siRNA is only one of the several mediators that can be used to achieve RNAi. In the next chapter, we will continue to discuss RNAi, but will introduce another type of small RNA molecules, namely miRNA, for intervention design and development.

Directions for intervention development

When one's proposed intervention involves gene silencing, the following steps can be taken to move the intervention closer to practice:

1. Determine the gene sequence to be targeted.
2. Design an siRNA duplex to act on the selected sequence.
3. Construct the siRNA duplex.
4. Evaluate the performance (in terms of various parameters, including the gene silencing efficiency, off-target effects, cytotoxicity, and stability) of the duplex in vitro and in vivo.
5. Consider the need of structurally modifying the duplex to enhance the performance.
6. Select a carrier, if appropriate, to facilitate the delivery process.
7. Reexamine and optimize the performance of the duplex in vitro and in vivo.
8. Optimize the dose and the administration method, if necessary, to maximize the efficacy of the proposed intervention.

References

[1] W.J. Ma, S. Cheng, C. Campbell, A. Wright, H. Furneaux, Cloning and characterization of HuR, a ubiquitously expressed Elav-like protein, J. Biol. Chem. 271 (1996) 8144–8151.

[2] P.J. Good, A conserved family of elav-like genes in vertebrates, Proc. Natl Acad. Sci. U.S.A. 92 (1995) 4557–4561.

[3] W.J. Ma, S. Chung, H. Furneaux, The Elav-like proteins bind to AU-rich elements and to the poly(A) tail of mRNA, Nucleic Acids Res. 25 (1997) 3564–3569.

[4] W. Wang, H. Furneaux, H. Cheng, M.C. Caldwell, D. Hutter, Y. Liu, et al., HuR regulates p21 mRNA stabilization by UV light, Mol. Cell. Biol. 20 (2000) 760–769.

[5] X.C. Fan, J.A. Steitz, Overexpression of HuR, a nuclear-cytoplasmic shuttling protein, increases the in vivo stability of ARE-containing mRNAs, EMBO J. 17 (1998) 3448–3460.

[6] M. Gorospe, X. Wang, K.Z. Guyton, N.J. Holbrook, Protective role of p21(Waf1/Cip1) against prostaglandin A2-mediated apoptosis of human colorectal carcinoma cells, Mol. Cell. Biol. 16 (1996) 6654–6660.

[7] W. Wang, J. Fan, X. Yang, S. Furer-Galban, I. Lopez de Silanes, C. von Kobbe, et al., AMP-activated kinase regulates cytoplasmic HuR, Mol. Cell. Biol. 22 (2002) 3425–3436.

[8] Z.H. Beg, D.W. Allmann, D.M. Gibson, Modulation of 3-hydroxy-3-methylglutaryl coenzyme A reductase activity with cAMP and wth protein fractions of rat liver cytosol, Biochem. Biophys. Res. Commun. 54 (1973) 1362–1369.

[9] C.A. Carlson, K.H. Kim, Regulation of hepatic acetyl coenzyme A carboxylase by phosphorylation and dephosphorylation, J. Biol. Chem. 248 (1973) 378–380.

[10] D.G. Hardie, S.A. Hawley, AMP-activated protein kinase: the energy charge hypothesis revisited, Bioessays 23 (2001) 1112–1119.

[11] G. Brewer, Messenger RNA decay during aging and development, Ageing Res. Rev. 1 (2002) 607–625.

[12] W. Wang, X. Yang, V.J. Cristofalo, N.J. Holbrook, M. Gorospe, Loss of HuR is linked to reduced expression of proliferative genes during replicative senescence, Mol. Cell. Biol. 21 (2001) 5889–5898.

[13] M. Gupta, G. Brewer, MicroRNAs: new players in an old game, Proc. Natl Acad. Sci. U.S.A. 103 (2006) 3951–3952.

[14] S.S. Lee, Whole genome RNAi screens for increased longevity: important new insights but not the whole story, Exp. Gerontol. 41 (2006) 968–973.

[15] T.W. Day, A.R. Safa, RNA interference in cancer: targeting the anti-apoptotic protein c-FLIP for drug discovery, Mini Rev. Med. Chem. 9 (2009) 741–748.

[16] M. Folini, M. Pennati, N. Zaffaroni, RNA interference-mediated validation of genes involved in telomere maintenance and evasion of apoptosis as cancer therapeutic targets, Methods Mol. Biol. 487 (2009) 303–330.

[17] S. He, D. Zhang, F. Cheng, F. Gong, Y. Guo, Applications of RNA interference in cancer therapeutics as a powerful tool for suppressing gene expression, Mol. Biol. Rep. 36 (2009) 2153–2163.

[18] K. Kirby, J. Hu, A.J. Hilliker, J.P. Phillips, RNA interference-mediated silencing of Sod2 in Drosophila leads to early adult-onset mortality and elevated endogenous oxidative stress, Proc. Natl Acad. Sci. U.S.A. 99 (2002) 16162–16167.

[19] A. Duttaroy, A. Paul, M. Kundu, A. Belton, A Sod2 null mutation confers severely reduced adult life span in Drosophila, Genetics 165 (2003) 2295–2299.

[20] B.K. Kennedy, N.R. Austriaco Jr., J. Zhang, L. Guarente, Mutation in the silencing gene SIR4 can delay aging in S. cerevisiae, Cell 80 (1995) 485–496.

[21] B.K. Kennedy, E.D. Smith, M. Kaeberlein, The enigmatic role of Sir2 in aging, Cell 123 (2005) 548–550.

[22] S. Kusama, R. Ueda, T. Suda, S. Nishihara, E.T. Matsuura, Involvement of Drosophila Sir2-like genes in the regulation of life span, Genes. Genet. Syst. 81 (2006) 341–348.

[23] N. Minois, P. Sykacek, B. Godsey, D.P. Kreil, RNA interference in ageing research - a mini-review, Gerontology 56 (2010) 496–506.

[24] A. Li, Z. Xie, Y. Dong, K.M. McKay, M.L. McKee, R.E. Tanzi, Isolation and characterization of the Drosophila ubiquilin ortholog dUbqln: in vivo interaction with early-onset Alzheimer disease genes, Hum. Mol. Genet. 16 (2007) 2626–2639.

[25] J.M. Copeland, J. Cho, T. Lo Jr., J.H. Hur, S. Bahadorani, et al., Extension of Drosophila life span by RNAi of the mitochondrial respiratory chain, Curr. Biol. 19 (2009) 1591–1598.

[26] S. Choung, Y.J. Kim, S. Kim, H.O. Park, Y.C. Choi, Chemical modification of siRNAs to improve serum stability without loss of efficacy, Biochem. Biophys. Res. Commun. 342 (2006) 919–927.

[27] J.K. Watts, G.F. Deleavey, M.J. Damha, Chemically modified siRNA: tools and applications, Drug Discov. Today 13 (2008) 842–855.

[28] B.A. Kraynack, B.F. Baker, Small interfering RNAs containing full 2'-O-methylribonucleotide-modified sense strands display Argonaute2/eIF$_2$C$_2$-dependent activity, RNA 12 (2006) 163–176.

[29] F. Czauderna, M. Fechtner, S. Dames, H. Aygun, A. Klippel, G.J. Pronk, et al., Structural variations and stabilising modifications of synthetic siRNAs in mammalian cells, Nucleic Acids Res. 31 (2003) 2705–2716.

[30] S.M. Elbashir, J. Martinez, A. Patkaniowska, W. Lendeckel, T. Tuschl, Functional anatomy of siRNAs for mediating efficient RNAi in Drosophila melanogaster embryo lysate, EMBO J. 20 (2001) 6877–6888.

[31] D.A. Braasch, S. Jensen, Y. Liu, K. Kaur, K. Arar, M.A. White, et al., RNA interference in mammalian cells by chemically-modified RNA, Biochemistry 42 (2003) 7967–7975.

[32] Y.L. Chiu, T.M. Rana, siRNA function in RNAi: a chemical modification analysis, RNA. 9 (2003) 1034–1048.

[33] J.M. Layzer, A.P. McCaffrey, A.K. Tanner, Z. Huang, M.A. Kay, B.A. Sullenger, In vivo activity of nuclease-resistant siRNAs, RNA. 10 (2004) 766–771.

[34] D.S. Schwarz, Y. Tomari, P.D. Zamore, The RNA-induced silencing complex is a Mg^{2+}-dependent endonuclease, Curr. Biol. 14 (2004) 787–791.

[35] M. Amarzguioui, T. Holen, E. Babaie, H. Prydz, Tolerance for mutations and chemical modifications in a siRNA, Nucleic Acids Res. 31 (2003) 589–595.

[36] A.H. Hall, J. Wan, E.E. Shaughnessy, B. Ramsay Shaw, K.A. Alexander, RNA interference using boranophosphate siRNAs: structure-activity relationships, Nucleic Acids Res. 32 (2004) 5991–6000.

[37] J. Xia, A. Noronha, I. Toudjarska, F. Li, A. Akinc, R. Braich, et al., Gene silencing activity of siRNAs with a ribo-difluorotoluyl nucleotide, ACS Chem. Biol. 1 (2006) 176–183.

[38] J.B. Bramsen, M.B. Laursen, C.K. Damgaard, S.W. Lena, B.R. Babu, J. Wengel, et al., Improved silencing properties using small internally segmented interfering RNAs, Nucleic Acids Res. 35 (2007) 5886–5897.

[39] D.H. Kim, M.A. Behlke, S.D. Rose, M.S. Chang, S. Choi, J.J. Rossi, Synthetic dsRNA Dicer substrates enhance RNAi potency and efficacy, Nat. Biotechnol. 23 (2005) 222–226.

[40] M.A. Minks, D.K. West, S. Benvin, C. Baglioni, Structural requirements of double-stranded RNA for the activation of 2',5'-oligo(A) polymerase and protein kinase of interferon-treated HeLa cells, J. Biol. Chem. 254 (1979) 10180–10183.

[41] L. Aagaard, J.J. Rossi, RNAi therapeutics: principles, prospects and challenges, Adv. Drug Deliv. Rev. 59 (2007) 75–86.

[42] C.X. Li, A. Parker, E. Menocal, S. Xiang, L. Borodyansky, J.H. Fruehauf, Delivery of RNA interference, Cell Cycle 5 (2006) 2103–2109.

[43] W.F. Lai, M.C. Lin, Nucleic acid delivery with chitosan and its derivatives, J. Control Release 134 (2009) 158–168.

[44] H. Arima, M. Arizono, T. Higashi, A. Yoshimatsu, H. Ikeda, K. Motoyama, et al., Potential use of folate-polyethylene glycol (PEG)-appended dendrimer (G3) conjugate with alpha-cyclodextrin as DNA carriers to tumor cells, Cancer Gene Ther. 19 (2012) 358–366.

[45] H. Arima, A. Yoshimatsu, H. Ikeda, A. Ohyama, K. Motoyama, T. Higashi, et al., Folate-PEG-appended dendrimer conjugate with alpha-cyclodextrin as a novel cancer cell-selective siRNA delivery carrier, Mol. Pharmaceut. 9 (2012) 2591–2604.

[46] S.H. Lee, Y.Y. Kang, H.E. Jang, H. Mok, Current preclinical small interfering RNA (siRNA)-based conjugate systems for RNA therapeutics, Adv. Drug Deliv. Rev. 104 (2016) 78–92.

[47] D.B. Rozema, D.L. Lewis, D.H. Wakefield, S.C. Wong, J.J. Klein, P.L. Roesch, et al., Dynamic PolyConjugates for targeted in vivo delivery of siRNA to hepatocytes, Proc. Natl Acad. Sci. U. S.A. 104 (2007) 12982–12987.

[48] J. Soutschek, A. Akinc, B. Bramlage, K. Charisse, R. Constien, M. Donoghue, et al., Therapeutic silencing of an endogenous gene by systemic administration of modified siRNAs, Nature 432 (2004) 173–178.

[49] C. Lorenz, P. Hadwiger, M. John, H.P. Vornlocher, C. Unverzagt, Steroid and lipid conjugates of siRNAs to enhance cellular uptake and gene silencing in liver cells, Bioorg. Med. Chem. Lett. 14 (2004) 4975–4977.

[50] A. Muratovska, M.R. Eccles, Conjugate for efficient delivery of short interfering RNA (siRNA) into mammalian cells, FEBS Lett. 558 (2004) 63–68.

[51] J.O. McNamara 2nd, E.R. Andrechek, Y. Wang, K.D. Viles, R.E. Rempel, et al., Cell type-specific delivery of siRNAs with aptamer-siRNA chimeras, Nat. Biotechnol. 24 (2006) 1005–1015.

[52] J.P. Dassie, X.Y. Liu, G.S. Thomas, R.M. Whitaker, K.W. Thiel, K.R. Stockdale, et al., Systemic administration of optimized aptamer-siRNA chimeras promotes regression of PSMA-expressing tumors, Nat. Biotechnol. 27 (2009) 839–849.

[53] E. Song, P. Zhu, S.K. Lee, D. Chowdhury, S. Kussman, D.M. Dykxhoorn, et al., Antibody mediated in vivo delivery of small interfering RNAs via cell-surface receptors, Nat. Biotechnol. 23 (2005) 709–717.

[54] M.E. Kleinman, K. Yamada, A. Takeda, V. Chandrasekaran, M. Nozaki, J.Z. Baffi, et al., Sequence- and target-independent angiogenesis suppression by siRNA via TLR3, Nature 452 (2008) 591–597.

[55] M.E. Davis, The first targeted delivery of siRNA in humans via a self-assembling, cyclodextrin polymer-based nanoparticle: from concept to clinic, Mol. Pharmaceut. 6 (2009) 659–668.

[56] M.E. Davis, J.E. Zuckerman, C.H. Choi, D. Seligson, A. Tolcher, C.A. Alabi, et al., Evidence of RNAi in humans from systemically administered siRNA via targeted nanoparticles, Nature 464 (2010) 1067–1070.

[57] D. Strumberg, B. Schultheis, U. Traugott, C. Vank, A. Santel, O. Keil, et al., Phase I clinical development of Atu027, a siRNA formulation targeting PKN3 in patients with advanced solid tumors, Int. J. Clin. Pharm. Th. 50 (2012) 76–78.

[58] M. Aleku, P. Schulz, O. Keil, A. Santel, U. Schaeper, B. Dieckhoff, et al., Atu027, a liposomal small interfering RNA formulation targeting protein kinase N3, inhibits cancer progression, Cancer Res. 68 (2008) 9788–9798.

[59] F. Kemble, Conduction in the normal adult median nerve: the different effect of ageing in men and women, Electromyography 7 (1967) 275–287.

[60] B.E. Tomlinson, G. Henderson, Some quantitative cerebral findings in normal and demented old people, in: R.D. Terry, S. Gershon (Eds.), Neurobiology of Aging, Raven Press, New York, 1976, pp. 83–227.

[61] T.C. Hall, A.K.H. Miller, J.A.N. Corsellis, Variations in the human purkinje cell population according to age and sex, Neuropathol. Appl. Neurobiol. 1 (2010) 267–292.

[62] T. Ikura, V.V. Ogryzko, Chromatin dynamics and DNA repair, Front. Biosci. 8 (2003) s149–s155.

[63] M.V. Chao, Neurotrophins and their receptors: a convergence point for many signalling pathways, Nat. Rev. Neurosci. 4 (2003) 299–309.

[64] S.R. Burzynski, Gene silencing - a new theory of aging, Med. Hypotheses 60 (2003) 578–583.

[65] A.M. da Silva, S.L. Payao, B. Borsatto, P.H. Bertolucci, M.A. Smith, Quantitative evaluation of the rRNA in Alzheimer's disease, Mech. Ageing Dev. 120 (2000) 57–64.

[66] M. Ohno, Genetic and pharmacological basis for therapeutic inhibition of beta- and gamma-secretases in mouse models of Alzheimer's memory deficits, Rev. Neurosci. 17 (2006) 429–454.

[67] B. Nawrot, Targeting BACE with small inhibitory nucleic acids - a future for Alzheimer's disease therapy? Acta Biochim. Pol. 51 (2004) 431–444.

[68] P. Nilsson, N. Iwata, S. Muramatsu, L.O. Tjernberg, B. Winblad, T.C. Saido, Gene therapy in Alzheimer's disease - potential for disease modification, J. Cell. Mol. Med. 14 (2010) 741–757.

[69] L. Alvarez-Erviti, Y. Seow, H. Yin, C. Betts, S. Lakhal, M.J. Wood, Delivery of siRNA to the mouse brain by systemic injection of targeted exosomes, Nat. Biotechnol. 29 (2011) 341–345.

[70] Y.Y. Liu, X.Y. Yang, Z. Li, Z.L. Liu, D. Cheng, Y. Wang, et al., Characterization of polyethylene glycol-polyethyleneimine as a vector for alpha-synuclein siRNA delivery to PC12 cells for Parkinson's disease, CNS Neurosci. Ther. 20 (2014) 76–85.

[71] T.A. Shtam, R.A. Kovalev, E.Y. Varfolomeeva, E.M. Makarov, Y.V. Kil, M.V. Filatov, Exosomes are natural carriers of exogenous siRNA to human cells in vitro, Cell Commun. Signal. 11 (2013) 88.

[72] S.K. Limoni, M.F. Moghadam, S.M. Moazzeni, H. Gomari, F. Salimi, Engineered exosomes for targeted transfer of siRNA to HER2 positive breast cancer cells, Appl. Biochem. Biotechnol. (2018). Available from: https://doi.org/10.1007/s12010-018-2813-4.

[73] Z. Ju, J. Ma, C. Wang, J. Yu, Y. Qiao, F. Hei, Exosomes from iPSCs delivering siRNA attenuate intracellular adhesion molecule-1 expression and neutrophils adhesion in pulmonary microvascular endothelial cells, Inflammation 40 (2017) 486–496.

[74] H. Ma, X. Shi, X. Yuan, Y. Ding, IL-1β siRNA adenovirus benefits liver regeneration by improving mesenchymal stem cells survival after acute liver failure, Ann. Hepatol. 15 (2016) 260–270.

[75] X. Du, W. Li, X. Gao, M.B. West, W.M. Saltzman, C.J. Cheng, et al., Regeneration of mammalian cochlear and vestibular hair cells through Hes1/Hes5 modulation with siRNA, Hearing Res. 304 (2013) 91–110.

[76] Z.K. Cui, J.A. Sun, J.J. Baljon, J. Fan, S. Kim, B.M. Wu, et al., Simultaneous delivery of hydrophobic small molecules and siRNA using Sterosomes to direct mesenchymal stem cell differentiation for bone repair, Acta Biomater. 58 (2017) 214–224.

Chapter 9

Use of delivery technologies to manipulate miRNA expression

Introduction

The occurrence of aging is not only mediated by the accumulation of mutations that undermine the functions of normal genes, but is also caused by the occurrence of changes (spanning from chromatin remodeling and histone modification to DNA methylation) at the genetic level. To combat these changes, over the years extensive efforts have been made to manipulate genes by either gene overexpression or gene silencing. In the previous chapter, we have introduced RNAi technologies mediated by using siRNA. In fact, RNAi can also be mediated by using miRNA and small hairpin RNA (shRNA). These RNA molecules, in the end, necessitate the involvement of the RISC for mRNA degradation and posttranscriptional gene silencing. Because shRNA is only an RNA sequence, possessing a tight hairpin turn, encoded within an expression vector, its delivery can be mediated simply by using the technologies for DNA delivery, as discussed in Part II of this book. The objective of this chapter will lie on the possible use of miRNA for intervention design.

Since the turn of the last century, studies on miRNA have been greatly facilitated by technological advances. Traditionally, discovery of miRNA molecules and their targets rely on hybridization-based array methods. The efficiency is highly limited. The situation has been improved by the increased affordability and accessibility of commercial platforms (e.g., Roche's 454/FLX system, Illumina's Genome Analyzer, and ABI's SOLiD) for parallel sequencing, which enables high-resolution views of miRNA expression to be attained [1,2]. Identification of molecular targets of miRNA molecules has also been enhanced by using HITS-CLIP (i.e., high-throughput sequencing of RNA isolated by crosslinking immunoprecipitation) [3–5] and PARCLIP (i.e., photoactivatable-ribonucleoside-enhanced cross-linking and immunoprecipitation) [6,7]. The great expansion of knowledge of miRNA has already opened a new era in biomedicine. To exploit the feasibility and technical demands of turning the expanding knowledge of miRNA into antiaging therapies, in this chapter we will provide an overview of the current understanding of the roles of miRNA in modulating the aging network, and will offer insights into the opportunities and challenges of bench-to-clinic translation.

Mechanisms of action of miRNA

MiRNA is short and noncoding single-stranded RNA, generally consisting of 18–25 nucleotides. It plays regulatory roles in diverse biological processes, ranging from cell proliferation and apoptosis to development and differentiation [8–13]. At most of the time, miRNA functions at the posttranscriptional level by complementarily matching with the coding region or 3′-UTR of the target mRNA molecule, though there are also few recent reports on the capacity of miRNA to upregulate translation [11,14]. During action, the primary precursor of miRNA, namely pri-miRNA, is first transcribed as a capped, polyadenylated RNA strand, which subsequently forms a double-stranded stem-loop structure (Fig. 9.1). This structure is then processed in the nucleus by DGCR8 and Drosha, forming pre-miRNA, which is a hairpin structure consisting of approximately 70–100 nucleotides. Pre-miRNA undergoes nuclear export to the cytoplasm through the help of a RanGTP-dependent dsRNA-binding protein, namely Exportin-5, for further processing by Dicer [16,17]. Finally, the duplex binds with the miRNA-induced silencing complex (miRISC), and is unwound into a mature strand as well as a passenger strand [18]. The mature strand is retained in the miRISC; whereas the passenger strand is released and degraded.

In fact, interactions with the miRISC have been regarded as the major mechanism for the action of miRNA [19]. With the help of the mature miRNA strand,

Delivery of Therapeutics for Biogerontological Interventions. DOI: https://doi.org/10.1016/B978-0-12-816485-3.00009-X

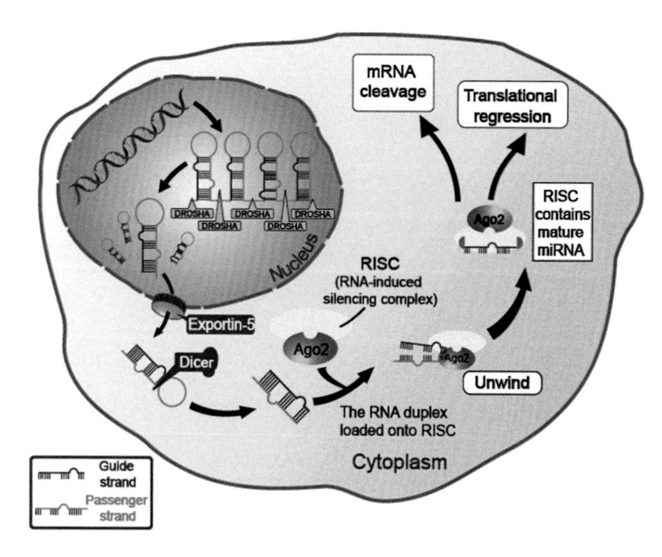

FIGURE 9.1 A schematic diagram showing the process of miRNA generation and gene silencing. *Reproduced from Y. Chen, D.Y. Gao, L. Huang, In vivo delivery of miRNAs for cancer therapy: challenges and strategies, Adv. Drug Deliv. Rev. 81 (2015) 128–141 with permission from Elsevier B.V., [15].*

the miRISC can silence the expression of a gene by inhibiting mRNA translation or even causing degradation of the target mRNA transcript [20,21]. When miRNA is involved in gene silencing, the activity of miRNA is similar to that of siRNA; however, perfect pairing is not required when the miRISC complex binds to the target mRNA molecule. This explains why one miRNA strand can recognize multiple mRNA targets. It is this promiscuity that makes miRNA a more versatile tool than siRNA for gene expression modulation.

Roles in regulating the aging network

To combat aging, cellular senescence, which causes a cell to be locked into a state of permanent growth arrest in response to cellular stress, can be a target to approach because it is one of the fundamental processes accounting for the occurrence of aging and age-related diseases. One major cause of cellular senescence is telomere shortening. An example of miRNA molecules that can be manipulated for telomere lengthening is miR-138, whose downregulation can enhance the endogenous level of human telomerase reverse transcriptase (hTERT) [22], possibly helping cells to overcome the Hayflick limit. A low level of miR-138, however, has also been found in anaplastic thyroid carcinoma. This has raised concerns on the safety of developing interventions that use miR-138 as a target. Apart from telomere shortening, cellular senescence can be caused by oxidative stress. By examining alternations in miRNA expression profiles associated with stress-induced premature senescence (SIPS) in human diploid fibroblasts (HDF) and human trabecular meshwork (HTM) cells, downregulation of five members of the miR-106b family (miR-17-5p, miR-18a, miR-20a, miR-106a, and miR-106b) and four members of the miR-15 family (miR-15a, miR-15b, miR-16, and miR-195), as well as upregulation of two miRNA molecules (miR-182 and miR-183) from the miR-183-96-182 cluster, have been linked with SIPS [21]. Further studies have revealed that SIPS-associated up-regulation of $p21^{CDKN1A}$ has been inhibited by transfection with a miR-106a mimic, which has promoted cell proliferation [21]. This has demonstrated the feasibility of manipulating the process of cellular senescence by using miRNA as a modulator.

Biochemically, the target of rapamycin (TOR) pathway and the insulin/IGF pathway are two major pathways that are known to have a potent impact on organismal aging and life span determination. The involvement of the TOR pathway in aging has been demonstrated by the observation that *C. elegans* having loss-of-function mutations of the TOR orthologs (*daf-15* and *let-363*) and raptor have a more extended life span [23]. Examples of miRNA molecules that modulate the PTEN/Akt/mTOR signaling pathway include miR-146b [24], miR-616-3p [25], miR-21 [26], miR-101 [27], miR-181 [28], miR-206 [29], miR-494 [30], and miR-126 [31]. For the insulin/IGF pathway, it is an important mechanism regulating protein synthesis and glucose homeostasis. The effect of downregulating the pathway in extending the life span has been verified in *D. melanogaster* [32], *C. elegans* [33], mice [34], and humans [35]. Over the years, different miRNA molecules have been reported to involve in regulating insulin/IGF signaling. For example, miR-1, miR-18a, miR-320, and miR-206 have been reported to target IGF-1 [36–40]. The insulin receptor substrate-1 (IRS-1) and the insulin-like growth factor binding protein (IGFBP)-5 have also been found to be targeted by miR-145 and miR-140-5p, respectively [41,42]. Until now, there have been at least two miRNA molecules explicitly shown to affect the life span of *C. elegans* [43]. One is miR-71. Mutants of which have displayed upregulation of PI3K and PDK1, causing an elevation in the activity of the insulin/IGF pathway. These mutants have been reported to be short-lived. Another one is miR-239. Mutants of which have shown a reduction in the insulin/IGF signaling activity and an increase in stress resistance, possessing a more extended life span.

Structural design for an intervention

Generally speaking, partly due to the degradation mediated by nucleases, the half-life of naked DNA or RNA in blood is very short [44]. This explains the poor efficiency and stability of the first generation antisense phosphodiester ODNs in therapeutic use [45]. To enhance the efficiency and stability, various chemical derivatives of nucleic acid therapeutics have been reported. One example is phosphorothioate ODNs, in which one of the nonbridging oxygen atoms in the phosphate group is replaced by sulfur to enhance the ODN stability [46]. Compared to the unmodified counterparts, phosphorothioate ODNs have shown enhanced nuclease resistance for parenteral administration, and have undergone effective cellular uptake [47,48]. Unfortunately, the half-lives of many of them are still too short for any practical use. For instance, half-lives of pentadecamers have been reported to be less than 10 hours in serum, and less than 1 day in the rabbit reticulocyte lysate and the rat cerebrospinal fluid [46]. Along with their nonspecific activity on cell growth inhibition and their low affinity with target RNA molecules, phosphorothioate ODNs are needed to be modified before clinical use [49]. Over the years, various modification strategies have been proposed. One example is the 2′-*O*-methoxyethyl modification of the ODN backbone to reduce the clearance of oligoribonucleotides from tissues [47]. Another example is the introduction of 2′-*O*-methyl groups to the ribose moiety in a phosphorothioate ODN [50]. By using this method, a 20-mer ODN, which has displayed less nonspecific inhibitory effects on cell growth, has been generated in a previous study to target to sites 109 and 277 of *bcl-2* mRNA [50].

Besides phosphorothioate ODNs, other examples of chemically modified nucleic acid therapeutics include locked nucleic acid (LNA) oligonucleotides [51] and peptide nucleic acids (PNAs) [52]. LNA oligonucleotides are RNA analogs in which the ribose moiety is locked using a bridge that connects the 4′-carbon and 2′-oxygen in a RNA-mimicking N-type (C3′-endo) conformation [53]. Their potential in clinical use has been corroborated by the case of Miravirsen (Santaris Pharma, Hørsholm, Denmark), which is an LNA-based therapeutic that has proceeded into clinical trials for treatment of hepatitis C virus (HCV) infection [54,55]. For PNAs, they are uncharged oligonucleotide analogs with the sugar-phosphodiester backbone being replaced by an achiral structure that comprises *N*-(2-aminoethyl)-glycine units. These molecules show high stability, and can be easily incorporated with different functional moieties (e.g., targeting ligands, PEG, and peptides) using standard bioconjugation techniques to enhance their tissue specificity and half-lives [56–58].

Strategies for intervention execution

When miRNA-based therapies are designed for antiaging purposes, two directions are available. The first one is to use miRNA as a tool to modulate gene expression regardless of the "normal" miRNA expression profile. In this case, miRNA is used to silence specific genes whose silencing is deemed to bring benefits toward life span extension or aging retardation. The second approach is to take miRNA as the therapeutic target. In this case, there are two ways to go. One is to inhibit the expression of the miRNA molecule, and the other one is to elevate it. In fact, changes in miRNA expression profiles have not only been implicated for age-associated diseases [59–61], but are also related to aging and longevity per se [62–68]. By using peripheral blood mononuclear cells (PBMCs) as a model, a cohort of 21 miRNA molecules have been found to be upregulated during human aging, whereas 144 miRNA molecules have been shown to be suppressed [20]. This has suggested the possibility of intervening in the aging process by modulating miRNA expression.

One approach to compensate for the reduction in the level of a specific endogenous miRNA molecule is to provide an oligonucleotide mimic that on one hand has a sequence the same as that of the mature endogenous one and on the other hand possesses the capacity of entering the RISC. Theoretically, the mimic can be designed as a single-stranded RNA molecule; however, the potency of a double-stranded mimic, which possesses a guide strand as well as a passenger strand, has been found to be higher than that of the single-stranded counterpart [69]. To downregulate the activity of an overexpressed miRNA molecule, one may use the anti-miRNA oligonucleotide (AMO), which is complementary to the miRNA mature strand and can inhibit the interactions of the miRISC with either the miRNA molecule or the target mRNA transcript. The ultimate goal of this approach is to prevent the mRNA transcript from degradation mediated by the overexpressed miRNA molecule. More recently, miRNA sponges have emerged as a new tool to manipulate the activity of endogenous miRNA molecules. These sponges are actually RNA molecules that possess repeated miRNA antisense sequences that are capable of sequestering miRNA molecules away from their endogenous targets [70]. They cannot only be applicable to long-term loss-of-function studies, but may also function as a decoy to modulate the miRNA activity.

Technologies for delivery of miRNA

To translate the theory of miRNA-based therapies into practice, apart from enhancing the potency of the miRNA therapeutics per se, effective delivery of the therapeutics into the target site is pivotal. Practically, to deliver miRNA, two approaches can be adopted. One approach is to deliver miRNA expression vectors, which contain DNA sequences that are later transcribed into pre-miRNA. To succeed, the vectors have to overcome various barriers (e.g., cellular internalization and endolysosomal escape) imposed to nucleic acid transfer, to undergo effective nuclear localization, and finally to have

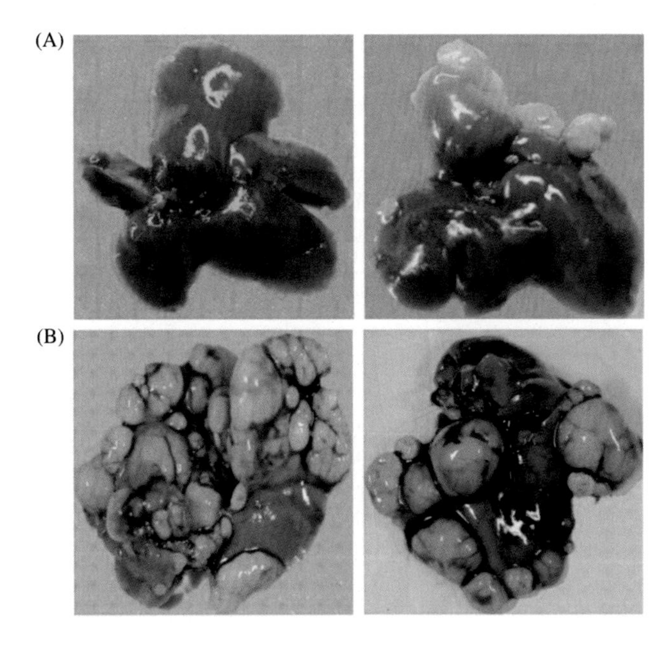

FIGURE 9.2 Representative images of livers from the (A) miR-26a-treated mice and (B) control mice. *Reproduced from J. Kota, R.R. Chivukula, K.A. O'Donnell, E.A. Wentzel, C.L. Montgomery, H.W. Hwang, et al., Therapeutic microRNA delivery suppresses tumorigenesis in a murine liver cancer model, Cell 137 (2009) 1005–1017 with permission from Elsevier B.V., [73].*

the transcribed mRNA molecule to be processed properly by the cellular RNA machinery [71]. Besides delivering expression vectors, directly delivering functional miRNA molecules, whose activity can be elicited as soon as the molecules reach the cytosol, can be an alternative.

Similar to the case of DNA delivery, delivery of miRNA can be achieved by using viral vectors (e.g., adenoviruses, lentiviruses, and AAVs), which can deliver miRNA expression vectors into nuclei for transcription. Previously, in a de novo New Zealand Black (NZB) mouse model, systemic administration of lentiviral vectors expressing miR-15a/16 has been found to exhibit little systemic toxicity but has remarkably increased the expression of miR-15a/16 in transduced cells [72]. Using AAVs systemically to deliver miR-26a in vivo to hepatocellular carcinoma (HCC) cells, which have displayed a reduced level of miR-26a expression, has successfully inhibited the tumor growth and has induced apoptosis (Fig. 9.2) [73]. Recently, nano-sized lipid vesicles, namely exosomes, generated by virus-infected cells have been employed to carry miRNA mimics or inhibitors. In an earlier study, exosomes loaded with a miRNA-155 mimic have elevated the miRNA-155 level in primary mouse hepatocytes and in the liver of miRNA-155 knockout mice [74]. Delivery of a miRNA-155 inhibitor to RAW macrophages using exosomes has also effectively reduced lipopolysaccharide (LPS)-induced TNFα production and has partially inhibited the LPS-induced decrease in the *SOCS1* mRNA level [74].

Example protocols for experimental design

The method below is an example protocol for preparing liposomes in form of large unilamellar vesicles. This protocol is based on the one previously reported by Bender and coworkers [75].

1. Prepare a solvent by mixing 5 mL of chloroform with 5 mL of methanol.
2. Mix 2 nmol of DOTAP and 2 nmol of cholesterol into 10 mL of the solvent.
3. Use N_2 gas to evaporate 9 mL of the solvent.
4. Evaporate the last 1 mL of the solvent in a fume hood overnight to obtain a thin lipid film.
5. Dissolve 1 g of sucrose in 10 mL of distilled water.
6. Heat the sucrose solution to 55°C.
7. Pour the heated sucrose solution onto the thin lipid film at a rate of 1 mL/min under gentle swirling.
8. Incubate the solution mixture at 55°C for 2 hours.
9. Assemble an extruder.
10. Size the liposome suspension using the extruder according to manufacturer's instructions.

Apart from viruses, synthetic carriers are available for miRNA delivery. In an earlier study, *N*-[1-(2,3-dioleyloxy)propyl]-*N*,*N*,*N*-trimethylammonium chloride (DOTMA): cholesterol: D-α-tocopheryl PEG 1000 succinate (TPGS) lipoplexes have been generated by mixing pre-miR-133b with empty liposomes [76]. In A549 nonsmall cell lung cancer (NSCLC) cells, the transfection efficiency of the lipoplexes has been found to be much higher than that of the commercially available siPORT NeoFX transfection agent, leading to significant *MCL1* knockdown [76]. In a recent study, solid lipid nanoparticles (SLNs) containing dimethyldioctadecylammonium bromide (DDAB) have also been fabricated using a film-ultrasonic method, and have been used to deliver miR-34a mimics into the cancer stem cell (CSC) [77], leading to an increase in the survival rate of CSC-bearing mice. Notwithstanding these advances, wide applications of systems containing cationic lipids have been impeded by several practical problems, including nonspecific cellular uptake, type I and type II interferon induction, and toxicity [78–80]. While the nonspecific cellular uptake can be tackled by ligand conjugation to liposomes, other problems are partly caused by the positive surface charge, which is the inherent property of cationic lipids [78]. To address these problems, neutral lipids have been exploited for miRNA delivery. Upon administration of miR-34a into NSCLC xenograft mouse models using a neutral lipid-based carrier, no significant change in the serum levels of cytokines and liver enzymes has been observed, but miRNA-mediated gene silencing has been detected in tumor tissues [81]. Despite this, compared with cationic liposomes, those generated from neutral lipids generally have lower efficiency in transfection and nucleic acid condensation [15]. More efforts are required to develop an effective delivery system with low toxicity for delivery of RNA therapeutics.

Example protocols for experimental design

The method below is an example protocol for preparing PEI polyplexes:

1. Dissolve bPEI in distilled water to reach a desired concentration.
2. Select an appropriate polymer/RNA mass-to-mass ratio.
3. Mix the PEI solution with an miRNA solution to reach the selected polymer/RNA ratio.
4. Vortex the mixture for 30 seconds.
5. Incubate the mixture at ambient conditions for 15 minutes.
6. Use the polyplexes immediately for transfection.

In addition to lipid-based carriers, polymers have been widely used for miRNA delivery. For example, cationic amphiphilic copolymers have been adopted to deliver miR-34a [82], which has been found to involve in aortic senescence in streptozotocin-induced diabetic mice and in life span determination in *C. elegans* under stress conditions [83]. Reducible hyperbranched polymers have also been used to complex with pre-miRNA to generate polyplexes, which have been shown to undergo redox-activated disassembly in the presence of dithiothreitol (DTT) [84]. In an EGFP-expressing cancer cell line, the polyplexes have been internalized into cells effectively to elicit EGFP gene silencing, with no toxicity observed [84]. To enhance the efficiency in endosomal escape, chloroquine-containing 2-(dimethylamino)ethyl methacrylate copolymers have been generated via reversible addition-fragmentation chain transfer polymerization, with the polyplexes formed with miR-210 found to inhibit the migration of cancer cells [85]. Furthermore, by using a cationic gene vector possessing matrix metalloproteinase-2 (MMP2)-cleavable substrate peptides (which target tumor sites that show a high level of MMP2), miR-34a has been successfully delivered to target cells for cancer treatment [86]. While polycations can complex with miRNA to generate polyplexes, oppositely charged polymers can also interact with each other to generate degradable complexes for miRNA delivery. A good example is the hyaluronic acid/protamine sulfate interpolyelectrolyte complexes, which have been used to deliver miR-34a to breast cancer cells or tissues [87]. By targeting CD44 and Notch-1, the miRNA molecule has effectively induced cell apoptosis and has suppressed the migration, proliferation, and growth of breast cancer cells [87].

Example protocols for experimental design

The method below is an example protocol for determining the ability of a polycation to condense RNA:

1. Prepare a 1% agarose gel containing a fluorescent nucleic acid dye right before the experiment.
2. Add 2 μL of an RNA solution (1 μg/μL) into a microcentrifuge tube.
3. Dilute the RNA solution with 2 μL of distilled water.
4. Add 2 μL of the polymer solution (with the concentration decided by the polymer/RNA mass-to-mass ratio required) into another microcentrifuge tube.
5. Dilute the polymer solution with 2 μL of distilled water.
6. Add the diluted polymer solution into the diluted RNA solution.
7. Pipette the mixture up and down vigorously 10 times.
8. Vortex the mixture for 15 seconds.
9. Incubate the mixture at ambient conditions for 15 minutes.
10. Add 2 μL of a loading dye into the mixture.
11. Pipette the mixture up and down vigorously 10 times.
12. Vortex the mixture for 10 seconds.
13. Load the samples into the wells of the gel.
14. Run the gel at a voltage of 100 V for 40 minutes, or until the dye line is around 80 % of the way down the gel.
15. Turn off the power supply.
16. Remove the gel from the gel box.
17. Examine the bands in the gel under UV illumination.

Multifunctionality in miRNA delivery

To enhance the versatility of miRNA delivery, optically active systems have been developed. For example, by capping the particle surface with PEI, CdSe/ZnS QDs have been incorporated with the capacity of condensing the miR-26a expression vector, and have delivered the vector to HepG2 cells to induce cell cycle arrest and to inhibit cell proliferation [88]. Furthermore, due to the strong red luminescence given by the QDs, imaging of living cells has been made possible [88]. These modified QDs have shown the potential to be exploited as a theranostic carrier for the execution of an miRNA-mediated intervention. More recently, a biodegradable photoluminescent polymer has been generated by granting poly(1,8-octanedio-citric acid)-co-PEG (POCG) with PEI [89]. During the synthetic process, POCG has first been generated using PEG, citric acid, and 1,8-octanediol through a facile melt-derived polymerization process (Fig. 9.3). Amidation reactions between the polymer and PEI have subsequently been performed to generate POCG-PEI.

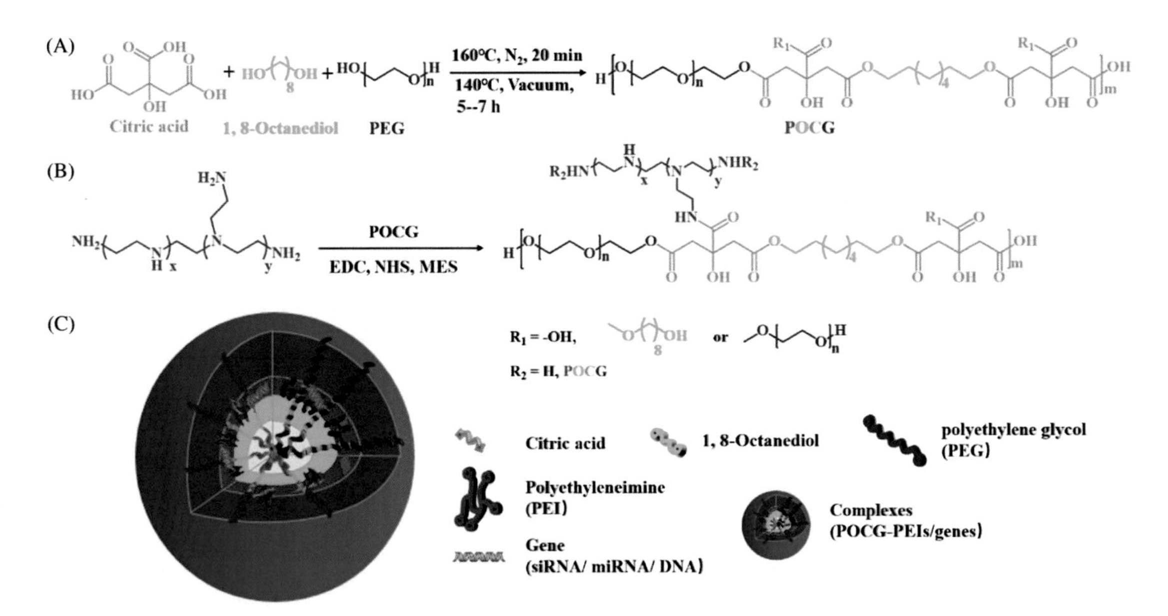

FIGURE 9.3 A schematic diagram showing the generation of (A) POCG, (B) POCG-PEI, and (C) the polyplexes of POCG-PEI with nucleic acids. *EDC*, 1-(3-dimethylaminopropyl)-3-ethylcarbodiimide hydrochloride; *NHS*, *N*-hydroxysuccinimide; *MES*, 2-(*N*-morpholino)ethanesulfonic acid. *Reproduced from M. Wang, Y. Guo, M. Yu, P.X. Ma, C. Mao, B. Lei, Photoluminescent and biodegradable polycitrate-polyethylene glycol- polyethyleneimine polymers as highly biocompatible and efficient vectors for bioimaging-guided siRNA and miRNA delivery, Acta Biomater. 54 (2017) 69–80 with permission from Elsevier B.V., [89].*

This polymer has not only condensed nucleic acids successfully [89], but has also been found to be photostable, thereby potentially enabling real-time tracking during miRNA delivery [89].

Apart from delivering miRNA alone, some carriers can deliver RNA and other agents concomitantly. Previously, methoxy PEG-block-poly(2-methyl-2-carboxyl-propylene carbonate-graft-dodecanol-graft-tetraethylenepentamine), miR-29b1, and a small molecule hedgehog inhibitor (viz., GDC-0449) have been coformulated into micelles, which have then been administered into common bile duct ligation (CBDL) mice [90]. High concentrations of both GDC-0449 and miR-29b1 have been detected in liver cells [90]. In a recent study, a core-shell nanocarrier coated with cationic albumin has also been adopted to codeliver miRNA-34a and docetaxel for treatment of metastatic breast cancer [91]. The size of the nanocarrier has been approximated to be around 180 nm. The nanocarrier has protected miRNA from enzymatic degradation [91]. Importantly, it has been shown in vitro to suppress the expression of antiapoptosis gene *Bcl-2* [91], leading to the inhibition of tumor cell migration and the induction of cancer cell death. In addition to small molecule compounds, carriers can codeliver contrast agents for imaging purposes. This has been exemplified by the case of PEG-modified liposomes, which have been used to entrap ultrasound contrast gas while delivering miRNA. They have not only enabled detection using diagnostic ultrasound [92], but have also delivered miR-126 to ischemic hindlimbs in vivo to inhibit negative regulators of VEGF signaling to promote angiogenesis and to improve the blood flow [92].

Summary

In this chapter, we have exploited the possible use of miRNA for intervention development to combat aging. Here it is worth mentioning that right now most of the studies on the miRNA biology of aging have been performed on invertebrate models such as *C. elegans*. No doubt, these models have been providing an accessible platform for us to elucidate the mechanisms of action of pro-longevity compounds [93] and to develop interventions to combat aging and related diseases. Yet, owing to the genetic differences between humans and these invertebrate models, directed research into the manipulation of the aging network in mammalian models based on miRNA is needed, or the practical potential of miRNA-based antiaging therapies can hardly be fairly verified. Moving beyond invertebrate models should, therefore, be the path to go for future bench-to-clinic translation. Despite this, manipulation of miRNA activity has already been attained technically by using antisense oligonucleotides, LNAs, and antagomirs [94–96]. This success, along with the rapid technological advances in nucleic acid delivery [97–101], has established a platform for the development of miRNA-based antiaging therapies.

Directions for intervention development

If miRNA is selected to be the target of one's proposed intervention, the following steps can be taken to move the intervention closer to practice:

1. Select one miRNA target (or an array of miRNA targets) to be manipulated.
2. Decide whether upregulation or downregulation of the miRNA target(s) should be exercised.
3. Select an appropriate therapeutic (e.g., an miRNA mimic or an analog) for the proposed manipulation.
4. Modify the chemical structure of the therapeutic to maximize the performance.
5. Evaluate potential off-target effects caused by the therapeutic.
6. Examine adverse effects led by the manipulation of the miRNA target(s) in vitro and in vivo.
7. Optimize the dose, the administration method, and the delivery system, if necessary, to enhance the efficacy of the proposed intervention.

References

[1] J. Wittmann, H.M. Jack, New surprises from the deep-the family of small regulatory RNAs increases, ScientificWorldJournal 10 (2010) 1239–1243.

[2] K.P. McCormick, M.R. Willmann, B.C. Meyers, Experimental design, preprocessing, normalization and differential expression analysis of small RNA sequencing experiments, Silence 2 (2011) 2.

[3] R.B. Darnell, HITS-CLIP: panoramic views of protein-RNA regulation in living cells, Wiley Interdiscip. Rev. RNA 1 (2010) 266–286.

[4] T. Maurin, K. Lebrigand, S. Castagnola, A. Paquet, M. Jarjat, A. Popa, et al., HITS-CLIP in various brain areas reveals new targets and new modalities of RNA binding by fragile X mental retardation protein, Nucleic Acids Res. 46 (2018) 6344–6355.

[5] P.N. Grozdanov, C.C. Macdonald, High-throughput sequencing of RNA isolated by cross-linking and immunoprecipitation (HITS-CLIP) to determine sites of binding of CstF-64 on nascent RNAs, Methods Mol. Biol. 1125 (2014) 187–208.

[6] H. Maatz, M. Kolinski, N. Hubner, M. Landthaler, Transcriptome-wide identification of RNA-binding protein binding sites using photoactivatable-ribonucleoside-enhanced crosslinking immunoprecipitation (PAR-CLIP), Curr. Protoc. Mol. Biol. 118 (2017). 27 26 21-27 26 19.

[7] J. Spitzer, M. Hafner, M. Landthaler, M. Ascano, T. Farazi, G. Wardle, et al., PAR-CLIP (photoactivatable ribonucleoside-enhanced crosslinking and immunoprecipitation): a step-by-step protocol to the transcriptome-wide identification of binding sites of RNA-binding proteins, Methods Enzymol. 539 (2014) 113–161.

[8] C. Zhao, J. Jiang, Y.L. Wang, Y.Q. Wu, Overexpression of microRNA-590-3p promotes the proliferation of and inhibits the apoptosis of myocardial cells through inhibition of the NF-κB signaling pathway by binding to RIPK1, J. Cell. Biochem. (2018). Available from: https://doi.org/10.1002/jcb.27633.

[9] S. Yao, Y. Liu, Z. Yao, Y. Zhao, H. Wang, Y. Xu, et al., MicroRNA-376a regulates cell proliferation and apoptosis by targeting forkhead box protein P2 in lymphoma, Oncol. Lett. 16 (2018) 3169–3176.

[10] J. Du, P. Zhang, M. Gan, X. Zhao, Y. Xu, Q. Li, et al., MicroRNA-204-5p regulates 3T3-L1 preadipocyte proliferation, apoptosis and differentiation, Gene 668 (2018) 1–7.

[11] U.A. Orom, F.C. Nielsen, A.H. Lund, MiRNA-10a binds the 5'UTR of ribosomal protein mRNAs and enhances their translation, Mol. Cell 30 (2008) 460–471.

[12] D. Fernandez-Perez, M.A. Brieno-Enriquez, J. Isoler-Alcaraz, E. Larriba, J. Del Mazo, MicroRNA dynamics at the onset of primordial germ and somatic cell sex differentiation during mouse embryonic gonad development, RNA 24 (2018) 287–303.

[13] H. Cao, J. Shi, J. Du, K. Chen, C. Dong, D. Jiang, et al., MicroRNA-194 regulates the development and differentiation of sensory patches and statoacoustic ganglion of inner ear by fgf4, Med. Sci. Monit. 24 (2018) 1712–1723.

[14] S. Vasudevan, Y. Tong, J.A. Steitz, Switching from repression to activation: microRNAs can up-regulate translation, Science 318 (2007) 1931–1934.

[15] Y. Chen, D.Y. Gao, L. Huang, In vivo delivery of miRNAs for cancer therapy: challenges and strategies, Adv. Drug Deliv. Rev. 81 (2015) 128–141.

[16] R. Yi, Y. Qin, I.G. Macara, B.R. Cullen, Exportin-5 mediates the nuclear export of pre-microRNAs and short hairpin RNAs, Genes Dev. 17 (2003) 3011–3016.

[17] Y. Lee, C. Ahn, J. Han, H. Choi, J. Kim, J. Yim, et al., The nuclear RNase III Drosha initiates microRNA processing, Nature 425 (2003) 415–419.

[18] A.M. Denli, B.B. Tops, R.H. Plasterk, R.F. Ketting, G.J. Hannon, Processing of primary microRNAs by the microprocessor complex, Nature 432 (2004) 231–235.

[19] V. Ambros, The functions of animal microRNAs, Nature 431 (2004) 350–355.

[20] N. Noren Hooten, K. Abdelmohsen, M. Gorospe, N. Ejiogu, A.B. Zonderman, M.K. Evans, MicroRNA expression patterns reveal differential expression of target genes with age, PLoS One 5 (2010) e10724.

[21] G.R. Li, C. Luna, J.M. Qiu, D.L. Epstein, P. Gonzalez, Alterations in microRNA expression in stress-induced cellular senescence, Mech. Ageing Dev. 130 (2009) 731–741.

[22] S. Mitomo, C. Maesawa, S. Ogasawara, T. Iwaya, M. Shibazaki, A. Yashima-Abo, et al., Downregulation of miR-138 is associated with overexpression of human telomerase reverse transcriptase protein in human anaplastic thyroid carcinoma cell lines, Cancer Sci. 99 (2008) 280–286.

[23] K. Jia, D. Chen, D.L. Riddle, The TOR pathway interacts with the insulin signaling pathway to regulate C. elegans larval development, metabolism and life span, Development 131 (2004) 3897–3906.

[24] S. Gao, Z. Zhao, R. Wu, L. Wu, X. Tian, Z. Zhang, MiR-146b inhibits autophagy in prostate cancer by targeting the PTEN/Akt/mTOR signaling pathway, Aging 10 (2018) 2113–2121.

[25] Z.H. Wu, C. Lin, C.C. Liu, W.W. Jiang, M.Z. Huang, X. Liu, et al., MiR-616-3p promotes angiogenesis and EMT in gastric cancer via the PTEN/AKT/mTOR pathway, Biochem. Biophys. Res. Commun. 501 (2018) 1068–1073.

[26] W.J. Wang, W. Yang, Z.H. Ouyang, J.B. Xue, X.L. Li, J. Zhang, et al., MiR-21 promotes ECM degradation through inhibiting autophagy via the PTEN/akt/mTOR signaling pathway in human degenerated NP cells, Biomed. Pharmacother. 99 (2018) 725–734.

[27] S. Zhang, M. Wang, Q. Li, P. Zhu, MiR-101 reduces cell proliferation and invasion and enhances apoptosis in endometrial cancer via regulating PI3K/Akt/mTOR, Cancer Biomark. 21 (2017) 179–186.

[28] J. Liu, Y. Xing, L. Rong, MiR-181 regulates cisplatin-resistant non-small cell lung cancer via downregulation of autophagy through the PTEN/PI3K/AKT pathway, Oncol. Rep. 39 (2018) 1631–1639.

[29] F. Liu, X. Zhao, Y. Qian, J. Zhang, Y. Zhang, R. Yin, MiR-206 inhibits head and neck squamous cell carcinoma cell progression by targeting HDAC6 via PTEN/AKT/mTOR pathway, Biomed. Pharmacother. 96 (2017) 229–237.

[30] H. Zhu, R. Xie, X. Liu, J. Shou, W. Gu, S. Gu, et al., MicroRNA-494 improves functional recovery and inhibits apoptosis by modulating PTEN/AKT/mTOR pathway in rats after spinal cord injury, Biomed. Pharmacother. 92 (2017) 879–887.

[31] F. Tang, T.L. Yang, MicroRNA-126 alleviates endothelial cells injury in atherosclerosis by restoring autophagic flux via inhibiting of PI3K/Akt/mTOR pathway, Biochem. Biophys. Res. Commun. 495 (2018) 1482–1489.

[32] M. Tatar, A. Kopelman, D. Epstein, M.P. Tu, C.M. Yin, R.S. Garofalo, A mutant Drosophila insulin receptor homolog that extends life-span and impairs neuroendocrine function, Science 292 (2001) 107–110.

[33] K.D. Kimura, H.A. Tissenbaum, Y. Liu, G. Ruvkun, daf-2, an insulin receptor-like gene that regulates longevity and diapause in Caenorhabditis elegans, Science 277 (1997) 942–946.

[34] M. Bluher, B.B. Kahn, C.R. Kahn, Extended longevity in mice lacking the insulin receptor in adipose tissue, Science 299 (2003) 572–574.

[35] D. van Heemst, M. Beekman, S.P. Mooijaart, B.T. Heijmans, B. W. Brandt, B.J. Zwaan, et al., Reduced insulin/IGF-1 signalling and human longevity, Aging Cell 4 (2005) 79–85.

[36] L. Liang, J. Wang, Y. Yuan, Y. Zhang, H. Liu, C. Wu, et al., MicRNA-320 facilitates the brain parenchyma injury via regulating IGF-1 during cerebral I/R injury in mice, Biomed. Pharmacother. 102 (2018) 86–93.

[37] Z.X. Shan, Q.X. Lin, Y.H. Fu, C.Y. Deng, Z.L. Zhou, J.N. Zhu, et al., Upregulated expression of miR-1/miR-206 in a rat model of myocardial infarction, Biochem. Biophys. Res. Commun. 381 (2009) 597–601.

[38] X.Y. Yu, Y.H. Song, Y.J. Geng, Q.X. Lin, Z.X. Shan, S.G. Lin, et al., Glucose induces apoptosis of cardiomyocytes via microRNA-1 and IGF-1, Biochem. Biophys. Res. Commun. 376 (2008) 548–552.

[39] W. Hu, T. Li, L. Wu, M. Li, X. Meng, Identification of microRNA-18a as a novel regulator of the insulin-like growth factor-1 in the proliferation and regeneration of deer antler, Biotechnol. Lett. 36 (2014) 703–710.

[40] L. Elia, R. Contu, M. Quintavalle, F. Varrone, C. Chimenti, M.A. Russo, et al., Reciprocal regulation of microRNA-1 and insulin-like growth factor-1 signal transduction cascade in cardiac and skeletal muscle in physiological and pathological conditions, Circulation 120 (2009) 2377–2385.

[41] L. Yu, Y. Lu, X. Han, W. Zhao, J. Li, J. Mao, et al., MicroRNA-140-5p inhibits colorectal cancer invasion and metastasis by targeting ADAMTS5 and IGFBP5, Stem Cell Res. Ther. 7 (2016) 180.

[42] Y. Guo, Y. Chen, Y. Zhang, Y. Zhang, L. Chen, D. Mo, Up-regulated miR-145 expression inhibits porcine preadipocytes differentiation by targeting IRS1, Int. J. Biol. Sci. 8 (2012) 1408–1417.

[43] A. de Lencastre, Z. Pincus, K. Zhou, M. Kato, S.S. Lee, F.J. Slack, MicroRNAs both promote and antagonize longevity in C. elegans, Curr. Biol. 20 (2010) 2159–2168.

[44] F. Czauderna, M. Fechtner, S. Dames, H. Aygun, A. Klippel, G.J. Pronk, et al., Structural variations and stabilising modifications of synthetic siRNAs in mammalian cells, Nucleic Acids Res. 31 (2003) 2705–2716.

[45] J. Krutzfeldt, N. Rajewsky, R. Braich, K.G. Rajeev, T. Tuschl, M. Manoharan, et al., Silencing of microRNAs in vivo with 'antagomirs', Nature 438 (2005) 685–689.

[46] J.M. Campbell, T.A. Bacon, E. Wickstrom, Oligodeoxynucleoside phosphorothioate stability in subcellular extracts, culture media, sera and cerebrospinal fluid, J. Biochem. Biophys. Methods 20 (1990) 259–267.

[47] S.T. Crooke, M.J. Graham, J.E. Zuckerman, D. Brooks, B.S. Conklin, L.L. Cummins, et al., Pharmacokinetic properties of several novel oligonucleotide analogs in mice, J. Pharmacol. Exp. Ther. 277 (1996) 923–937.

[48] S.D. Patil, D.G. Rhodes, Influence of divalent cations on the conformation of phosphorothioate oligodeoxynucleotides: a circular dichroism study, Nucleic Acids Res. 28 (2000) 2439–2445.

[49] T.P. Prakash, A.M. Kawasaki, E.V. Wancewicz, L. Shen, B.P. Monia, B.S. Ross, et al., Comparing in vitro and in vivo activity of 2'-O-[2-(methylamino)-2-oxoethyl]-and 2'-O-methoxyethyl-modified antisense oligonucleotides, J. Med. Chem. 51 (2008) 2766–2776.

[50] B.H. Yoo, E. Bochkareva, A. Bochkarev, T.C. Mou, D.M. Gray, 2'-O-methyl-modified phosphorothioate antisense oligonucleotides have reduced non-specific effects in vitro, Nucleic Acids Res. 32 (2004) 2008–2016.

[51] C. Wahlestedt, P. Salmi, L. Good, J. Kela, T. Johnsson, T. Hokfelt, et al., Potent and nontoxic antisense oligonucleotides containing locked nucleic acids, Proc. Natl. Acad. Sci. U.S.A. 97 (2000) 5633–5638.

[52] B. Hyrup, P.E. Nielsen, Peptide nucleic acids (PNA): synthesis, properties and potential applications, Bioorg. Med. Chem. 4 (1996) 5–23.

[53] J. Elmen, M. Lindow, S. Schutz, M. Lawrence, A. Petri, S. Obad, et al., LNA-mediated microRNA silencing in non-human primates, Nature 452 (2008) 896–899.

[54] H.L. Janssen, H.W. Reesink, E.J. Lawitz, S. Zeuzem, M. Rodriguez-Torres, K. Patel, et al., Treatment of HCV infection by targeting microRNA, N. Engl. J. Med. 368 (2013) 1685–1694.

[55] R.E. Lanford, E.S. Hildebrandt-Eriksen, A. Petri, R. Persson, M. Lindow, M.E. Munk, et al., Therapeutic silencing of microRNA-122 in primates with chronic hepatitis C virus infection, Science 327 (2010) 198–201.

[56] A.S. Ricciardi, E. Quijano, R. Putman, W.M. Saltzman, P.M. Glazer, Peptide nucleic acids as a tool for site-specific gene editing, Molecules 23 (2018).

[57] E. Quijano, R. Bahal, A. Ricciardi, W.M. Saltzman, P.M. Glazer, Therapeutic peptide nucleic acids: principles, limitations, and qpportunities, Yale J. Biol. Med. 90 (2017) 583–598.

[58] A. Gupta, A. Mishra, N. Puri, Peptide nucleic acids: advanced tools for biomedical applications, J. Biotechnol. 259 (2017) 148–159.

[59] E. Schraml, J. Grillari, From cellular senescence to age-associated diseases: the miRNA connection, Longev. Healthspan 1 (2012) 10.

[60] I.A. Zaporozhchenko, E.S. Morozkin, A.A. Ponomaryova, E.Y. Rykova, N.V. Cherdyntseva, A.A. Zheravin, et al., Profiling of 179 miRNA expression in blood plasma of lung cancer patients and cancer-free individuals, Sci. Rep. 8 (2018) 6348.
[61] P. Xu, J. Wang, B. Sun, Z. Xiao, Comprehensive analysis of miRNAs expression profiles revealed potential key miRNA/mRNAs regulating colorectal cancer stem cell self-renewal, Gene 656 (2018) 30–39.
[62] S. Park, S. Kang, K.H. Min, K. Woo Hwang, H. Min, Age-associated changes in microRNA expression in bone marrow derived dendritic cells, Immunol. Invest. 42 (2013) 179–190.
[63] X. Zhang, G. Azhar, E.D. Williams, S.C. Rogers, J.Y. Wei, MicroRNA clusters in the adult mouse heart: age-associated changes, Biomed Res. Int. 2015 (2015) 732397.
[64] T. Huan, G. Chen, C. Liu, A. Bhattacharya, J. Rong, B.H. Chen, et al., Age-associated microRNA expression in human peripheral blood is associated with all-cause mortality and age-related traits, Aging Cell (2018) 17.
[65] P. Vilmos, A. Bujna, M. Szuperak, Z. Havelda, E. Varallyay, J. Szabad, et al., Viability, longevity, and egg production of Drosophila melanogaster are regulated by the miR-282 microRNA, Genetics 195 (2013) 469–480.
[66] A. Kogure, M. Uno, T. Ikeda, E. Nishida, The microRNA machinery regulates fasting-induced changes in gene expression and longevity in Caenorhabditis elegans, J. Biol. Chem. 292 (2017) 11300–11309.
[67] J.M. Debernardi, M.A. Mecchia, L. Vercruyssen, C. Smaczniak, K. Kaufmann, D. Inze, et al., Post-transcriptional control of GRF transcription factors by microRNA miR396 and GIF co-activator affects leaf size and longevity, Plant J. 79 (2014) 413–426.
[68] S. Inukai, A. de Lencastre, M. Turner, F. Slack, Novel microRNAs differentially expressed during aging in the mouse brain, PLoS One 7 (2012) e40028.
[69] A.G. Bader, D. Brown, J. Stoudemire, P. Lammers, Developing therapeutic microRNAs for cancer, Gene Therapy 18 (2011) 1121–1126.
[70] J. Kluiver, I. Slezak-Prochazka, K. Smigielska-Czepiel, N. Halsema, B.J. Kroesen, A. van den Berg, Generation of miRNA sponge constructs, Methods 58 (2012) 113–117.
[71] W.F. Lai, W.T. Wong, Design of polymeric gene carriers for effective intracellular delivery, Trends Biotechnol. 36 (2018) 713–728.
[72] S. Kasar, E. Salerno, Y. Yuan, C. Underbayev, D. Vollenweider, M.F. Laurindo, et al., Systemic in vivo lentiviral delivery of miR-15a/16 reduces malignancy in the NZB de novo mouse model of chronic lymphocytic leukemia, Genes Immun. 13 (2012) 109–119.
[73] J. Kota, R.R. Chivukula, K.A. O'Donnell, E.A. Wentzel, C.L. Montgomery, H.W. Hwang, et al., Therapeutic microRNA delivery suppresses tumorigenesis in a murine liver cancer model, Cell 137 (2009) 1005–1017.
[74] F. Momen-Heravi, S. Bala, T. Bukong, G. Szabo, Exosome-mediated delivery of functionally active miRNA-155 inhibitor to macrophages, Nanomedicine 10 (2014) 1517–1527.
[75] H.R. Bender, S. Kane, M.D. Zabel, Delivery of therapeutic siRNA to the CNS using cationic and anionic liposomes, J. Vis. Exp. (2016). Available from: https://doi.org/10.3791/54106.
[76] Y. Wu, M. Crawford, B. Yu, Y. Mao, S.P. Nana-Sinkam, L.J. Lee, MicroRNA delivery by cationic lipoplexes for lung cancer therapy, Mol. Pharm. 8 (2011) 1381–1389.
[77] S. Shi, L. Han, T. Gong, Z. Zhang, X. Sun, Systemic delivery of microRNA-34a for cancer stem cell therapy, Angew. Chem. Int. Ed. Engl. 52 (2013) 3901–3905.
[78] H. Lv, S. Zhang, B. Wang, S. Cui, J. Yan, Toxicity of cationic lipids and cationic polymers in gene delivery, J. Control. Release 114 (2006) 100–109.
[79] D.M. Pereira, P.M. Rodrigues, P.M. Borralho, C.M. Rodrigues, Delivering the promise of miRNA cancer therapeutics, Drug Discov. Today 18 (2013) 282–289.
[80] C.V. Pecot, G.A. Calin, R.L. Coleman, G. Lopez-Berestein, A.K. Sood, RNA interference in the clinic: challenges and future directions, Nat. Rev. Cancer 11 (2011) 59–67.
[81] J.F. Wiggins, L. Ruffino, K. Kelnar, M. Omotola, L. Patrawala, D. Brown, et al., Development of a lung cancer therapeutic based on the tumor suppressor microRNA-34, Cancer Res. 70 (2010) 5923–5930.
[82] S. Sharma, S. Mazumdar, K.S. Italiya, T. Date, R.I. Mahato, A. Mittal, et al., Cholesterol and morpholine grafted cationic amphiphilic copolymers for miRNA-34a delivery, Mol. Pharm 15 (2018) 2391–2402.
[83] J. Chao, Y. Guo, P. Li, L. Chao, Role of kallistatin treatment in aging and cancer by modulating miR-34a and miR-21 expression, Oxid. Med. Cell. Longev. 2017 (2017) 5025610.
[84] U.L. Rahbek, A.F. Nielsen, M. Dong, Y. You, A. Chauchereau, D. Oupicky, et al., Bioresponsive hyperbranched polymers for siRNA and miRNA delivery, J. Drug Target. 18 (2010) 812–820.
[85] Y. Xie, F. Yu, W. Tang, B.O. Alade, Z.H. Peng, Y. Wang, et al., Synthesis and evaluation of chloroquine-containing DMAEMA copolymers as efficient anti-miRNA delivery vectors with improved endosomal escape and antimigratory activity in cancer cells, Macromol. Biosci. 18 (2018).
[86] Y. Zeng, Z. Zhou, M. Fan, T. Gong, Z. Zhang, X. Sun, PEGylated cationic vectors containing a protease-sensitive peptide as a miRNA delivery system for treating breast cancer, Mol. Pharm. 14 (2017) 81–92.
[87] S. Wang, M. Cao, X. Deng, X. Xiao, Z. Yin, Q. Hu, et al., Degradable hyaluronic acid/protamine sulfate interpolyelectrolyte complexes as miRNA-delivery nanocapsules for triple-negative breast cancer therapy, Adv. Healthc. Mater. 4 (2015) 281–290.
[88] G. Liang, Y. Li, W. Feng, X. Wang, A. Jing, J. Li, et al., Polyethyleneimine-coated quantum dots for miRNA delivery and its enhanced suppression in HepG2 cells, Int. J. Nanomedicine 11 (2016) 6079–6088.
[89] M. Wang, Y. Guo, M. Yu, P.X. Ma, C. Mao, B. Lei, Photoluminescent and biodegradable polycitrate-polyethylene glycol-polyethyleneimine polymers as highly biocompatible and efficient vectors for bioimaging-guided siRNA and miRNA delivery, Acta Biomater. 54 (2017) 69–80.
[90] V. Kumar, G. Mondal, R. Dutta, R.I. Mahato, Co-delivery of small molecule hedgehog inhibitor and miRNA for treating liver fibrosis, Biomaterials 76 (2016) 144–156.
[91] L. Zhang, X. Yang, Y. Lv, X. Xin, C. Qin, X. Han, et al., Cytosolic co-delivery of miRNA-34a and docetaxel with core-shell nanocarriers via caveolae-mediated pathway for the treatment of metastatic breast cancer, Sci. Rep. 7 (2017) 46186.

[92] Y. Endo-Takahashi, Y. Negishi, A. Nakamura, S. Ukai, K. Ooaku, Y. Oda, et al., Systemic delivery of miR-126 by miRNA-loaded bubble liposomes for the treatment of hindlimb ischemia, Sci. Rep. 4 (2014) 3883.
[93] M. Lucanic, G.J. Lithgow, S. Alavez, Pharmacological lifespan extension ofinvertebrates, Ageing Res. Rev. 12 (2013) 445–458.
[94] S.J. Lee, S.J. Kim, H.H. Seo, S.P. Shin, D. Kim, C.S. Park, et al., Over-expression of miR-145 enhances the effectiveness of HSVtk gene therapy for malignant glioma, Cancer Lett. 320 (2012) 72–80.
[95] M. Brock, V.J. Samillan, M. Trenkmann, C. Schwarzwald, S. Ulrich, R.E. Gay, et al., AntagomiR directed against miR-20a restores functional BMPR2 signalling and prevents vascular remodelling in hypoxia-induced pulmonary hypertension, Eur. Heart J. 35 (2014) 3203–3211.
[96] A. Selvamani, P. Sathyan, R.C. Miranda, F. Sohrabji, An antagomir to microRNA Let7f promotes neuroprotection in an ischemic stroke model, PLoS One 7 (2012) e32662.
[97] W.F. Lai, In vivo nucleic acid delivery with PEI and its derivatives: current status and perspectives, Expert Rev. Med. Devices 8 (2011) 173–185.
[98] W.F. Lai, Nucleic acid therapy for lifespan prolongation: present and future, J. Biosci. 36 (2011) 725–729.
[99] W.F. Lai, Nucleic acid delivery: roles in biogerontological interventions, Ageing Res. Rev. 12 (2013) 310–315.
[100] W.F. Lai, Cyclodextrins in non-viral gene delivery, Biomaterials 35 (2014) 401–411.
[101] W.F. Lai, M.C. Lin, Nucleic acid delivery with chitosan and its derivatives, J. Control. Release 134 (2009) 158–168.

Chapter 10

Use of delivery technologies to modulate protein kinase activity

Introduction

Proteins involve in diverse physiological processes, ranging from cell maintenance to aging. One example is the protein deacetylases, Sir2 proteins, which have been associated, albeit controversially, with calorie restriction-mediated longevity [1,2]. Other examples are TSC1/2 [3], Rheb [4], ribosomal RNA processing factors [3,5], Raptor [6], and S6 kinase 1 [7]. These proteins constitute the signaling core or the signaling output of the TOR pathway. Here it is worth noting that, as far as "proteins" and "peptides" are concerned, a clear boundary between the two terms is absent at the moment, with no consent reached among the scientific community on the use of these two terms. For discussions in this book, the upper limit of routine peptide synthesis in the solid phase is taken as the guide. A "peptide" is, therefore, thought to be one that contains no more than 50 amino acids in length; whereas a "protein" is one that has a larger size. Because proteins are a very large group of macromolecules with highly diverse 3D structures, covering all proteins exclusively in one chapter is impossible. We will, therefore, use protein kinases as an example to illustrate how interventions can be designed for physiological modulation.

Physiologically, protein kinases are enzymes that catalyze the covalent attachment of γ phosphate from adenosine triphosphate (ATP) to the side chain of threonine, serine, or tyrosine of the substrate protein. Based on the structure of the catalytic domain, protein kinases can be classified into several groups, including AGC kinases, tyrosine kinase-like kinases, tyrosine kinases, calcium/calmodulin-dependent protein kinases (CAMKs), type I casein kinases, CMGC kinases, and STE kinases [8]. In addition, according to the amino acid they phosphorylate, protein kinases can be roughly divided into two groups: serine/threonine kinases and tyrosine kinases [8]. Because protein kinases can act on various substrate proteins, ranging from transcription factors (e.g., c-Jun) to other enzymes (e.g., glycogen synthase), they can modulate the properties (e.g., subcellular localization and enzymatic activities) of protein players in different signal cascades. Because of this, protein kinases play important roles in gene expression and many other physiological processes (e.g., energy metabolism, cell cycle progression, and apoptosis). They can serve as targets to be manipulated for antiaging purposes.

Roles of protein kinases in aging

Over the years, disrupted protein kinase activity has not only been attributed to the cause of gastrointestinal stromal tumors, chronic myelogenous leukemia, and different nonmalignant disorders [8], but has also been linked with various age-associated diseases. For instance, the adult-onset inhibition of the α catalytic subunit of AMPK, which is a highly conserved protein that involves in systemic and cellular energy homeostasis [9], in muscles has been found to increase the sensitivity of fruit flies to stress, leading to the shortening of the life span [10]. Impairment in AMPK activity has also been associated with the onset of age-induced myocardial contractile dysfunctions [11], causing an increase in the generation of ROS and a decrease in the mitochondrial membrane potential in cells.

PKC, which is a serine/threonine kinase possessing four conserved domains (C1–C4) separated by five variable regions (V1–V5) [12], is another example of kinases whose disruption in activity may lead to aging symptoms. This protein is a key player in neuronal processes (such as synaptic plasticity and neurotransmitter release) [13]. Its activity with the prefrontal cortex is inversely correlated with the working memory performance [13]. During signal transduction, PKC is first activated by 3-phosphoinositide-dependent protein kinase-1 (PDK-1),

Delivery of Therapeutics for Biogerontological Interventions. DOI: https://doi.org/10.1016/B978-0-12-816485-3.00010-6

which phosphorylates the threonine residues in the C4 domain of PKC. This results in autophosphorylation of PKC [12]. The possibility of manipulating protein activity for antiaging purposes has been demonstrated by the success of chelerythrine-mediated inhibition of PKC in rescuing the compromised integrity of neuronal dendrites in aged rats [13]. In fact, AMPK and PKC are only a few of the examples. Many other kinases, whose impaired activity links with the aging phenotype, have been identified. For example, memory deficits and altered synaptic plasticity during aging have been attributed partly to age-associated oxidation of calcium/CaMKII [14]. IGF-1 receptor kinase and phosphatidylinositol 3-kinase (PI3K) have also been found to involve in insulin/IGF-1 signaling, playing a conserved role in aging across species [15,16].

Protein kinases as targets for interventions

The feasibility of intervening in aging in living organisms has been corroborated in *C. elegans*, in which overexpression of *aak-2* (which encodes the α-subunit of AMPK) mediates life span extension led by daf-2/insulin-like signaling mutations [17]. In yeasts, the life span has been found to be increased upon either the incorporation of mutations into *TPK1*, *TPK2*, and *TPK3* [which encode homologs of the catalytic subunits of protein kinase A (PKA) in mammals] [18] or the inhibition of kinase activity of the target of rapamycin complex 1 (TORC 1) and Sch9 [19]. Although the causal relationship between the onset of aging and the impairment in protein kinase activity still has not been fully elucidated at the moment, the technical possibility of manipulating aging by modulating kinase activity has already been documented in literature [17,19].

To modulate protein activity, manipulating the expression of kinase-encoding genes is the most direct method (Fig. 10.1A) [20]. This can be achieved by using RNA duplexes such as miRNA and siRNA as depicted in the last two chapters. The success in regulating protein kinase activity via genetic manipulation has been demonstrated in plants, in which RNAi has been adopted to silence the gene encoding the protein kinase, namely AtMPK6, to enhance the sensitivity of the plants to ozone [21]. Recently, the inhibition of the migration of MM-RU human metastatic melanoma cells has been achieved by silencing the expression of the gene for PKCα in cells [22]. The advances in techniques for genetic manipulation, along with the diversity of systems for nucleic acid delivery as depicted in Part II of this book, have made direct manipulation of metabolic pathways technically viable. Despite this, the success of genetic manipulation in reversing aging symptoms relies largely on the identification of genes whose expression changes with age. Unfortunately, not all genes show changes in expression even though the activity of the corresponding protein may be disrupted during aging. One example is AMPKα. Although its activity has been found to decline in aged rats, leading to the development of insulin resistance, the expression of the corresponding gene has not been affected by the aging process [23]. Under this circumstance, genetic manipulation targeting the gene for AMPKα may no longer be useful. Instead, other methods that can directly modulate the activity of the protein should be applied.

One method is to modulate the activity of the protein directly by using small-molecule compounds (Fig. 10.1B)

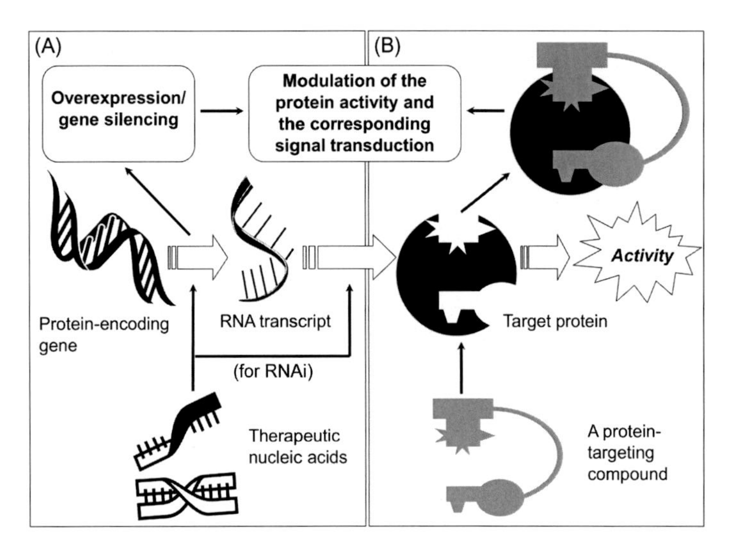

FIGURE 10.1 A schematic diagram showing the two major approaches to target proteins for antiaging interventions: (A) modification of gene expression and (B) modulation of protein activity. *Reproduced from W.F. Lai, Protein kinases as targets for interventive biogerontology: overview and perspectives, Exp. Gerontol. 47 (2012) 290–294 with permission from Elsevier B. V., [20].*

[20]. In an earlier study, the defective PKC signal transduction machinery in aged rats has been restored upon administration of dehydroepiandrosterone [24]. More recently, a cytokine-suppressive antiinflammatory drug, namely SB203580, has been shown to inhibit the action of p38 mitogen-activated protein kinases (MAPKs), and to prevent Werner syndrome (WS) fibroblasts from undergoing accelerated aging [25]. As a matter of fact, although genes are the fundamental units showing an impact on all levels of biology, sometimes genetic manipulation is still not sufficient to reverse the situation. Fortunately, advances in technologies for delivery of nongenetic materials have already been made over the last several decades [26–29], and have enabled the need for delivery of both genetic and nongenetic agents to be addressed. The translation of these advances into interventions has been further facilitated recently by the success in combining computational technologies (e.g., in silico protein–ligand docking) with information about structures and chemical microenvironments (e.g., molecular symmetry, hydrophobicity, and molecular dynamics) to reconstruct the 3D configuration of a protein/ligand complex and to identify ligands showing a high affinity with specific proteins [30,31]. This has made the design of interventions to modulate protein activity in a nongenetic manner more technically plausible.

Selection of delivery methods

The physiological roles played by proteins, as well as the extensive research on the development of protein pharmaceuticals, have made proteins and peptides an important group of biomacromolecules to be considered during the design and execution of an intervention. When a method for delivery of proteins or peptides is selected, several areas have to be considered. These include the efficiency in retaining protein activity, the route of administration, the target population, the duration of the treatment, the pharmacology of the protein, and the desired exposure profile in the patient (e.g., pulsatile or sustained exposure). Apart from this, patient convenience has to be considered during the selection and design of a delivery method. This is especially important when the intervention targets the elderly population, where frequent and repeated injections via parental routes may not be desirable. Compared to systemic administration, local administration of protein pharmaceuticals sometimes is more preferable because this can enhance the availability of the pharmaceuticals in the pathological site. For example, pulmonary administration has been adopted to deliver recombinant human deoxyribonuclease directly to the lungs of patients suffering from cystic fibrosis [32]. Growth factors [e.g., BMPs [33], fibroblast growth factor (FGF) [34], and VEGF [35]] are also delivered to target tissues in a nonsystemic manner in general to avoid systemic toxicity. In addition to the aforementioned, if the intervention to be developed is to be commercialized, market competition and other factors (e.g., production costs, and scalability of the manufacturing process) have to be considered. Finally, the impact of the processes involved in the production (or administration) of the protein-loaded delivery system to the integrity of the loaded protein has to be evaluated. Adverse side effects or a loss of potency may result if the protein is degraded at the time of administration.

Over the years, different technologies have been developed for protein delivery. For instance, previously a transdermal patch containing polymer microneedles has been developed for the administration of protein pharmaceuticals [36]. To fabricate the patch, photolithography is first adopted to produce a master structure from a polymeric photoresist epoxy (viz., SU-8). The master structure is then used to produce reverse molds out of PDMS. From these molds, polyvinylpyrrolidone (PVP) microneedles are fabricated via room temperature UV-induced photopolymerization of monomeric vinyl pyrrolidone, with the encapsulation of protein pharmaceutics performed at the same time. After insertion into the skin, the polymer needles can be dissolved, and the encapsulated proteins can be released into the body. More recently, a range of polymers have been reported for protein delivery. One example is poly(*N*-vinyl imidazole) (PVI), which has been used to form a hydrogel with xanthan gum for encapsulation of bovine serum albumin (BSA) [37] (Fig. 10.2). The loading efficiency and encapsulation efficiency of the hydrogel have been found to be as high as 59.50% and 99.17%, respectively [37]. These values have been successfully tuned by changing the length of the gelation time, the concentration of the loaded BSA, and the total polymer concentration in the system [37]. Other examples of applications of polymers as protein carriers include the use of poly(*N*-isopropylacrylamide) (PNIPAAm) and the poly(β-amino ester) (PAE) pentablock copolymer for delivery of insulin [38,39] and the use of mPEG–PLGA–mPEG for delivery of calcitonin [40].

To enhance the sustainability of protein release, one can design an affinity-controlled release system by making use of the possible interactions between a peptide moiety and its binding partner. For instance, Delplace et al. [41] have modified a thermogel of hyaluronan and methylcellulose with Src homology 3 (SH3) binding peptides (Fig. 10.3). The modified

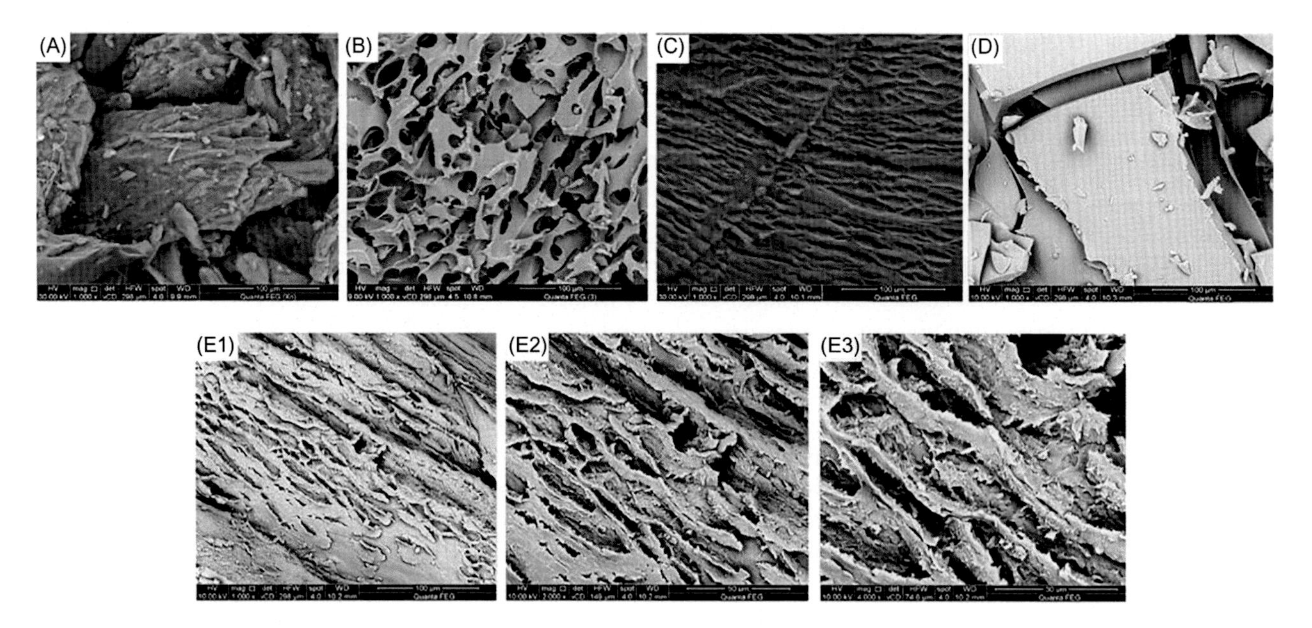

FIGURE 10.2 Scanning electron micrographs of (A) xanthan gum, (B) poly(N-vinyl imidazole), (C) the hydrogel, (D) BSA, and (E1–E3) the BSA-loaded hydrogel matrix. The magnification factor in (E1), (E2), and (E3) is ×1000, ×2000, and ×4000, respectively. *Reproduced from M.W. Sabaa, D.H. Hanna, M.H. Abu Elella, R.R. Mohamed, Encapsulation of bovine serum albumin within novel xanthan gum based hydrogel for protein delivery, Mater. Sci. Eng. C Mater. Biol. Appl. 94 (2019) 1044–1055 with permission from Elsevier B.V., [37].*

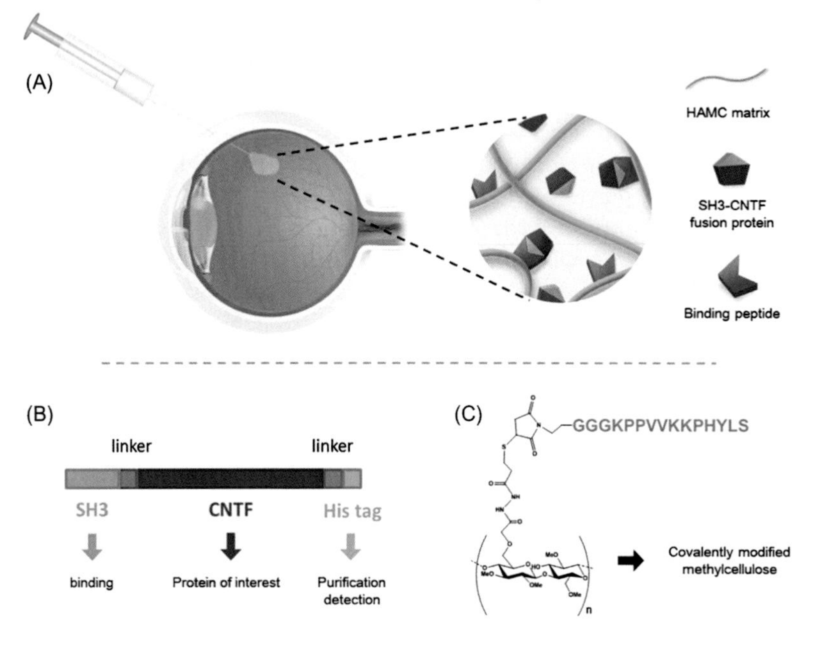

FIGURE 10.3 (A) A schematic illustration depicting the use of the thermogel for intravitreal protein delivery. (B) The structure of the CNTF–SH3 fusion protein and (C) the binding peptide. The fusion protein has been given a His tag to facilitate purification; whereas the binding peptide has been incorporated with a maleimide moiety for subsequent grafting onto thiolated methylcellulose prior to hydrogel preparation. *Reproduced from V. Delplace, A. Ortin-Martinez, E.L.S. Tsai, A.N. Amin, V. Wallace, M.S. Shoichet, Controlled release strategy designed for intravitreal protein delivery to the retina, J. Control Release 293 (2018) 10–20 with permission from Elsevier B.V., [41].*

thermogel has been used for sustained release of CNTF that has been expressed as a fusion protein with SH3. Upon an intravitreal injection of the CNTF-loaded thermogel, successful delivery of CNTF to the mouse retina has been achieved, leading to upregulation of *STAT1* and *STAT3* expression and downregulation of phototransduction genes [41]. Along with the high biocompatibility of the thermogel as demonstrated by immunohistochemical analysis of the retinal tissues of the treated eyes [41], the system has shown high potential to be exploited as an injectable carrier for sustained release of proteins.

Example protocols for experimental design

The method below is an example protocol for evaluating the concentration of a protein in a solution. This protocol is especially useful when determining the loading efficiency or encapsulation efficiency of a protein carrier, or when evaluating the release profile of a protein delivery system.

1. Prepare a series of solutions with different concentrations of a specific protein.
2. Pipet 10 μL of the solution to each well of a 96-well plate.
3. Add 200 μL of 1X Bradford Reagent into each well.
4. Incubate the plate in darkness at room temperature for 15 minutes.
5. Record the absorbance at 595 nm.
6. Repeat the measurement for at least three times.
7. Plot a calibration curve based on the recorded data.
8. Pipet 10 μL of a solution with an unknown protein concentration into one well of a 96-well plate.
9. Add 200 μL of 1X Bradford Reagent into the well.
10. Incubate the plate in darkness at room temperature for 15 minutes.
11. Measure absorbance at 595 nm.
12. Dilute the sample solution and repeat the measurement if the absorbance is beyond the linear region of the calibration curve.
13. Determine the protein concentration in the solution using the calibration curve.
14. Repeat the measurement for at least three times.
15. Take the average of the three measurements as the protein concentration of the solution.

Use of stimuli-responsive polymers as carriers

Polymeric carriers can be developed to respond to various types of stimuli, including chemical stimuli (e.g., alternations in pH, or changes in the concentration of a specific chemical/ion), physical stimuli (e.g., temperature), or biochemical stimuli (e.g., changes in the concentration of a specific biological agent). Among different polymeric carriers, phase-sensitive carriers are an important group of stimuli-responsive systems for protein delivery. Generation of a phase-sensitive system usually involves the use of a water-insoluble polymer dissolved in an organic solvent. Examples of polymers used for this purpose include poly(e-caprolactone), PLGA, and poly(D,L-lactide) (PLA); whereas glycofurol, triacetin, benzyl alcohol, *N*-methyl-2-pyrrolidone (NMP), ethyl acetate, benzyl benzoate (BB), and DMSO are some of the widely used organic solvents [42,43]. After the administration of the phase-sensitive system into a body site, dissipation of the organic solvent into the aqueous phase occurs, with water molecules also penetrating into the polymer matrix. The polymer then precipitates to form a network to entrap the protein pharmaceutical, generating a depot for sustained release of the protein. With the use of this concept, previously a system for controlled delivery of insulin has been developed [44]. The system has been generated simply by dissolving PLA in triacetin. By changing the polymer concentration, the release sustainability of the system has been tuned. In addition, in vivo evaluation has demonstrated that, after a subcutaneous injection, the reduction in the blood glucose level has been sustained in rats for around 1 month. This study has evidenced the possible use of a sustained release system for attaining systemic effects after local administration.

Another important group of stimuli-responsive systems for protein delivery is temperature-responsive carriers. These carriers are often fabricated by using a temperature-responsive polymer, which may help control the rate of protein release and may undergo reversible sol–gel transitions near the body temperature. In general, a temperature-responsive polymeric delivery system exists as a monophasic system before experiencing a temperature trigger, but dehydration of solvated polymer chains occurs when the temperature is elevated to or above the lower critical solution temperature (LCST), leading to phase separation. The process of phase separation is affected by the polymer/solvent interactions and by the ratio of hydrophobic and hydrophilic moieties possessed by the polymer [45,46]. Pluronics, Tetronics, PDEAAm, poly(*N*-isopropylacrylamide) (poly(NIPAAm)), PLGA–PEG–PLGA, and PLA–PEG–PLA are some examples of commonly used polymers that show temperature-responsive properties [47–49]. In a previous study, CS–zinc–insulin complexes have been incorporated into PLA–PEG–PLA to generate a delivery system for controlled release of insulin [47]. Results have shown that the system has preserved the conformational and structural stability of the loaded protein, with the aggregation of the loaded insulin prevented during the entire period of protein release and storage.

Example protocols for experimental design

The method below is an example protocol for determining the activity of the loaded protein in a hydrogel system.

1. Load *Bacillus licheniformis* α-amylase (BLA) into a selected hydrogel system.
2. Put the hydrogel into a mortar.
3. Add an appropriate amount of PBS.
4. Crush the hydrogel thoroughly with a pestle.

(*Continued*)

(Continued)

5. Removing the debris by filtration to obtain a sample solution in which the extracted protein is present.
6. Mix 100 mg of soluble starch and 100 mg of povidone-iodine with 50 mL of distilled water to form a substrate solution.
7. Add 0.5 mL of the sample solution to 2 mL of the substrate solution.
8. Measure the absorbance of the mixture at 580 nm using a UV-visible spectrophotometer at preset time intervals.
9. Determine the activity of the extracted protein based on the initial linear portion of the graph constructed.

In addition to the polymers mentioned above, polymeric systems responsive to other stimuli have been developed for protein delivery. For example, an earlier study has encapsulated a protein, during in situ interfacial polymerization, into a positively charged polymeric shell interconnected by disulfide-containing cross-linkers [50]. The polymeric shell has been found to be redox-responsive by being able to be dissolved under a reducing environment [50]. By taking advantage of the LbL self-assembly process, glucose-sensitive polyelectrolyte nanocapsules have also been generated as protein carriers. The nanocapsules have been fabricated by alternatively coating the CS−*N*-acetyl-L-cysteine conjugate and the random glycopolymer poly(D-gluconamidoethyl methacrylate-*r*-3-acrylamidophenylboronic acid) onto amino-functionalized silica nanospheres [51]. The nanospheres have subsequently been removed by treatment with an NH_4F/HF solution to generate hollow capsules. By using dynamic light scattering (DLS), the size of the nanocapsules has been found to change reversibly in response to changes in the concentration of glucose in the surrounding medium [51]. More recently, biodegradable silica-iron oxide hybrid nanovectors with large mesopores have been generated for protein delivery to cancer cells [52] (Fig. 10.4). The surface of the nanovectors has been functionalized with aminopropyltriethoxysilane to enable electrostatic interactions between the nanovectors and the negatively charged protein molecules [52]. Because half of the content of the nanovectors is composed of iron oxide nanophases, the nanovectors have shown responsiveness to the magnetic field, enabling more precise control of the release of the protein using magnetic stimuli [52]. In fact, with the incorporation of stimuli-responsiveness into the design of a protein carrier, the flexibility and versatility of the delivery process can be enhanced. This enables the execution of interventions to be more controllable.

Direct protein modification for protein delivery

Apart from using an external carrier, some studies have devoted to modifying the protein structure per se by incorporating the protein with arginine-rich cell-penetrating peptides (RPPs), which can cross the plasma membrane and transport the cargo effectively to the cytosol without causing significant membrane damage [53,54], to enhance the delivery efficiency. For instance, a previous study has modified the surface of purified transcription activator-like effector nucleases (TALENs) with the Cys (Npys)-(D-Arg)$_9$ peptide. Upon modification, TALENs have been internalized into cells effectively to induce gene knockout in transformed human cell lines without

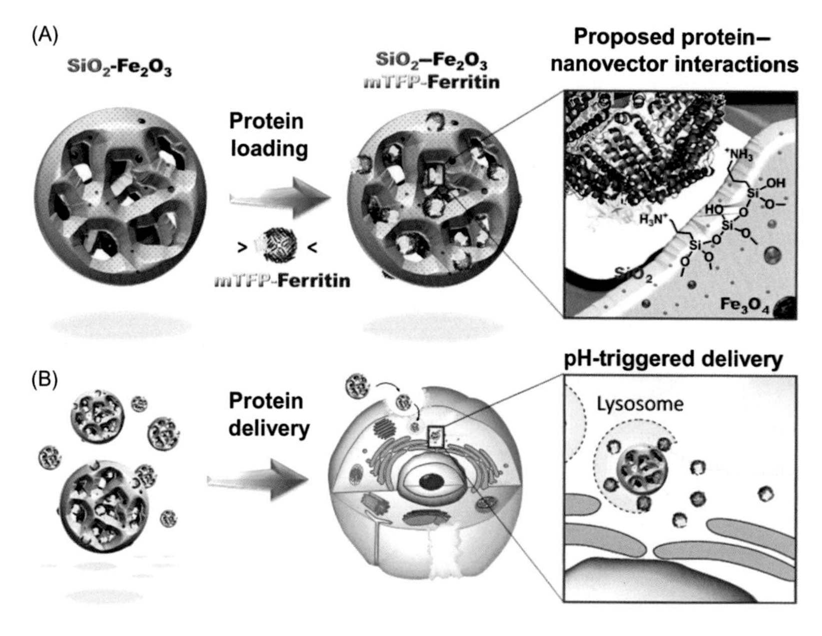

FIGURE 10.4 A schematic diagram showing (A) the loading of a protein model (viz., mTFP-Ferritin) into the nanovector and (B) the subsequent protein delivery process at the cellular level. *Reproduced from H. Omar, J.G. Croissant, K. Alamoudi, S. Alsaiari, I. Alradwan, M.A. Majrashi, et al., Biodegradable magnetic silica@iron oxide nanovectors with ultra-large mesopores for high protein loading, magnetothermal release, and delivery, J. Control Release 259 (2017) 187−194 with permission from Elsevier B.V., [52].*

causing significant toxicity [55]. A similar strategy has been adopted by Ramakrishna et al. [56], who have modified Cas9 with 4-maleimidobutyryl-GGGRRRRRRRRRLLLL. Meanwhile, they have complexed a guide RNA with the C3G9R4LC peptide to generate positively charged nanoparticles. In HEK293T cells, cell treatment with the modified Cas9 protein and the complexed guide RNA has successfully led to gene editing [56]. This work has shed light on the feasibility of engineering the genome via direct delivery of protein- or peptide-based effectors. Such feasibility has been corroborated by a recent study, which has generated genome-edited rats via electroporation of intact rat zygotes with the guide RNA and Cas9 [57]. Despite the promising potential mentioned above, the effective use of RPPs is partly affected by the size of the cargo [58]. RPPs can, therefore, be used only to enhance the delivery of small peptides. Transportation of large proteins using RPPs may result in endosomal trapping rather than nonendocytic delivery [58].

Example protocols for experimental design

The method below is an example protocol for evaluating the membrane integrity of a cell after treatment with a protein carrier

1. Cultivate the cells to be tested in a 12-well culture plate at 37°C under a humidified atmosphere with 5% CO_2 until a confluence of 70%–80% is obtained.
2. Prepare a solution with a known concentration of the carrier using the HEPES-buffered Krebs Ringer (HKR) solution.
3. Dilute Triton X-100 in the HKR solution to reach a concentration of 0.1%.
4. Aspirate the cell culture medium from each well.
5. Rinse the well twice, each with 0.5 mL of the HKR solution.
6. Add 0.5 mL of the solution prepared in Step 2 to each well. Reserve one well as the control, to which the solution prepared in Step 3 is added.
7. Incubate the plate at 37°C under a humidified atmosphere with 5% CO_2 for 30 minutes.
8. Transfer 100 μL of the solution from each well to the corresponding well in a black polypropylene FluoroNunc plate.
9. Assess the membrane integrity using the CytoTox-ONE Homogeneous Membrane Integrity Assay or other commercially available kits.

To improve the delivery performance of the cell-penetrating peptide, cyclization of the peptide can be performed [59]. The success of this has been demonstrated by Nischan et al. [60], who have adopted cyclic TAT as a covalent transporter for transporting green fluorescent protein (GFP) into the cytosol. GFP has been selected as the protein model because its intracellular localization in living cells can be easily detected by using confocal microscopy. Moreover, the correct folding of GFP can be readily determined simply based on the intensity of the emitted fluorescence [60]. Here the generation of the cyclic TAT–GFP conjugate has been achieved by using copper-catalyzed azide-alkyne cycloaddition, in which the cyclic azido-TAT peptide has been conjugated to the GFP mutant (which has been incorporated with homopropargyl glycine at the N terminus). By examining the cellular uptake and intracellular distribution of the conjugate, the cyclic TAT has been found to facilitate cellular internalization of the delivered protein to lead to immediate cytosolic and nuclear availability [60].

Summary

Nucleic acids function to create, encode, and store biological information in cells, and serve to transmit and express that information inside and outside the nucleus. Because of their important roles, during the design of an intervention, sometimes it is easy to be tempted to think that genetic manipulation is an omnipotent method of manipulating biochemical pathways. In this chapter, we have used the case of protein kinases to demonstrate both the opportunities and limitations possessed by genetic manipulation during intervention development. Although the content of this chapter has focused on protein kinases for illustration purposes, the concepts and approaches described can be applicable to other proteins that involve in the aging network. In fact, interventions can be designed to manipulate the physiology of an organism not only at the molecular level but also at the subcellular level. The latter will be discussed in the next chapter by using mitochondrial metabolism as a case study.

Directions for intervention development

Development of an intervention that targets proteins can be performed in two ways. The first is to use proteins or peptides as effectors. This can be achieved by following the steps below:

1. Select a biological target in a body.
2. Decide the protein/peptide to be delivered into the target site as an effector.
3. Select an appropriate administration route.
4. Design an appropriate system to mediate the delivery process.
5. Modify the effector (or the carrier) with the cell-penetrating peptide, if necessary, to enhance the cellular internalization efficiency.
6. Evaluate the performance of the protein-loaded carrier in vitro and in vivo.
7. Optimize the performance of the protein-loaded carrier for intervention execution.

(Continued)

(Continued)

Besides functioning as effectors, proteins or peptides can serve as biological targets. This can be achieved by following the steps below:

1. Choose an endogenous protein whose activity is associated with the aging process.
2. Select an effector (e.g., ligands, competitors, or therapeutics that can modulate the expression of the gene for the selected protein) to be delivered to the target site.
3. Select a carrier to deliver the effector for intervention execution.
4. Study changes in the activity of the target protein after administration of the intervention.
5. Evaluate the short-term and long-term effect of the intervention.
6. Examine the occurrence of off-target effects.
7. Optimize the intervention to maximize the antiaging efficiency.

References

[1] J. Couzin-Frankel, Aging genes: the sirtuin story unravels, Science 334 (2011) 1194–1198.

[2] B. Rogina, S.L. Helfand, Sir2 mediates longevity in the fly through a pathway related to calorie restriction, Proc. Natl. Acad. Sci. U.S.A. 101 (2004) 15998–16003.

[3] M. Kaeberlein, R.W. Powers 3rd, K.K. Steffen, E.A. Westman, D. Hu, N. Dang, et al., Regulation of yeast replicative life span by TOR and Sch9 in response to nutrients, Science 310 (2005) 1193–1196.

[4] S. Honjoh, T. Yamamoto, M. Uno, E. Nishida, Signalling through RHEB-1 mediates intermittent fasting-induced longevity in C. elegans, Nature 457 (2009) 726–730.

[5] K.K. Steffen, V.L. MacKay, E.O. Kerr, M. Tsuchiya, D. Hu, L.A. Fox, et al., Yeast life span extension by depletion of 60s ribosomal subunits is mediated by Gcn4, Cell 133 (2008) 292–302.

[6] K. Jia, D. Chen, D.L. Riddle, The TOR pathway interacts with the insulin signaling pathway to regulate C. elegans larval development, metabolism and life span, Development 131 (2004) 3897–3906.

[7] G.M. Thomas, G.R. Rumbaugh, D.B. Harrar, R.L. Huganir, Ribosomal S6 kinase 2 interacts with and phosphorylates PDZ domain-containing proteins and regulates AMPA receptor transmission, Proc. Natl. Acad. Sci. U.S.A. 102 (2005) 15006–15011.

[8] I. Shchemelinin, L. Sefc, E. Necas, Protein kinases, their function and implication in cancer and other diseases, Folia Biol. 52 (2006) 81–100.

[9] A.A. Gonzalez, R. Kumar, J.D. Mulligan, A.J. Davis, K.W. Saupe, Effects of aging on cardiac and skeletal muscle AMPK activity: basal activity, allosteric activation, and response to in vivo hypoxemia in mice, Am. J. Physiol. Regul. Integr. Comp. Physiol. 287 (2004) R1270–R1275.

[10] D. Tohyama, A. Yamaguchi, A critical role of SNF1A/dAMPKα (Drosophila AMP-activated protein kinase α) in muscle on longevity and stress resistance in Drosophila melanogaster, Biochem. Biophys. Res. Commun. 394 (2010) 112–118.

[11] S. Turdi, X. Fan, J. Li, J. Zhao, A.F. Huff, M. Du, et al., AMP-activated protein kinase deficiency exacerbates aging-induced myocardial contractile dysfunction, Aging Cell 9 (2010) 592–606.

[12] A. Pascale, M. Amadio, S. Govoni, F. Battaini, The aging brain, a key target for the future: the protein kinase C involvement, Pharmacol. Res. 55 (2007) 560–569.

[13] A.R. Brennan, P. Yuan, D.L. Dickstein, A.B. Rocher, P.R. Hof, H. Manji, et al., Protein kinase C activity is associated with prefrontal cortical decline in aging, Neurobiol. Aging 30 (2009) 782–792.

[14] K. Bodhinathan, A. Kumar, T.C. Foster, Intracellular redox state alters NMDA receptor response during aging through Ca^{2+}/calmodulin-dependent protein kinase II, J. Neurosci. 30 (2010) 1914–1924.

[15] M. Holzenberger, J. Dupont, B. Ducos, P. Leneuve, A. Geloen, P. C. Even, et al., IGF-1 receptor regulates lifespan and resistance to oxidative stress in mice, Nature 421 (2003) 182–187.

[16] H. Watanabe, H. Saito, P.G. Rychahou, T. Uchida, B.M. Evers, Aging is associated with decreased pancreatic acinar cell regeneration and phosphatidylinositol 3-kinase/Akt activation, Gastroenterology 128 (2005) 1391–1404.

[17] R. Curtis, G. O'Connor, P.S. DiStefano, Aging networks in Caenorhabditis elegans: AMP-activated protein kinase (aak-2) links multiple aging and metabolism pathways, Aging Cell 5 (2006) 119–126.

[18] S.J. Lin, P.A. Defossez, L. Guarente, Requirement of NAD and SIR2 for life-span extension by calorie restriction in Saccharomyces cerevisiae, Science 289 (2000) 2126–2128.

[19] V. Wanke, E. Cameroni, A. Uotila, M. Piccolis, J. Urban, R. Loewith, et al., Caffeine extends yeast lifespan by targeting TORC1, Mol. Microbiol. 69 (2008) 277–285.

[20] W.F. Lai, Protein kinases as targets for interventive biogerontology: overview and perspectives, Exp. Gerontol. 47 (2012) 290–294.

[21] G.P. Miles, M.A. Samuel, Y. Zhang, B.E. Ellis, RNA interference-based (RNAi) suppression of AtMPK6, an Arabidopsis mitogen-activated protein kinase, results in hypersensitivity to ozone and misregulation of AtMPK3, Environ. Pollut. 138 (2005) 230–237.

[22] H.R. Byers, S.J. Boissel, C. Tu, H.Y. Park, RNAi-mediated knockdown of protein kinase C-alpha inhibits cell migration in MM-RU human metastatic melanoma cell line, Melanoma Res. 20 (2010) 171–178.

[23] W. Qiang, K. Weiqiang, Z. Qing, Z. Pengju, L. Yi, Aging impairs insulin-stimulated glucose uptake in rat skeletal muscle via suppressing AMPKα, Exp. Mol. Med. 39 (2007) 535–543.

[24] M. Racchi, S. Govoni, S.B. Solerte, C.L. Galli, E. Corsini, Dehydroepiandrosterone and the relationship with aging and memory: a possible link with protein kinase C functional machinery, Brain Res. Brain Res. Rev. 37 (2001) 287–293.

[25] T. Davis, D.M. Baird, M.F. Haughton, C.J. Jones, D. Kipling, Prevention of accelerated cell aging in Werner syndrome using a p38 mitogen-activated protein kinase inhibitor, J. Gerontol. A Biol. Sci. Med. Sci. 60 (2005) 1386–1393.

[26] W.F. Lai, A.S. Susha, A.L. Rogach, Multicompartment microgel beads for co-delivery of multiple drugs at individual release rates, ACS Appl. Mater. Interfaces 8 (2016) 871–880.

[27] W.F. Lai, H.C. Shum, Hypromellose-graft-chitosan and Its polyelectrolyte complex as novel systems for sustained drug delivery, ACS Appl. Mater. Interfaces 7 (2015) 10501–10510.

[28] W.F. Lai, A.L. Rogach, Hydrogel-based materials for delivery of herbal medicines, ACS Appl. Mater. Interfaces 9 (2017) 11309–11320.
[29] W.F. Lai, Z.D. He, Design and fabrication of hydrogel-based nanoparticulate systems for in vivo drug delivery, J. Control Release 243 (2016) 269–282.
[30] Y. Fukunishi, Post processing of protein-compound docking for fragment-based drug discovery (FBDD): in-silico structure-based drug screening and ligand-binding pose prediction, Curr. Top. Med. Chem. 10 (2010) 680–694.
[31] Y. Fukunishi, Y. Mizukoshi, K. Takeuchi, I. Shimada, H. Takahashi, H. Nakamura, Protein-ligand docking guided by ligand pharmacophore-mapping experiment by NMR, J. Mol. Graph. Model. 31 (2011) 20–27.
[32] F. Christopher, D. Chase, K. Stein, R. Milne, rhDNase therapy for the treatment of cystic fibrosis patients with mild to moderate lung disease, J. Clin. Pharm. Ther. 24 (1999) 415–426.
[33] C.A. Kirker-Head, Potential applications and delivery strategies for bone morphogenetic proteins, Adv. Drug Deliv. Rev. 43 (2000) 65–92.
[34] R.J. Laham, F.W. Sellke, E.R. Edelman, J.D. Pearlman, J.A. Ware, D.L. Brown, et al., Local perivascular delivery of basic fibroblast growth factor in patients undergoing coronary bypass surgery: results of a phase I randomized, double-blind, placebo-controlled trial, Circulation 100 (1999) 1865–1871.
[35] J.L. Cleland, E.T. Duenas, A. Park, A. Daugherty, J. Kahn, J. Kowalski, et al., Development of poly-(D,L-lactide-coglycolide) microsphere formulations containing recombinant human vascular endothelial growth factor to promote local angiogenesis, J. Control Release 72 (2001) 13–24.
[36] S.P. Sullivan, N. Murthy, M.R. Prausnitz, Minimally invasive protein delivery with rapidly dissolving polymer microneedles, Adv. Mater. 20 (2008) 933–938.
[37] M.W. Sabaa, D.H. Hanna, M.H. Abu Elella, R.R. Mohamed, Encapsulation of bovine serum albumin within novel xanthan gum based hydrogel for protein delivery, Mater. Sci. Eng. C Mater. Biol. Appl. 94 (2019) 1044–1055.
[38] J.Y. Wu, S.Q. Liu, P.W. Heng, Y.Y. Yang, Evaluating proteins release from, and their interactions with, thermosensitive poly(N-isopropylacrylamide) hydrogels, J. Control Release 102 (2005) 361–372.
[39] D.P. Huynh, M.K. Nguyen, B.S. Pi, M.S. Kim, S.Y. Chae, K.C. Lee, et al., Functionalized injectable hydrogels for controlled insulin delivery, Biomaterials 29 (2008) 2527–2534.
[40] Y. Tang, J. Singh, Thermosensitive drug delivery system of salmon calcitonin: in vitro release, in vivo absorption, bioactivity and therapeutic efficacies, Pharm. Res. 27 (2010) 272–284.
[41] V. Delplace, A. Ortin-Martinez, E.L.S. Tsai, A.N. Amin, V. Wallace, M.S. Shoichet, Controlled release strategy designed for intravitreal protein delivery to the retina, J. Control Release 293 (2018) 10–20.
[42] K. Al-Tahami, J. Singh, Smart polymer based delivery systems for peptides and proteins, Recent Pat. Drug Deliv. Formul. 1 (2007) 65–71.
[43] S. Singh, J. Singh, Controlled release of a model protein lysozyme from phase sensitive smart polymer systems, Int. J. Pharm. 271 (2004) 189–196.
[44] K. Al-Tahami, M. Oak, J. Singh, Controlled delivery of basal insulin from phase-sensitive polymeric systems after subcutaneous administration: in vitro release, stability, biocompatibility, in vivo absorption, and bioactivity of insulin, J. Pharm. Sci. 100 (2011) 2161–2171.
[45] D. Schmaljohann, Thermo- and pH-responsive polymers in drug delivery, Adv. Drug Deliv. Rev. 58 (2006) 1655–1670.
[46] E. Ruel-Gariepy, J.C. Leroux, In situ-forming hydrogels–review of temperature-sensitive systems, Eur. J. Pharm. Biopharm. 58 (2004) 409–426.
[47] M. Oak, J. Singh, Controlled delivery of basal level of insulin from chitosan-zinc-insulin-complex-loaded thermosensitive copolymer, J. Pharm. Sci. 101 (2012) 1079–1096.
[48] S. Choi, M. Baudys, S.W. Kim, Control of blood glucose by novel GLP-1 delivery using biodegradable triblock copolymer of PLGA-PEG-PLGA in type 2 diabetic rats, Pharm. Res. 21 (2004) 827–831.
[49] Y. Qiu, K. Park, Environment-sensitive hydrogels for drug delivery, Adv. Drug Deliv. Rev. 53 (2001) 321–339.
[50] M. Zhao, A. Biswas, B. Hu, K.I. Joo, P. Wang, Z. Gu, et al., Redox-responsive nanocapsules for intracellular protein delivery, Biomaterials 32 (2011) 5223–5230.
[51] H. Guo, Q. Guo, T. Chu, X. Zhang, Z. Wu, D. Yu, Glucose-sensitive polyelectrolyte nanocapsules based on layer-by-layer technique for protein drug delivery, J. Mater. Sci. Mater. Med. 25 (2014) 121–129.
[52] H. Omar, J.G. Croissant, K. Alamoudi, S. Alsaiari, I. Alradwan, M.A. Majrashi, et al., Biodegradable magnetic silica@iron oxide nanovectors with ultra-large mesopores for high protein loading, magnetothermal release, and delivery, J. Control Release 259 (2017) 187–194.
[53] H.D. Herce, A.E. Garcia, M.C. Cardoso, Fundamental molecular mechanism for the cellular uptake of guanidinium-rich molecules, J. Am. Chem. Soc. 136 (2014) 17459–17467.
[54] G. Ter-Avetisyan, G. Tunnemann, D. Nowak, M. Nitschke, A. Herrmann, M. Drab, et al., Cell entry of arginine-rich peptides is independent of endocytosis, J. Biol. Chem. 284 (2009) 3370–3378.
[55] J. Liu, T. Gaj, J.T. Patterson, S.J. Sirk, C.F. Barbas 3rd, Cell-penetrating peptide-mediated delivery of TALEN proteins via bioconjugation for genome engineering, PLoS One 9 (2014) e85755.
[56] S. Ramakrishna, A.B. Kwaku Dad, J. Beloor, R. Gopalappa, S.K. Lee, H. Kim, Gene disruption by cell-penetrating peptide-mediated delivery of Cas9 protein and guide RNA, Genome Res. 24 (2014) 1020–1027.
[57] S. Remy, V. Chenouard, L. Tesson, C. Usal, S. Menoret, L. Brusselle, et al., Generation of gene-edited rats by delivery of CRISPR/Cas9 protein and donor DNA into intact zygotes using electroporation, Sci. Rep. 7 (2017) 16554.
[58] G. Tunnemann, R.M. Martin, S. Haupt, C. Patsch, F. Edenhofer, M.C. Cardoso, Cargo-dependent mode of uptake and bioavailability of TAT-containing proteins and peptides in living cells, FASEB J. 20 (2006) 1775–1784.
[59] G. Lattig-Tunnemann, M. Prinz, D. Hoffmann, J. Behlke, C. Palm-Apergi, I. Morano, et al., Backbone rigidity and static presentation of guanidinium groups increases cellular uptake of arginine-rich cell-penetrating peptides, Nat. Commun. 2 (2011) 453.
[60] N. Nischan, H.D. Herce, F. Natale, N. Bohlke, N. Budisa, M.C. Cardoso, et al., Covalent attachment of cyclic TAT peptides to GFP results in protein delivery into live cells with immediate bioavailability, Angew. Chem. 54 (2015) 1950–1953.

Chapter 11

Use of delivery technologies to manipulate mitochondrial metabolism

Introduction

The natural nucleotide, ATP, whose expenditure is tightly regulated by ATP/adenosine diphosphate (ADP) monitors, such as mitochondrial ATP-sensitive potassium [mitoK (ATP)] channels, is ubiquitous in mammalian cells [1]. Physiologically, ATP plays an important role in intracellular energy metabolism, and involves in diverse extracellular functions (e.g., bone metabolism, neurotransmission, hepatic glycogen metabolism, muscle contraction, cardiac function, and vasodilation) [2–5]. ATP also exhibits immunomodulating effects through adenosine (which is one of the breakdown products of ATP, and can modulate inflammation and immune responses [6,7]). Due to the physiological roles played by ATP, proper maintenance of ATP-related metabolism is vital to the normal functioning of an organism.

In this chapter, we will explore how interventions can be developed to target mitochondrial metabolism for anti-aging purposes. The objective of this chapter is to use mitochondrial bioenergenesis, which has been found to be disrupted by aging [8], as an example to illustrate how interventions can be designed, by using either the geriatric approach or the gerontological approach as described in Chapter 1, to combat aging.

Relationships between ATP and aging

ATP involves in different biological processes, ranging from energy storage and signal transduction [9–12] to nucleic acid synthesis [13], in mammalian cells. Once the metabolic activity associated with ATP is compromised, biological risks result. Based on the knowledge accumulated over the past few decades, links between ATP-related metabolism and age-associated symptoms have already been identified. For example, overactive bladder (OAB), which is a common disease in the elderly population, results from detrusor overactivity, which is caused by age-related changes in ATP release in the detrusor [14]. By examining the age-associated changes in the electric reaction of the intact rat aorta endothelium, Tkachenko et al. [15] have also noted that aged rats fail to elicit a typical course of reactions in response to acetylcholine and ATP. This failure has been thought to lead to functional clusterization of endothelial cells and altered electric properties of the endothelium in aged rats, ultimately causing the age-associated reduction in the release of the vasodilator, namely nitric oxide [15].

In mammals, over 90% of endogenous ATP molecules are produced by oxidative phosphorylation (OXPHOS), in which five membrane-associated electron transport chain (ETC) complexes are involved. While ubiquinone oxidoreductase (complex II) has all of its 4 subunits being encoded by the nuclear genome, the other four complexes [i.e., NADH:ubiquinone oxidoreductase (complex I), panthenol:cytochrome c oxidoreductase (complex III), cytochrome c oxidase (complex IV), and ATP synthase (complex V)] are encoded by both mitochondrial and nuclear DNA [16,17]. During OXPHOS, electrons generated by catabolism of oxidizable metabolic substrates (e.g., glucose) are transferred to molecular oxygen in the mitochondrial matrix. Meanwhile, hydrogen ions on the inner mitochondrial membrane are pumped from the mitochondrial matrix to the membrane gap. This generates an electrochemical gradient [18]. This proton gradient drives ATP synthase. The membrane electrochemical potential, the presence/absence of ATPase inhibitory peptides, and allosteric regulation are some of the factors that can modulate the activity of OXPHOS [1]. By measuring the activity of cytochrome c oxidase in skin fibroblasts obtained from donors of different ages (0–97 years old), the activity of OXPHOS has been found by an earlier study to decline with age [19]. An age-associated decline in mitochondrial transcripts encoding cytochrome c oxidase and other OXPHOS components has also been observed in *D. melanogaster* [20]. All these have revealed the impact of aging on the OXPHOS activity.

Delivery of Therapeutics for Biogerontological Interventions. DOI: https://doi.org/10.1016/B978-0-12-816485-3.00011-8

Over the years, increasing efforts have been devoted to elucidating the relevance of the OXPHOS activity to age-associated diseases. For example, Petrosillo et al. [21] have examined the impact of aging on various parameters (e.g., the complex I activity, oxygen consumption, membrane potential, ROS generation, and cardiolipin oxidation) that have been associated with mitochondrial bioenergetics in the rat heart. Results have shown that during aging, the level of oxidized cardiolipin increases; while the mitochondrial content of cardiolipin (which is a phospholipid that serves as an important component of the inner mitochondrial membrane) decreases. Based on these results, ROS-induced cardiolipin peroxidation has been proposed as one of the causes of age-associated changes in the cardiolipin content and hence the mitochondrial complex I deficiency in the heart. Recently, Tatarkova et al. [22] have studied the ETC complex activity and the oxidative damage in cardiac mitochondria obtained from rats with different ages and have attributed lipid peroxidation and mitochondrial oxidative damage to the reduction in the activity of different ETC complexes in aged rats. In fact, oxidative damage not only causes mutations in the mitochondrial genome, but also damages the structures of ETC complexes [23], thereby jeopardizing the bioenergetic functions of mitochondria [24]. In humans, mtDNA encodes 22 tRNA molecules, 2 ribosomal RNA molecules, and 13 protein subunits of the respiratory chain apparatus [23]. The rate of mutations in mtDNA is usually much higher than that in chromosomal DNA [25,26], partly because of the lack of systems for nucleotide excision and mismatch repair in mitochondria. The close relationship between aging and mtDNA mutations has been suggested by an earlier study [8] in which impairment in the mitochondrial bioenergetic function in skeletal muscles in mice has been found to accumulate with age. This observation has suggested that the damage of tissue mitochondria can be one of the underlying mechanisms responsible for the age-associated disruption of bioenergetic processes.

Implications for intervention development

Until now, directed research on the use of ATP as a target for the development of biogerontological interventions has been limited. Due to the roles played by ATP in maintaining proper biological functions, ATP-related metabolic components (e.g., ETC complexes) and the mitochondrial apparatus may be used as intervention points. The plausibility of intervening in the ATP-related metabolic activity to modulate the aging network has been documented in an earlier study, which has found that, upon activation of mitoK(ATP) channels, the susceptibility of an aged subject to neurodegenerative diseases (such as Parkinson's disease) has been escalated due to an increase in angiotensin-induced oxidative stress, neuroinflammation, and dopaminergic neuron degeneration [27]. This has revealed the possibility of manipulating the activity of mitoK(ATP) channels to tackle age-associated neurodegeneration.

In addition to targeting the ATP-related metabolic components, the respiratory chain apparatus may be manipulated to minimize ROS generation. The viability of this has been shown in some plants, protozoa, and fungi, which possess respiratory enzymes other than those OXPHOS components found in eukaryotes [28,29]. An earlier study on the filamentous fungus, *Podospora anserine* (*P. anserine*), has found that the fungus displays a series of degenerative changes after a prolonged vegetative state, as contrary to many other fungi that exhibit indefinite growth [30]. Some of these changes include a reduction in the growth rate, cell death at the tips of hyphae, and alternations in the pigmentation of the mycelium [30]. By examining some of the long-lived *P. anserine* mutants displaying impaired complex IV activity, the mutants have been found to possess an alternative oxidase [30]. That oxidase can directly transport electrons from ubiquinol to molecular oxygen while bypassing those electrons from being conveyed to complexes III and IV [31], thereby reducing ROS generation in mitochondria [32,33]. This finding has provided insights into the possibility of minimizing oxidative stress in mitochondria by incorporating alternative mechanisms into mammalian cells to replace the conventional OXPHOS pathway. Despite this, as mentioned in Chapter 1, re-designing a metabolic pathway to halt the emergence of intrinsic age-associated damages is difficult. This is because this kind of interventions necessitates comprehensive understanding of the high complexity of the intimately interwoven metabolic pathways. In addition, after introducing an alternative mechanism, the effects of potential cross-links between that mechanism and the conventional OXPHOS pathway are unknown. The physiological price involved may be too high to be paid by the recipient.

Strategies to target mitochondria

Over the years, different strategies have been adopted to target mitochondria during therapeutics delivery. Most of these strategies take advantage of the strong negative membrane potential and the high hydrophobicity of the inner mitochondrial membrane. One example of mitochondrial targeting moieties is dequalinium (DQA), which tends to accumulate in mitochondria in response to the electrochemical gradient across the mitochondrial membrane [34]. Another example is the derivatives of the TPP ion. These derivatives display high lipophilicity, which enables effective penetration into the phospholipid membrane. In addition, they show delocalized cationic properties, and hence can reduce the free energy change involved when they move from a hydrophilic environment to a hydrophobic area [34]. Unfortunately, because TPP derivatives may increase proton leakage across the inner mitochondrial membrane [35–37], gene carriers may become more cytotoxic after the incorporation of TTP derivatives.

Targeting mitochondria can also be achieved using mitochondria penetrating peptides or related mimics. One example is the cationic amphiphilic polyproline helix P11LRR [38], which consists of a polyproline scaffold structure functionalized with cationic and hydrophobic moieties. The polyproline backbone enables the production of the polyproline type II helix that can enhance mitochondrial import (Fig. 11.1) [38]. Based on the concentration of the peptide adopted, cell entry has been found to be mediated via an endosomal pathway at a low concentration (10 μM) and a direct transport mechanism of cell penetration, which leads to localization at mitochondria, at a high concentration (> 20 μM) [38]. More recently, P14LRR has been generated from P11LRR, with additional cationic and hydrophobic groups incorporated into the helix [39]. By monitoring the localization of the green fluorescence from fluorescein-labeled P14LRR in mitochondria (labeled with Mitotracker) and lysosomes (labeled with Lysotracker), significant mitochondrial localization of the peptide has been observed after the cells have been treated with the peptide at the concentration of only 5 μM (Fig. 11.2) [38,39]. This has confirmed the high efficiency of P14LRR in mediating mitochondrial localization.

Another example of mitochondria penetrating peptides is the Szeto-Schiller (SS) tetra-peptide, which is positively charged and lipophilic in nature. This peptide can target cardiolipin in the inner mitochondrial membrane [40,41]. It can also promote cardiolipin production

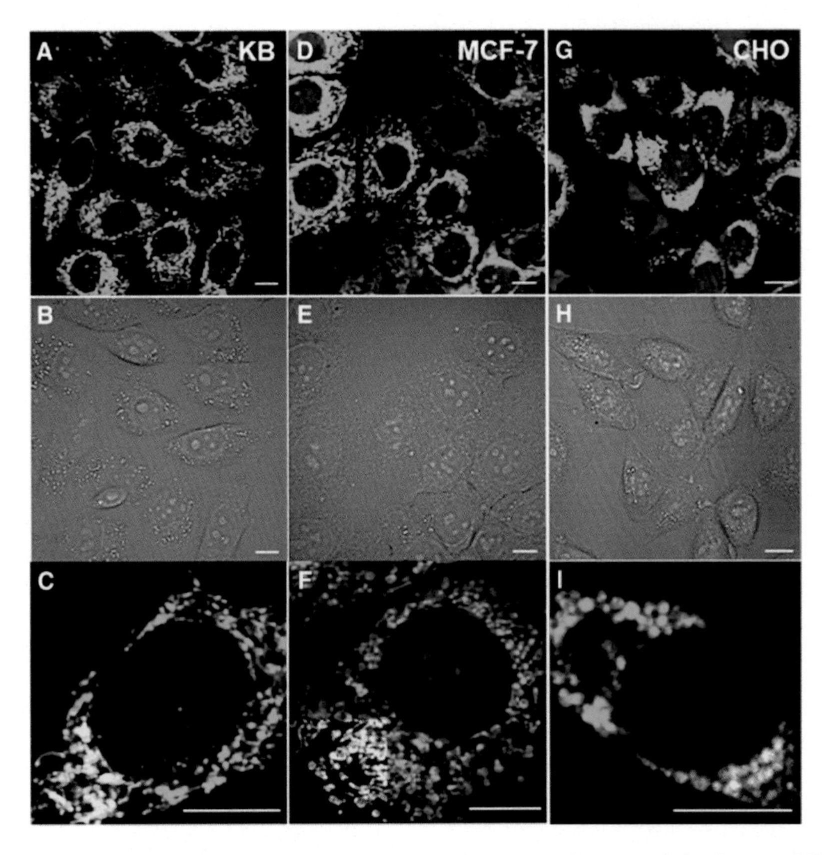

FIGURE 11.1 (A, C, D, F, G, I) Confocal fluorescence images and (B, E, H) the corresponding transmission images of (A-C) KB cells, (D-F) MCF-7 cells, and (G-I) CHO cells after treatment with P11LRR. Colocalization of P11LRR-Fl with Mitotracker Red, which is a mitochondria-specific dye, has suggested the localization of P11LRR-Fl in mitochondria. Scale bar = 10 μm. *Reproduced from L. Li, I. Geisler, J. Chmielewski, J.X. Cheng, Cationic amphiphilic polyproline helix P11LRR targets intracellular mitochondria, J. Control Release 142 (2010) 259–266 with permission from Elsevier B.V., [38].*

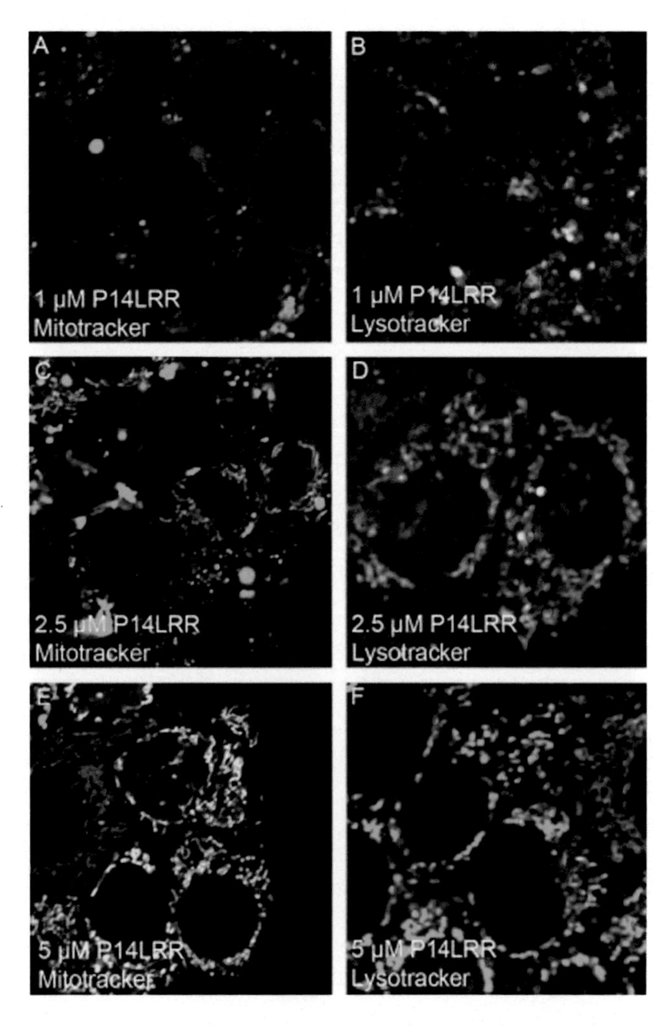

FIGURE 11.2 Subcellular localization of P14LRR in HeLa cells. Different concentrations ((A, B) 1 μM, (C, D) 2.5 μM, and (E, F) 5 μM) of P14LRR have been used for cell treatment. *Reproduced from D. Kalafut, T.N. Anderson, J. Chmielewski, Mitochondrial targeting of a cationic amphiphilic polyproline helix, Bioorg. Med. Chem. Lett. 22 (2012) 561–563 with permission from Elsevier B.V., [39].*

to repair mitochondrial cristae in aged mice [42]. This has been revealed in mice that, starting at the age of 24 months, have been subcutaneously administered with SS-31 daily at a dose of 1 mg/kg for 2 months [42]. As shown in Fig. 11.3, in young mice, cardiac mitochondria are localized in clusters beneath the sarcolemma or in longitudinal rows along myofibrils. Stacks of flat and thin lamellar cristae can be observed in interfibrillar mitochondria. On the contrary, a loss of cristae membranes and cristeolysis occur in aged mice, along with the collapse of the tubular cristae in retinal pigment epithelium (RPE) cells. This age-associated loss of mitochondrial cristae, however, has been successfully restored after treatment with SS-31. This peptide has the potential to be further exploited during the development of antiaging interventions that are mediated by mitochondrial manipulation.

Development of mitochondria-targeted carriers

To manipulate mitochondrial metabolism, we need a carrier that does not release the therapeutics during endocytosis, but can transport it to mitochondria [34]. In fact, if the carrier tends to accumulate in the mitochondria, off-target effects of the intervention can be minimized. Such preferential accumulation can be achieved by taking advantage of the available protein import machinery in mitochondria and the proton electrochemical potential gradient across the inner mitochondrial membrane. The former can be exploited for delivering proteins and peptides into mitochondria; whereas the latter can be used to enhance the mitochondrial entry of positively charged lipophilic cationic molecules.

Over the years, different carriers have been studied for mitochondria-targeted delivery. One example is the water-soluble amorphous silica nanocage [43], which can function as a mitochondria-targeted carrier for drug delivery and bioimaging. Cationic vesicles, namely DQAsomes, have also been reported for mitochondria-targeted gene delivery. These vesicles are made of 1,1′-decamethylene *bis*(4-aminoquinaldiniumchloride), which is also known as DQA. Structurally, DQA is a cationic bolaamphiphile consisting of two quinaldinium rings linked by ten methylene groups. The tendency of this compound to accumulate in mitochondria was reported in the late 1980s [44–46], with the mechanism elucidated in recent years through the use of the Fick–Nernst–Planck physicochemical model and the quantitative structure–activity relationship (QSAR) model [47]. As DQA shows diverse activities (ranging from selectively blocking K^+ channels [48] to causing mtDNA deletion in human cervical carcinoma cells [49]) that may induce cell damage, it is interesting to see that this "mitochondrion poison" turns out to be one of the extensively used agents for the development of mitochondria-targeted carriers. By taking advantage of the properties of DQA, nanosomes consisting of DQA–1,2-dioleoyl-3-trimethylammonium-propane (DOTAP)–1,2-dioleoyl-*sn*-glycero-3-phosphoethanolamine (DOPE), namely DQA80s, have been generated recently [50]. In vitro studies have shown that, after an incubation period of 12 hours, while DQAsomes/pDNA complexes have still been trapped inside the endolysosomes, pDNA complexes formed with DQA80s have

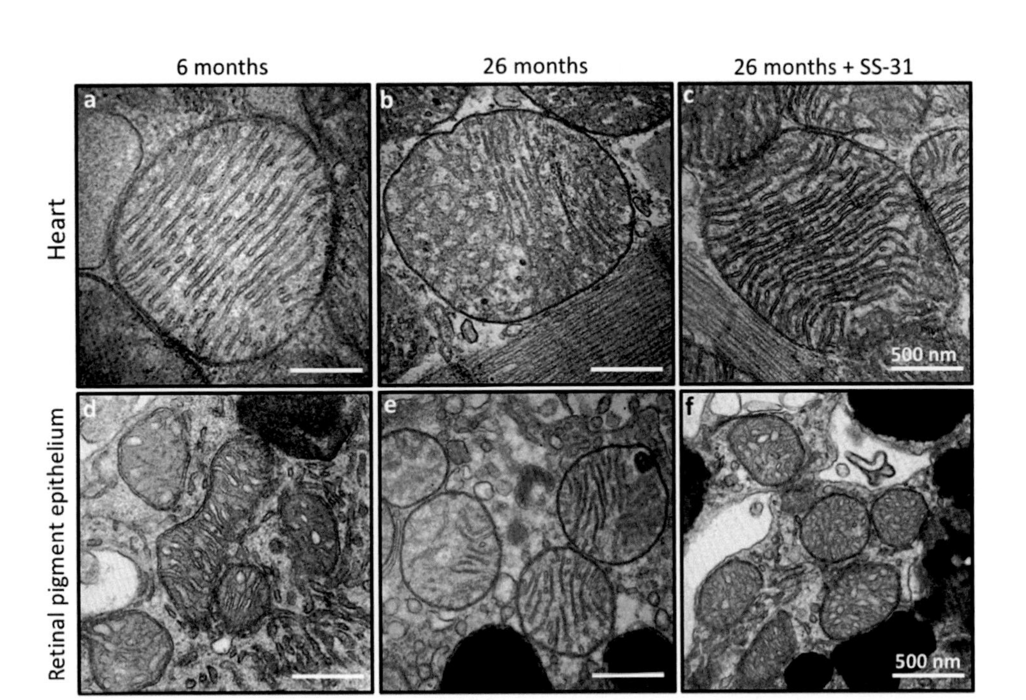

FIGURE 11.3 The effect of SS-31 on restoration of the mitochondrial cristae architecture in aged mice. The top panel and bottom panel show the transmission electron micrographs of the murine cardiac mitochondria and the murine retinal pigment epithelium mitochondria, respectively. These mitochondria have been obtained from the untreated mice at (a, d) 6 months old and (b, e) 26 months old, and from the (c, f) 26-month-old mice that have undergone 2 months of SS-31 treatment. *Reproduced from H.H. Szeto, S. Liu, Cardiolipin-targeted peptides rejuvenate mitochondrial function, remodel mitochondria, and promote tissue regeneration during aging, Arch. Biochem. Biophys. 660 (2018) 137–148 with permission from Elsevier B.V., [42].*

already undergone endosomal escape (Fig. 11.4). The nanosomes warrant further development as carriers for mitochondria-targeted therapy.

Exampleprotocols for experimental design

The method below is an example protocol for preparing DQAsomes. This protocol is based on the one previously reported by Weissig et al. [34].

1. Dissolve dequalinium chloride in methanol to reach a final concentration of 10 mM.
2. Evaporate the organic solvent using a rotary evaporator.
3. Resuspend the solvent-free DQA film in a HEPES buffer (5 mM, pH 7.4).
4. Sonicate the suspension for 1 hour.
5. Centrifuge the suspension at 10,000 rpm for 10 minutes.
6. Filter the generated DQAsomes using a 0.8 μm pore filter.

In addition to using carriers, mitochondria-targeted delivery can be achieved using physical techniques. One example is electroporation [51,52], which has been employed to introduce plasmids into isolated mitochondria for genetic manipulation. The field strength of 14 kV/cm has been reported to lead to maximal plasmid internalization, though mitochondrial destruction has been observed [52]. Another example is microprojectile bombardment [53], which enables direct delivery of foreign genes into mitochondria. The feasibility of using this method has been reported by Johnston et al. [54], who have bombarded a nonreverting yeast strain, which is respiratory deficient owing to the presence of a deletion in the mitochondrial *oxi3* gene, with tungsten microprojectiles that have been coated with DNA bearing sequences for correcting the *oxi3* deletion. Results have shown that the introduced DNA sequences have integrated into the homologous site in the mitochondrial genome of the recipient yeast cell, with the respiratory function of the strain restored [54]. This study has demonstrated that genetic manipulation can be applied not only to the nuclear genome, but also to organelle genomes. Despite this, thus far applications of these physical techniques have been limited to isolated mitochondria or simple in vivo models. There is still a long way to go before these techniques can be translated into interventions that can be executed in a mammalian body.

Replacement of mitochondria

Strategies to develop mitochondria-targeted interventions can be roughly divided into three main groups. One is to

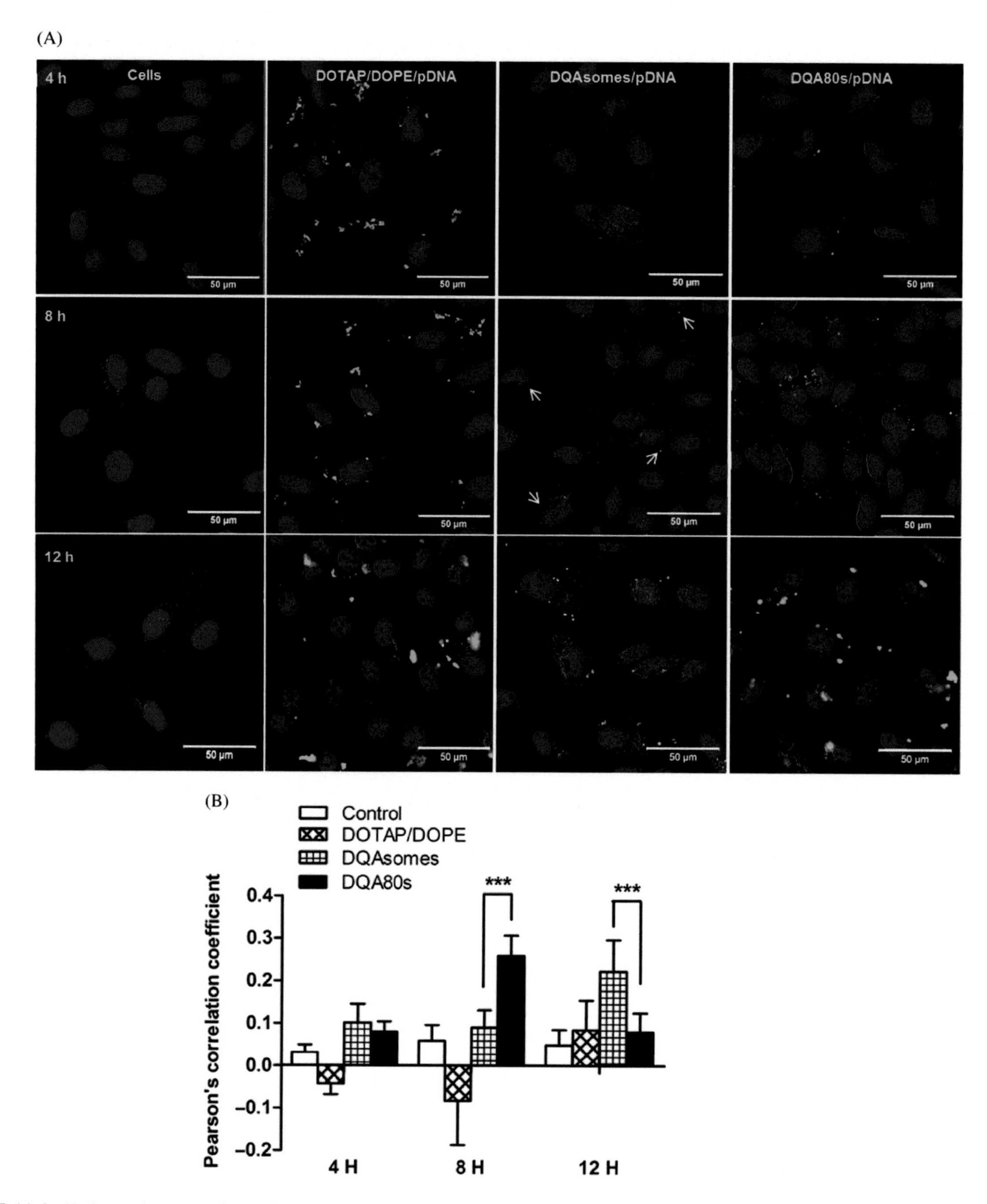

FIGURE 11.4 Endosomal escape of complexes with pDNA. (A) HeLa cells have been transfected with pDNA complexed with various carriers: DOTAP/DOPE, DQAsomes, or DQA80s. Nuclei have been stained with NucBlue Live Ready Probes. Lysosomes have been stained with Lysotracker Red. The carriers have been labeled with a green fluorescent ammonium salt. Arrows indicate the colocalization of lysosomes and DQAsomes/pDNA complexes. Scale bar = 50 μm. (B) Quantification of the colocalization of each of the complexes with Lysotracker Red. ***$P < .001$. *Reproduced from Y. Bae, M.K. Jung, S.J. Song, E.S. Green, S. Lee, H.S. Park, et al., Functional nanosome for enhanced mitochondria-targeted gene delivery and expression, Mitochondrion 37 (2017) 27–40 with permission from Elsevier B.V., [50].*

treat physiological dysfunctions caused by mitochondrial damage. This is a geriatric approach to antiaging medicine. The other approach is to re-program the mitochondria so that the occurrence of damage can be prevented. This is a gerontological approach that aims at manipulating or re-designing the inborn metabolic process. The last approach is to restore the cellular functions of cells by replacing the old mitochondria with healthy exogenous ones. This can be regarded as an engineering approach to the development of antiaging interventions. The viability of this concept has been supported by an earlier study, which has adopted centrifugation to transfer exogenous mitochondria (which have been isolated from human umbilical cord-derived MSCs using differential centrifugation) into mtDNA-deleted Rho^0 cells and dexamethasone-treated atrophic muscle cells [55]. Upon the incorporation of exogenous mitochondria, the declined production of ATP in the target cells has been restored [55]. In addition, the dysfunctional mitochondrial membrane potential, the mitochondrial ROS level, and the oxygen consumption rate of the target cells have been normalized [55]. Despite this success, mitochondrial transfer via centrifugation can only be possible for ex vivo cell manipulation. This method can hardly be translated into a practicable intervention for replacing damaged mitochondria in a living being bodywide.

Exampleprotocols for experimental design

The method below is an example protocol for studying the impact of an agent on the mitochondrial membrane potential in cells.

1. Cultivate the cells in a 6-well culture plate at 37°C under a humidified atmosphere with 5% CO_2 until a confluence of 70%–80% is obtained.
2. Prepare a solution with a known concentration of the agent to be tested using the fresh cell culture medium.
3. Dilute the solution using the fresh cell culture medium to obtain a series of solutions with different concentrations of the agent.
4. Add 3 mL of the solution to each well.
5. Incubate the plate at 37°C for 24 hours under a humidified atmosphere with 5% CO_2.
6. Remove the cell culture medium.
7. Wash the cells with PBS twice.
8. Dissolve TREM (tetramethylrhodamine, ethyl ester) and JC-1 (5,5′,6,6′-tetrachloro-1,1′,3,3′-tetraethylbenzimidazolyl-carbocyanine iodide) in the cell culture medium at a concentration of 100 ng/mL and 10 μg/mL, respectively.
9. Add 3 mL of the cell culture medium prepared in Step 8 to each well.
10. Incubate the plate at 37°C for 30 minutes in darkness under a humidified atmosphere with 5% CO_2.
11. Remove the medium from each well.
12. Wash the cells with PBS twice.
13. Resuspend the cells in each well in 1 mL of PBS.
14. Analyze fluorescence from cells using a flow cytometer.

Apart from centrifugation, various chemical methods have been used to deliver intact mitochondria. For example, a previous study has adopted Pep-1 (KETWWETWWTEWSQPKKKRKV-cysteamine), which is an amphipathic cell-penetrating peptide with three domains (viz., a spacer domain, a hydrophilic lysine-rich domain, and a hydrophobic tryptophan-rich motif), for direct mitochondrial transfer [56]. During the transfer process, mitochondria have first been isolated from parent 143B cells, followed by conjugation with Pep-1 and by incubation with a cybrid cell model of myoclonic epilepsy with ragged-red fibers (MERRF) syndrome. After direct mitochondrial transfer, the cells have displayed the normal spread-out morphology [56]. Flow cytometry analysis has revealed that around 75% of the mitochondria have been internalized by the MERRF cybrid cells [56]. By labeling both the endogenous and exogenous mitochondria in treated cells with fluorescent dyes, the exogenous mitochondria have been shown to come in direct contact with the host mitochondria after cellular internalization [56]. Importantly, after the uptake of exogenous mitochondria, the mitochondrial functions of the MERRF cybrid cells have been restored [56]. As changes in mitochondrial dynamics have been implicated in numerous age-associated diseases (including Alzheimer's disease [57] and Parkinson's disease [58]), the success of mitochondrial transfer has opened up a new avenue for the development of therapies to tackle tissue degeneration caused by the aging process.

Although Pep-1-conjugated mitochondria can be internalized into the host cells, the delivery process is not target specific. This problem may not be obvious in the in vitro context, in which the cell type is highly homogeneous; however, if the intervention is applied in the in vivo context in which different tissues and organs are involved, strategies for more target-specific mitochondrial transfer are desirable. This need can be fulfilled by using magnetomitotransfer. This process can be mediated by using anti-TOM22 magnetic bead-labeled mitochondria, which are generated by first incubating the lysate of donor cells with anti-TOM22 microbeads, followed by magnetic separation of mitochondria from the lysate [59]. The labeled mitochondria are then incubated with Lipofectamine at 37°C before adding to the recipient cells, and are finally pulled down into the cells using magnetic forces [59]. In MRC-5 fibroblasts and as compared to passive mitochondrial transfer, magnetomitotransfer has been found to increase the efficiency of exogenous mitochondria in being internalized into recipient cells, in which intracellular membrane vesicles containing membranous structured contents and microbeads have been found between day 2 and day 4 after magnetomitotransfer (Fig. 11.5) [59].

To ensure the applicability of direct mitochondrial transfer to a human body for antiaging purposes, few challenges have to be addressed. First, the impact of mitochondria from different origins (e.g., autologous, allogeneic, and xenogeneic mitochondria) on the function of the recipient cell should be fully elucidated. In addition, mtDNA is fragile and can be damaged easily due to the lack of effective systems in mitochondria for nucleotide excision and mismatch repair [25,26]. Because of this,

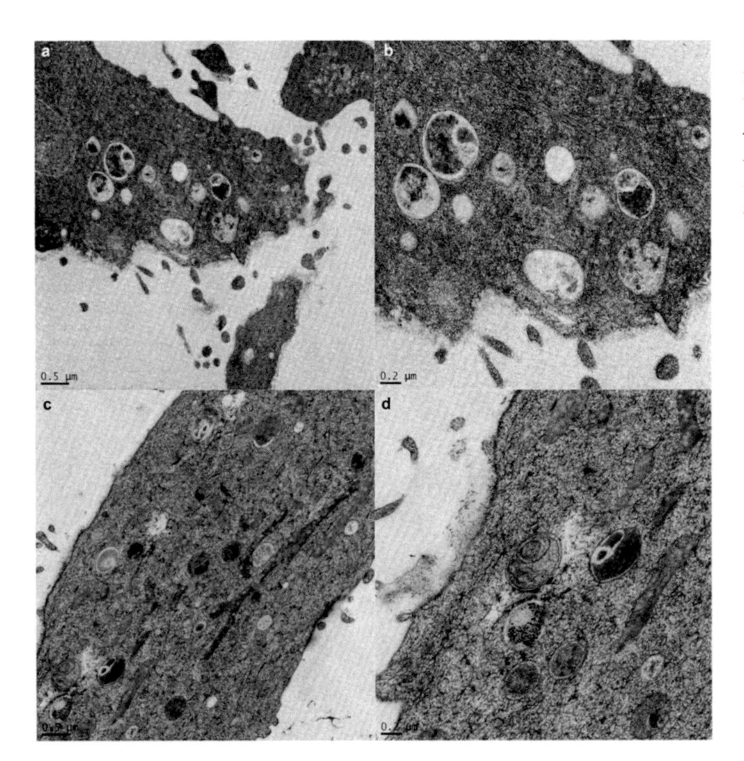

FIGURE 11.5 Transmission electron micrographs of MRC-5 fibroblasts (a, b) 2 days and (c, d) 4 days after magnetomitotransfer. *Reproduced from T. Macheiner, V.H. Fengler, M. Agreiter, T. Eisenberg, F. Madeo D. Kolb, et al., 2016. Magnetomitotransfer: an efficient way for direct mitochondria transfer into cultured human cells, Sci. Rep. 6, 35571 with permission from Springer Nature, [59].*

technologies to maintain mitochondria in a healthy state have to be made available before exogenous mitochondria delivery should be put into practice. Finally, the efficiency in direct mitochondrial transfer is partially affected by the membrane fluidity of the recipient cell. This has been suggested by the success in increasing the amount of mitochondria penetrating the plasma membrane after cell pre-treatment with PF-68, which is a nonionic surfactant that can enhance cell membrane permeability [55,60]. To enhance the efficiency in mitochondrial transfer, manipulating the cell permeability (e.g., via electroporation) may, therefore, be a direction that is worth pursuing.

Example protocols for experimental design

The method below is an example protocol for measuring the extent of mitochondrial ROS production in cells that have received exogenous mitochondria.

1. Cultivate the cells, which have undergone direct mitochondrial transfer, in a 24-well culture plate at 37°C under a humidified atmosphere with 5% CO_2 until a confluence of 80%–90% is obtained.
2. Wash the cells with Hank's balanced salt solution.
3. Dissolve MitoSOX Red in the cell culture medium to reach a concentration of 1 μM.
4. Add 1 mL of the medium into each well.
5. Incubate the plate at 37°C under a humidified atmosphere with 5% CO_2 for 30 minutes.
6. Determine the mitochondrial ROS level using a fluorescence microplate reader (excitation wavelength = 510 nm, emission wavelength = 528 nm).

Summary

Defects in mitochondrial metabolism are linked to diverse age-associated diseases. One example is type-2 diabetes, which is associated with the deficiency of subsarcolemmal mitochondria [61]. Another example is cancer. Numerous mitochondrial events, ranging from the swelling of mitochondria to the collapse of the intimal potential, can lead to apoptosis [62]; however, via overexpression of antiapoptotic proteins, cancer cells can inhibit the induction of mitochondrial-mediated apoptosis. Taking these pieces of evidence into account, proper functioning of the mitochondrial apparatus is essential to cell maintenance. In this chapter, we have presented the relevance of disrupted mitochondrial activities to the occurrence of age-associated symptoms, and have used the case of mitochondrial bioenergenesis as an example to show how interventions can be designed at the subcellular level. Along with the possibility of intervening in biological processes at the genetic and protein levels, as discussed in the proceeding chapters, it is hoped that a complete picture has been presented to illustrate the flexibility provided by delivery technologies to the development and execution of antiaging interventions.

Directions for intervention development

Age-associated damage to mitochondria has been linked to the occurrence of age-associated symptoms. The steps below can help to develop an intervention to replace defective endogenous mitochondria with exogenous ones.

(*Continued*)

(Continued)

1. Select appropriate donor cells.
2. Isolate healthy mitochondria from the selected cells.
3. Consider conjugating the isolated mitochondria (or the carrier for direct mitochondrial transfer) with cell-penetrating peptides to enhance the efficiency in cellular internalization.
4. Label the mitochondria so that their localization can be monitored after administration.
5. Deliver the isolated mitochondria (with or without using a carrier) to target cells in a body by either local or systemic administration.
6. Examine the integrity and biodistribution of the exogenous mitochondria after administration.
7. Study alterations in recipient cells before and after mitochondrial transfer
8. Optimize the delivery process to enhance the efficiency.

References

[1] Y. Kagawa, T. Hamamoto, H. Endo, M. Ichida, H. Shibui, M. Hayakawa, Genes of human ATP synthase: their roles in physiology and aging, Biosci. Rep. 17 (1997) 115–146.

[2] H.J. Agteresch, P.C. Dagnelie, J.W. van den Berg, J.H. Wilson, Adenosine triphosphate: established and potential clinical applications, Drugs 58 (1999) 211–232.

[3] M.J. Bours, E.L. Swennen, F. Di Virgilio, B.N. Cronstein, P.C. Dagnelie, Adenosine 5'-triphosphate and adenosine as endogenous signaling molecules in immunity and inflammation, Pharmacol. Ther. 112 (2006) 358–404.

[4] G. Burnstock, G.E. Knight, Cellular distribution and functions of P2 receptor subtypes in different systems, Int. Rev. Cytol. 240 (2004) 31–304.

[5] A. Hoebertz, T.R. Arnett, G. Burnstock, Regulation of bone resorption and formation by purines and pyrimidines, Trends Pharmacol. Sci. 24 (2003) 290–297.

[6] G. Gomez, M.V. Sitkovsky, Targeting G protein-coupled A2a adenosine receptors to engineer inflammation in vivo, Int. J. Biochem. Cell Biol. 35 (2003) 410–414.

[7] M.V. Sitkovsky, Use of the A_{2A} adenosine receptor as a physiological immunosuppressor and to engineer inflammation in vivo, Biochem. Pharmacol. 65 (2003) 493–501.

[8] P.A. Figueiredo, S.K. Powers, R.M. Ferreira, H.J. Appell, J.A. Duarte, Aging impairs skeletal muscle mitochondrial bioenergetic function, J. Gerontol. A. Biol. Sci. Med. Sci. 64 (2009) 21–33.

[9] P.H. Nunes, C. Calaza Kda, L.M. Albuquerque, L. Fragel-Madeira, A. Sholl-Franco, A.L. Ventura, Signal transduction pathways associated with ATP-induced proliferation of cell progenitors in the intact embryonic retina, Int. J. Dev. Neurosci. 25 (2007) 499–508.

[10] M. Stamatakis, N.V. Mantzaris, Modeling of ATP-mediated signal transduction and wave propagation in astrocytic cellular networks, J. Theor. Biol. 241 (2006) 649–668.

[11] L. van der Weyden, A.D. Conigrave, M.B. Morris, Signal transduction and white cell maturation via extracellular ATP and the P2Y11 receptor, Immunol. Cell Biol. 78 (2000) 369–374.

[12] R.S. Ostrom, C. Gregorian, P.A. Insel, Cellular release of and response to ATP as key determinants of the set-point of signal transduction pathways, J. Biol. Chem. 275 (2000) 11735–11739.

[13] E. Nakamura, Y. Uezono, K. Narusawa, I. Shibuya, Y. Oishi, M. Tanaka, et al., ATP activates DNA synthesis by acting on P2X receptors in human osteoblast-like MG-63 cells, Am. J. Physiol. Cell. Physiol. 279 (2000) C510–C519.

[14] M. Yoshida, K. Miyamae, H. Iwashita, M. Otani, A. Inadome, Management of detrusor dysfunction in the elderly: changes in acetylcholine and adenosine triphosphate release during aging, Urology 63 (2004) 17–23.

[15] M.M. Tkachenko, V.V. Iarots'kyi, S.M. Marchenko, V.F. Sahach, Effect of acetylcholine and adenosine triphosphate on membrane potential of intact rat aorta endothelium in aging, Fiziol. Zh. 48 (2002) 3–8.

[16] J.V. Leonard, A.H. Schapira, Mitochondrial respiratory chain disorders I: mitochondrial DNA defects, Lancet 355 (2000) 299–304.

[17] M.P. Murphy, How understanding the control of energy metabolism can help investigation of mitochondrial dysfunction, regulation and pharmacology, Biochim. Biophys. Acta 1504 (2001) 1–11.

[18] M.P. Fink, Ischemia and ischemia-reperfusion-induced organ injury, in: M.F. Newman, L.A. Fleisher, M.P. Fink (Eds.), Perioperative Medicine: Managing for Outcomes, Elsevier, New York, 2008, pp. 11–18.

[19] J. Hayashi, S. Ohta, Y. Kagawa, H. Kondo, H. Yonekawa, D. Takai, et al., Nuclear but not mitochondrial genome involvement in human age-related mitochondrial dysfunction, J. Biol. Chem. 269 (1994) 6878–6883.

[20] M. Calleja, P. Pena, C. Ugalde, C. Ferreiro, R. Marco, R. Garesse, Mitochondrial DNA remains intact during Drosophila aging, but the levels of mitochondrial transcripts are significantly reduced, J. Biol. Chem. 268 (1993) 18891–18897.

[21] G. Petrosillo, M. Matera, N. Moro, F.M. Ruggiero, G. Paradies, Mitochondrial complex I dysfunction in rat heart with aging: critical role of reactive oxygen species and cardiolipin, Free Radic. Biol. Med. 46 (2009) 88–94.

[22] Z. Tatarkova, S. Kuka, P. Racay, J. Lehotsky, D. Dobrota, D. Mistuna, et al., Effects of aging on activities of mitochondrial electron transport chain complexes and oxidative damage in rat heart, Physiol. Res. 60 (2011) 281–289.

[23] R.C. Scarpulla, Molecular biology of the OXPHOS system, in: J. A.M. Smeitink, R.C.A. Sengers, J.M. Frans Trijbels (Eds.), Oxidative Phosphorylation in Health and Diseases, Eurekah.com/ Landes Bioscience, New York, 2004, pp. 28–42.

[24] Y.S. Ma, S.B. Wu, W.Y. Lee, J.S. Cheng, Y.H. Wei, Response to the increase of oxidative stress and mutation of mitochondrial DNA in aging, Biochim. Biophys. Acta 1790 (2009) 1021–1029.

[25] A.W. Linnane, S. Marzuki, T. Ozawa, M. Tanaka, Mitochondrial DNA mutations as an important contributor to ageing and degenerative diseases, Lancet 1 (1989) 642–645.

[26] D.C. Wallace, J.H. Ye, S.N. Neckelmann, G. Singh, K.A. Webster, B.D. Greenberg, Sequence analysis of cDNAs for the human and bovine ATP synthase beta subunit: mitochondrial DNA genes sustain seventeen times more mutations, Curr. Genet. 12 (1987) 81–90.

[27] J. Rodriguez-Pallares, J.A. Parga, B. Joglar, M.J. Guerra, J.L. Labandeira-Garcia, Mitochondrial ATP-sensitive potassium

channels enhance angiotensin-induced oxidative damage and dopaminergic neuron degeneration. Relevance for aging-associated susceptibility to Parkinson's disease, Age (Dordr.) 34 (2012) 863–880.
[28] A.L. Moore, M.S. Albury, P.G. Crichton, C. Affourtit, Function of the alternative oxidase: is it still a scavenger? Trends Plant Sci. 7 (2002) 478–481.
[29] A.G. Rasmusson, K.L. Soole, T.E. Elthon, Alternative NAD(P)H dehydrogenases of plant mitochondria, Annu. Rev. Plant Biol. 55 (2004) 23–39.
[30] F. Krause, C.Q. Scheckhuber, A. Werner, S. Rexroth, N.H. Reifschneider, N.A. Dencher, et al., OXPHOS Supercomplexes: respiration and life-span control in the aging model Podospora anserina, Ann. N. Y. Acad. Sci. 1067 (2006) 106–115.
[31] H.D. Osiewacz, Genes, mitochondria and aging in filamentous fungi, Ageing Res. Rev. 1 (2002) 425–442.
[32] E. Dufour, J. Boulay, V. Rincheval, A. Sainsard-Chanet, A causal link between respiration and senescence in Podospora anserina, Proc. Natl. Acad. Sci. U.S.A. 97 (2000) 4138–4143.
[33] C. Borghouts, S. kerschner, H.D. Osiewacz, Copper-dependence of mitochondrial DNA rearrangements in Podospora anserina, Curr. Genet. 37 (2000) 268–275.
[34] V. Weissig, C. Lizano, V.P. Torchilin, Selective DNA release from DQAsome/DNA complexes at mitochondria-like membranes, Drug Deliv. 7 (2000) 1–5.
[35] S. Leo, G. Szabadkai, R. Rizzuto, The mitochondrial antioxidants $MitoE_2$ and $MitoQ_{10}$ increase mitochondrial Ca^{2+} load upon cell stimulation by inhibiting Ca^{2+} efflux from the organelle, Ann. N. Y. Acad. Sci. 1147 (2008) 264–274.
[36] T.A. Trendeleva, A.G. Rogov, D.A. Cherepanov, E.I. Sukhanova, T.M. Il'yasova, I.I. Severina, et al., Interaction of tetraphenylphosphonium and dodecyltriphenylphosphonium with lipid membranes and mitochondria, Biochemistry (Mosc.) 77 (2012) 1021–1028.
[37] B. Cunniff, K. Benson, J. Stumpff, K. Newick, P. Held, D. Taatjes, et al., Mitochondrial-targeted nitroxides disrupt mitochondrial architecture and inhibit expression of peroxiredoxin 3 and FOXM1 in malignant mesothelioma cells, J. Cell Physiol. 228 (2013) 835–845.
[38] L. Li, I. Geisler, J. Chmielewski, J.X. Cheng, Cationic amphiphilic polyproline helix P11LRR targets intracellular mitochondria, J. Control Release 142 (2010) 259–266.
[39] D. Kalafut, T.N. Anderson, J. Chmielewski, Mitochondrial targeting of a cationic amphiphilic polyproline helix, Bioorg. Med. Chem. Lett. 22 (2012) 561–563.
[40] Y. Pang, C. Wang, L. Yu, Mitochondria-targeted antioxidant SS-31 is a potential novel ophthalmic medication for neuroprotection in glaucoma, Med. Hypothesis Discov. Innov. Ophthalmol. 4 (2015) 120–126.
[41] S. Hao, J. Ji, H. Zhao, L. Shang, J. Wu, H. Li, et al., Mitochondrion-targeted peptide SS-31 inhibited oxidized low-density lipoproteins-induced foam cell formation through both ROS scavenging and inhibition of cholesterol influx in RAW264, Molecules 20 (2015) 21287–21297.
[42] H.H. Szeto, S. Liu, Cardiolipin-targeted peptides rejuvenate mitochondrial function, remodel mitochondria, and promote tissue regeneration during aging, Arch. Biochem. Biophys. 660 (2018) 137–148.
[43] L. Zhou, J.H. Liu, F. Ma, S.H. Wei, Y.Y. Feng, J.H. Zhou, et al., Mitochondria-targeting photosensitizer-encapsulated amorphous nanocage as a bimodal reagent for drug delivery and biodiagnose in vitro, Biomed. Microdevices. 12 (2010) 655–663.
[44] J.D. Steichen, M.J. Weiss, D.R. Elmaleh, R.L. Martuza, Enhanced in vitro uptake and retention of 3H-tetraphenylphosphonium by nervous system tumor cells, J. Neurosurg. 74 (1991) 116–122.
[45] J.E. Christman, D.S. Miller, P. Coward, L.H. Smith, N.N. Teng, Study of the selective cytotoxic properties of cationic, lipophilic mitochondrial-specific compounds in gynecologic malignancies, Gynecol. Oncol. 39 (1990) 72–79.
[46] M.J. Weiss, J.R. Wong, C.S. Ha, R. Bleday, R.R. Salem, G.D. Steele Jr., et al., Dequalinium, a topical antimicrobial agent, displays anticarcinoma activity based on selective mitochondrial accumulation, Proc. Natl. Acad. Sci. U.S.A. 84 (1987) 5444–5448.
[47] R.W. Horobin, S. Trapp, V. Weissig, Mitochondriotropics: a review of their mode of action, and their applications for drug and DNA delivery to mammalian mitochondria, J. Control Release 121 (2007) 125–136.
[48] D. Galanakis, C.A. Davis, B. Del Rey Herrero, C.R. Ganellin, P. M. Dunn, D.H. Jenkinson, Synthesis and structure-activity relationships of dequalinium analogues as K^+ channel blockers. Investigations on the role of the charged heterocycle, J. Med. Pharm. Chem. 38 (1995) 595–606.
[49] K.R. Schneider Berlin, C.V. Ammini, T.C. Rowe, Dequalinium induces a selective depletion of mitochondrial DNA from HeLa human cervical carcinoma cells, Exp. Cell Res. 245 (1998) 137–145.
[50] Y. Bae, M.K. Jung, S.J. Song, E.S. Green, S. Lee, H.S. Park, et al., Functional nanosome for enhanced mitochondria-targeted gene delivery and expression, Mitochondrion 37 (2017) 27–40.
[51] J.M. Gott, G.M. Naegele, S.J. Howell, Electroporation of DNA into physarum polycephalum mitochondria: effects on transcription and RNA editing in isolated organelles, Genes 7 (2016).
[52] J.M. Collombet, V.C. Wheeler, F. Vogel, C. Coutelle, Introduction of plasmid DNA into isolated mitochondria by electroporation. A novel approach toward gene correction for mitochondrial disorders, J. Biol. Chem. 272 (1997) 5342–5347.
[53] R.A. Butow, T.D. Fox, Organelle transformation: shoot first, ask questions later, Trends Biochem. Sci. 15 (1990) 465–468.
[54] S.A. Johnston, P.Q. Anziano, K. Shark, J.C. Sanford, R.A. Butow, Mitochondrial transformation in yeast by bombardment with microprojectiles, Science 240 (1988) 1538–1541.
[55] M.J. Kim, J.W. Hwang, C.K. Yun, Y. Lee, Y.S. Choi, Delivery of exogenous mitochondria via centrifugation enhances cellular metabolic function, Sci. Rep. 8 (2018) 3330.
[56] J.C. Chang, K.H. Liu, Y.C. Li, S.J. Kou, Y.H. Wei, C.S. Chuang, et al., Functional recovery of human cells harbouring the mitochondrial DNA mutation MERRF A8344G via peptide-mediated mitochondrial delivery, Neurosignals. 21 (2013) 160–173.
[57] E. Area-Gomez, A.J. de Groof, I. Boldogh, T.D. Bird, G.E. Gibson, C.M. Koehler, et al., Presenilins are enriched in endoplasmic reticulum membranes associated with mitochondria, Am. J. Pathol. 175 (2009) 1810–1816.
[58] D. Narendra, A. Tanaka, D.F. Suen, R.J. Youle, Parkin is recruited selectively to impaired mitochondria and promotes their autophagy, J. Cell Biol. 183 (2008) 795–803.

[59] T. Macheiner, V.H. Fengler, M. Agreiter, T. Eisenberg, F. Madeo, D. Kolb, et al., Magnetomitotransfer: an efficient way for direct mitochondria transfer into cultured human cells, Sci. Rep. 6 (2016) 35571.

[60] M.S. Clarke, M.A. Prendergast, A.V. Terry Jr., Plasma membrane ordering agent pluronic F-68 (PF-68) reduces neurotransmitter uptake and release and produces learning and memory deficits in rats, Learn. Mem. 6 (1999) 634–649.

[61] V.B. Ritov, E.V. Menshikova, J. He, R.E. Ferrell, B.H. Goodpaster, D.E. Kelley, Deficiency of subsarcolemmal mitochondria in obesity and type 2 diabetes, Diabetes 54 (2005) 8–14.

[62] J.E. Chipuk, L. Bouchier-Hayes, D.R. Green, Mitochondrial outer membrane permeabilization during apoptosis: the innocent bystander scenario, Cell Death Differ. 13 (2006) 1396–1402.

Chapter 12

Use of delivery technologies to mediate tissue regeneration and repair

Introduction

To combat aging, one strategy is to restore the function of a damaged body part by modulating or repairing the age-associated physiological changes. The other one is to replace the damaged part with a new one so that the body can go back to the normal state. In the last few chapters, we have focused extensively on the use of delivery technologies to transfer various agents, ranging from nucleic acids to proteins, to the body for antiaging purposes. In this chapter, we will discuss the use of delivery technologies to mediate tissue engineering, which aims at regenerating, or restoring the function of damaged tissues.

As far as tissue engineering is concerned, the conventional method is to generate a structure that can mimic the natural tissue for implantation; however, the success of tissue regeneration using this approach is affected by the ability of the tissue mimic to induce cellular proliferation, migration, and differentiation. Recently, genetic manipulation has been employed to modulate the physiology of implanted cells so as to augment the efficiency in tissue regeneration [1]. In this chapter, we will discuss major strategies for fabricating scaffolds, followed by a discussion of methods of enhancing the scaffold performance. Finally, by using the tooth as an example, the translation of the fundamental knowledge of tissue development into interventions will be demonstrated. The objective of this chapter is to illustrate how materials can be exploited for delivery of not only therapeutic agents but also living entities (such as cells) for tackling age-associated damage in the future.

Strategies for scaffold fabrication

Scaffolds used for cell seeding can be in the bulk size or micron size. Generation of the latter often involves the use of microfluidic technologies, which enable precise control of the production of cell-laden microgels [2,3]. Regardless of the size of the scaffold, the capacity of the scaffold to provide a 3D platform to mimic the physiological microenvironment experienced by cells is vital to tissue regeneration and repair. One method of offering a scaffold with the 3D architecture is lyophilization. By using this method, a highly porous polymeric structure with interconnected pores can be generated. This has been demonstrated by an earlier study [4], which has used this method to fabricate a 3D porous sponge containing pores in the range of 70–300 μm for seeding fibroblasts. To enhance the mechanical strength of the scaffold, the polymer chains involved can be further cross-linked by using cross-linkers (e.g., carbodiimide [5], glutaraldehyde [6], and citric acid [7]) or UV irradiation [8]. More recently, cross-linked porous gelatin hydrogels have been obtained by lyophilization for seeding corneal endothelial cells [5]. The Young's modulus and swelling ratio of the hydrogel have been found to be affected partly by the polymer concentration and the degree of cross-linking [5]. In addition, by tuning various parameters (e.g., the concentration of the polymer solution adopted, the size of the solvent crystals, or the freezing regime [4,6,9]), the pore size of the hydrogel has been successfully manipulated.

Example protocols for experimental design

The method below is an example protocol for evaluating the viability of cells within microgels.

1. Dissolve calcein acetoxymethyl ester (calcein-AM) in anhydrous DMSO to reach a concentration of 4 mM.
2. Dissolve ethidium homodimer-1 (EthD-1) in a 20% (v/v) DMSO solution to reach a concentration of 2 mM.
3. Collect the cell-laden microgels at a designated time point.
4. Add 20 μL of the EthD-1 solution to 10 mL of PBS.
5. Add 5 μL of the calcein-AM solution to 10 mL of the diluted EthD-1 solution.
6. Vortex for 30 seconds.
7. Add the solution mixture directly to the cell-laden microgels.

(Continued)

Delivery of Therapeutics for Biogerontological Interventions. DOI: https://doi.org/10.1016/B978-0-12-816485-3.00012-X

(Continued)

8. Treat the cells for 30–45 minutes in darkness at ambient conditions.
9. Visualize the live and dead cells using a fluorescence microscope.
10. Count the number of cells manually.
11. Obtain a percentage of live cells from the total number of cells.

Another method of scaffold fabrication is solvent casting and particulate leaching (SCPL), in which a polymer is first dissolved in a solvent followed by the addition of particles with specific dimensions. After the solvent is evaporated, a composite with uniformly distributed particles is generated. This composite is then immersed into another solvent to allow the particles to leach, leaving behind a highly porous scaffold. By using this method, the pore size can be tuned simply by choosing particles with different diameters. In an earlier study, a scaffold has been synthesized by cross-linking PEG with PPL around a salt-leached PLGA scaffold, which has later been dissolved and removed under basic conditions [10]. The construct generated has been shown to support the growth of endothelial cells and has enabled microvessel formation in vivo [10]. More recently, a block copolymer scaffold consisting of PEG and poly(ε-caprolactone) (PCL) has been generated for seeding chondrocytes [11]. The study has adopted sodium chloride as a porogen and DMSO as a solvent [11]. The mechanical strength of the scaffold has been manipulated by altering the ratio of PEG to PCL. This has demonstrated the tunability of the scaffold generated by SCPL. Nevertheless, if cytotoxic solvents are adopted during scaffold fabrication, multiple washing steps are required to remove the harmful agents before use.

The third method is gas foaming, which enables the generation of a porous scaffold with the pore size in the range of 100–500 μm [12]. Examples of commonly used foaming agents include nitrogen [13] and carbon dioxide [14]. Previously, gas foaming has been adopted to generate PLGA disks, which enable smooth muscle cells to be seeded and 3D tissue structures to be produced [15]. Another study has also used gas-foamed polyurethane as a template to fabricate a porous biphasic calcium phosphate scaffold [16], which has been found to be biocompatible and has shown good potential for applications in bone regeneration. As the generation of pores using the gas foaming method requires only the use of inert forming agents, the use of toxic agents can be avoided; however, due to the high polydispersity of the pores, an uneven distribution of seeded cells within the scaffold often results.

Since the turn of the last century, the development of bioprinting has made the generation of complex 3D structures, using a top-down approach, possible. There are two common methods of bioprinting. One is drop-based bioprinting. This technique has previously been adopted to generate fibrin scaffolds loaded with endothelial cells as microvascular structures [17]. Another method is extrusion bioprinting, in which a stage (or surface) is moved in a directed manner to control the spatial deposition of the bioink from the nozzle to generate a scaffold with the complex architecture [18]. This technique has been used to generate a poly(2-hydroxyethyl methacrylate) (PHEMA) hydrogel structure to seed primary hippocampal neurons (Fig. 12.1) [19], and an acrylamide hydrogel structure to cultivate fibroblasts [20]. Besides bioprinting, the advent of photolithography technologies has facilitated advances in tissue engineering. Photolithography enables the fabrication of 2D/3D structures for supporting

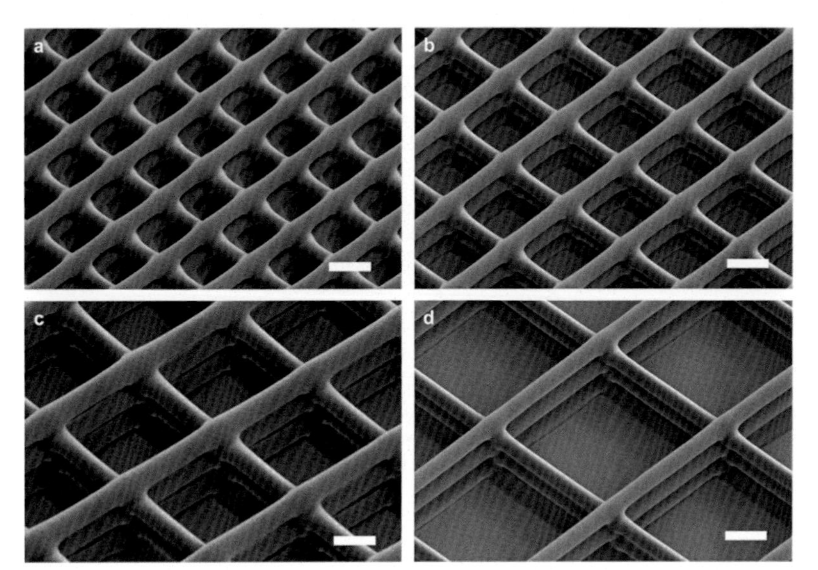

FIGURE 12.1 Scanning electron micrographs of scaffolds of varying architectures. Each scaffold consists of orthogonal arrays of cylindrical hydrogel filaments or rods, with the center-to-center distance between adjacent rods being (a) 30, (b) 40, (c) 60, and (d) 80 μm. Scale bar = 20 μm. *Reproduced from J.N. Hanson Shepherd, S.T. Parker, R.F. Shepherd, M.U. Gillette, J.A. Lewis, R.G. Nuzzo, 3D microperiodic hydrogel scaffolds for robust neuronal cultures, Adv. Funct. Mater. 21 (2011) 47–54 with permission from John Wiley & Sons, Inc., [19].*

the growth of various cell types [21,22], ranging from hepatocytes [23] and cardiac stem cells [24] to hippocampal neurons [25].

As far as cell growth in a scaffold is concerned, mammalian cells are known to be sensitive to scaffold properties (including roughness [26–28], stiffness [29,30], geometry [31,32], and topography [33–35]). By using mask-based photolithography, a scaffold possessing regions of soft and stiff hydrogels has been generated [36]. The soft hydrogel region has been found to direct MSCs to be differentiated into osteocytes; whereas the stiff region has enabled the stem cells to be differentiated into the adipogenic lineage. This work has evidenced the possible use of photolithography to control the stiffness gradient to modulate the fate of the seeded mammalian cells. Apart from manipulating the stiffness gradient, oligo(PEG-fumarate):PEGDA hydrogels with a high spatial resolution have been produced using photolithography [37]. The hydrogels have been seeded with tendon/ligament fibroblasts and marrow stromal cells, with high viability of the seeded cells maintained over a period of 14 days. More recently, photolithography has been adopted to generate PEGDA hydrogel–based biological robots, which have displayed spontaneous locomotion after rat cardiomyocytes have been seeded [38]. This technology may provide insights into the development of intelligent multicellular bio-bots for drug delivery, artificial immunity, or other biomedical applications. Here it is worth noting that, although photolithography enables precise temporal and spatial control of the reaction kinetics during scaffold fabrication [39], free radicals generated from a photoinitiator are often needed to be used to initiate chain reactions. These free radicals may cause cell death, particularly when polymerization reactions are carried out in the presence of cells [39,40].

Example protocols for experimental design

The method below is an example protocol for examining the rigidity of a spherical hydrogel scaffold that is in the micron size.

1. Place the sample stage of an atomic force microscope (AFM) on top of an optical microscope so that the sample can be monitored when indentation measurements are performed.
2. Set the spring constant of the cantilever for indentation as 0.20 N/m.
3. Position the cantilever at the center of the sample.
4. Compress at 3 μm/s to indent the sample.
5. Record the applied force (F) versus the indentation depth (nm).
6. Calculate the elastic modulus (E) using Hertz contact theory for the spherical elastic solid (where h is the indentation depth, R is the radius of the sample, and ν is the Poisson's ratio of the sample), with ν set as 0.5 by assuming that the sample follows the properties of an ideal rubber.

$$F = \frac{4}{3}\left(\frac{E}{1-\nu^2}\right)R^{1/2}h^{3/2} \quad (12.1)$$

Engineering scaffold properties for enhanced performance

To enhance the adhesion, proliferation, migration, differentiation, and maturation of the seeded cells, various methods of engineering the properties of a scaffold have been employed. One method is to conjugate bioactive molecules to the scaffold surface to make the scaffold more biocompatible. Another method is to immunologically modulate the biomaterial–host interactions to avoid the induction of inflammatory responses in a body, thereby minimizing the initiation of foreign body reactions (e.g., giant cell formation, fibrosis, and damage to the implant [41,42]). This can be achieved by manipulating the size and chemical properties of the scaffold. For instance, various triazole-containing analogs have been used to modify Alg to reduce foreign body reactions in vivo [43]. Macrophage accumulation on implanted spheres has also been found to be reduced by increasing the diameter of the spheres to 1.5 mm or above [44,45].

In addition to manipulating the biomaterial–host interactions, recent efforts incorporates biosensors into scaffolds to monitor the activity of the seeded cells in real time. This has been documented in a study reported by Tian et al. [46], who have integrated silicon-nanowire field-effect transistors into a scaffold. The generated scaffold can support the growth of the seeded cardiomyocytes while enabling the electrical activity of those cells to be monitored continuously. More recently, by incorporating carbon nanotubes (CNTs) into a gelatin methacryloyl (GelMA) hydrogel (Fig. 12.2), the spontaneous beating rate of the seeded neonatal rat ventricular myocytes has been found to be successfully enhanced, with the excitation threshold significantly reduced [47]. This success has illustrated the possibility of modulating cell growth by incorporating bioactive agents into the implant. This possibility is particularly important when the seeded cells are stem cells, which display multilineage differentiation potential. Failure of the stem cells to differentiate into appropriate cell lineages may lead to failure of tissue regeneration. One common approach to address this

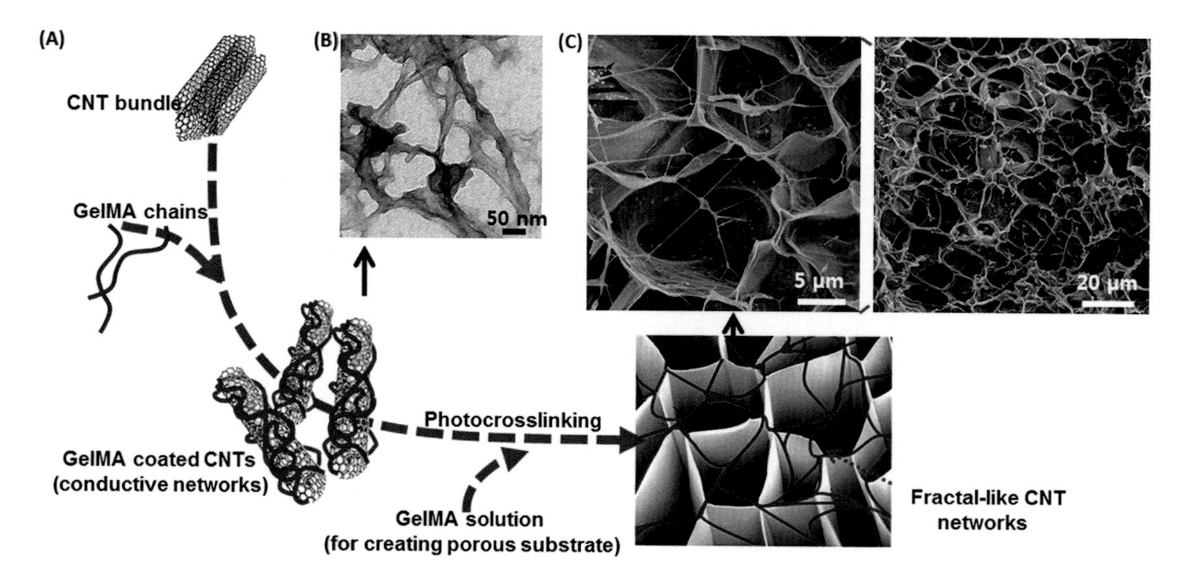

FIGURE 12.2 (A) The synthetic route for the production of fractal-like CNT networks embedded in a GelMA hydrogel. (B) Transmission electron micrographs of GelMA-coated CNTs. (C) Scanning electron micrographs showing the porous surface of a 1 mg/mL CNT–GelMA thin film. The magnified image shows the nanofibrous network across and inside the porous structure. *Reproduced from S.R. Shin, S.M. Jung, M. Zalabany, K. Kim, P. Zorlutuna, S.B. Kim, et al., Carbon-nanotube-embedded hydrogel sheets for engineering cardiac constructs and bioactuators, ACS Nano 7 (2013) 2369–2380 with permission from the American Chemical Society, [47].*

problem is to load the scaffold with bioactive agents (e.g., erythropoietin, transforming growth factor-β, and granulocyte colony-stimulating factor) to guide the proliferation and differentiation of the seeded stem and progenitor cells [48]. The possible leakage of the loaded agents, however, may lead to safety risks when those agents adversely affect the functioning of surrounding tissues. An alternative approach is to modify the cells genetically prior to cell seeding. Over the years, different technologies for gene delivery to stem cells have been developed. For instance, a recombinant adenoviral vector containing fiber proteins has been developed for stem cell transduction [49]. It effectively delivers transgenes into the MSCs, while displaying low cytotoxicity even at a dose of 5000 viral particles per cell. Gene delivery to stem cells can also be mediated by using nonviral vectors. One example is PEI–β-CD. Upon conjugation with a cell-penetrating peptide that possesses the protein transduction domain (PTD) of the HIV-1 TAT protein [50], the polymer can transfect MSCs while maintaining the phenotypic profile of the stem cells. These advances can potentially enable more precise control of the fate of the seeded stem cells for tissue regeneration.

Example protocols for experimental design

The method below is an example protocol for synthesizing GelMA.

1. Add 10 g of gelatin to 90 mL of DMSO.

(Continued)

(Continued)

2. Heat the solution to 50°C under constant stirring.
3. Add 0.5 g of 4-(dimethylamino)-pyridine (DMAP) to the solution in darkness under constant stirring until DMAP is completely dissolved.
4. Add 4 mL of glycidyl methacrylate dropwise to the solution.
5. Stir the reaction mixture at 50°C for 3 days.
6. Dialyze the reaction mixture (molecular weight cut-off = 12 kDa) against doubly deionized water for 2 days.
7. Lyophilize to obtain GelMA.

Translation into tissue regeneration and repair

To demonstrate tissue regeneration, teeth can be illustrative [51,52]. This is because, compared to other vertebrates, mammals have a limited capacity of tooth renewal. Humans have a diphyodont dentition and can only replace teeth maximally once. The capacity of overcoming this intrinsic limit to achieve tooth regeneration may provide insights into the regeneration of other damaged tissues. Furthermore, teeth are present in all vertebrates. They can be used as a model to examine the evolution and development of epithelial organs [53], which are derivatives of the epithelia and are also known as epithelial appendages [54]. The typical structure of a molar, as well as the strength of the compressive force across the whole length of the tooth, is presented in Fig. 12.3.

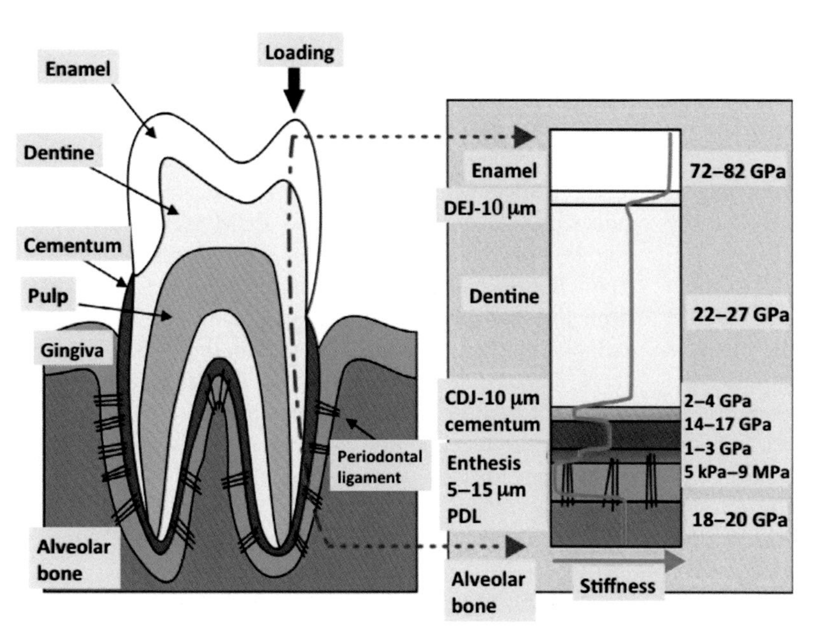

FIGURE 12.3 The typical structure of a molar, as well as the strength of the compressive force across the whole length of the tooth. *Reproduced from D.W. Green, W.F. Lai, H.S. Jung, Evolving marine biomimetics for regenerative dentistry, Mar. Drugs 12 (2014) 2877–2912 and J.W.C. Dunlop, R. Weinkamer, P. Fratzl, Artful interfaces within biological materials, Mater. Today14 (2011) 70–78 with permission from Elsevier B.V and MDPI, [51,52].*

Physiologically, the spatiotemporal expression and interactions of different signaling molecules (e.g., cytokines and adhesion molecules) regulate the development and morphogenesis of a tooth [55], whose macromorphological (e.g., the crown size and tooth length) and micromorphological (e.g., the number and position of cusps and roots) features define the tooth identity [56]. Due to the spatiotemporal complexity of tooth development, proper use of biodegradable scaffolds (which allow for interactions between epithelial and MSCs in a developing tooth bud) becomes pivotal [57]. The technical feasibility of artificially reconstructing dental structures from stem cells in vivo has been demonstrated by an earlier study, which has first adopted macroporous biphasic calcium phosphate (MBCP) as a scaffold for seeding stem cells [including human bone marrow–derived MSCs (BMMSCs), stem cells extracted from the periodontal ligament (PDL) of permanent teeth (pPDLSCs), and those extracted from the PDL of deciduous teeth (dPDLSCs)], followed by the transplantation of the cell-laden scaffold to the dorsal subcutaneous pocket of an immunocompromised mouse model [58]. Eight weeks after transplantation, the stem cells have successfully produced hard tissues at the periphery of the scaffold [58]. The BMMSC transplant has generated bone-like tissues with a lining of osteoblast-like cells; whereas the dPDLSC and pPDLSC transplants have produced cementum-like tissues [58]. Immunohistochemical analysis and quantitative reverse transcriptase polymerase chain reaction (RT-PCR) have revealed that the expression of genes associated with mineralization has been increased in the BMMSC transplant. On the other hand, genes associated with the cementum/PDL complex have been stimulated to be expressed in the pPDLSC and dPDLSC transplants [58]. This study has illustrated the feasibility of obtaining dental tissues by tissue engineering.

Over the last several decades, an increasing number of polymeric materials have been exploited for applications in tooth regeneration [59–64]. For instance, an earlier study has modified a fibrin gel with PEG, which can decelerate the degradation of fibrin [65], to generate a biocompatible, injectable, and tunable scaffold for supporting the proliferation of stem cells derived from the dental pulp or PDL. The alkaline phosphatase activity, as well as the expression of osteoblast-specific genes, in the seeded stem cells has been found to be increased upon osteogenic induction. More recently, platelet-rich fibrin (PRF) has been added to a fibrin glue, which has been seeded with dental bud cells (DBCs), to enrich the microenvironment with growth factors for tooth regeneration [64]. After the DBC–fibrin glue–PRF composite has been transplanted to the alveolar socket in a porcine model, one animal has developed a complete tooth with blood vessels [64] while another one has produced an unerupted tooth that has expressed the genes for DMP-1, cytokeratin 14, osteopontin, and VEGF [64]. Apart from fibrin, collagen can be used in tooth engineering because it enables the attachment of pulp cells and allows the attached cells to secrete the intensely mineralized extracellular matrix (ECM) [66]. Recently, a scaffold consisting of bioactive glass and gelatin has been found to enable the regeneration of bone tissues and to possess a porous and interconnected structure for cell proliferation and tissue ingrowth. Alkaline phosphatase staining and immunohistochemical analysis have revealed that the scaffold has supported the proliferation, osteogenic differentiation, and ECM deposition of the dental pulp stem cells [67]. The success in controlling cell physiology in a scaffold may provide insights into the transition from basic research on tooth development to interventions for tooth regeneration in the future.

Tooth regeneration is a one-off solution to confront tooth loss; however, its success relies on the availability of technologies for precise control of the tooth development processes in a biological body. Compared to tooth regeneration, tooth repair is more practicable at this moment. One of the factors mediating tooth repair is VEGF [68]. A previous study has observed that, after the introduction of dental trauma, the serum concentration of VEGF is closely associated with the revascularization rate during the healing process [69]. In addition, during the osseointegration of dental implants, VEGF has been found to enhance osteoblastic differentiation, vascularization, and bone regeneration [70]. To facilitate the effective use of VEGF in tooth repair, polymeric delivery systems may be adopted. One example of these systems is CS, which possesses primary amine groups with a pK_a value of around 6.5 [71–73]. In a previous study, titanium coupons have been coated with CS, which has been further loaded with VEGF to support the growth and matrix production of osteoblastic cells [70]. Compared to the cells seeded on uncoated titanium, those seeded on the coating have shown an increase in the alkaline phosphatase activity and the extent of calcium deposition [70]. This, along with the ability of CS to swirl across the membrane lipid bilayer [74] and to facilitate the paracellular transport of hydrophilic agents [75], has rendered CS applicable to restorative dentistry. To further enhance the efficiency of tooth repair, engineering techniques can be employed. This has been demonstrated by an earlier study [76], which has used a photolithographic method to fabricate patterned PEG hydrogels (which have been incorporated with VEGF) on the surface of silanized silicon wafers [76]. The micropatterns produced have been found to alter the behavior of osteoblasts, leading to membrane ruffling and cellular process formation [76]. This study has evidenced the feasibility of modulating cell–substrate interactions by manipulating the surface structures of a scaffold.

In addition to teeth, tissue engineering can be used to regenerate and repair other tissues such as cartilages, skin, vascular tissues, bones, and nerves [77–82]. For instance, one study has used laser irradiation to enhance the proliferation, intracellular ROS synthesis, and neuronal differentiation of neural stem cells seeded on a 3D-printed scaffold [83], demonstrating the possible use of low-level light therapy (LLLT) to facilitate the rehabilitation of degenerative nerves [83]. Another study has achieved osteochondral tissue regeneration by using a bilayered scaffold, in which the chondral layer and the bone scaffold layer have supported the adhesion, proliferation, and differentiation of the seeded stem cells into chondrocytes and osteoblasts, respectively [84]. Finally, with the use of emulsion electrospinning, basic FGF (bFGF) has been imbedded into ultrafine core-sheath fibers [85]. The bFGF-loaded fibers have been found to enhance cell adhesion, cell proliferation, and ECM secretion [85]. In the in vivo context, the fibers have facilitated the re-epithelialization and regeneration of skin appendages, with the gradual release of bFGF being able to enhance collagen deposition and ECM remodeling [85]. All these have demonstrated the versatility of tissue engineering and the future potential of using related technologies to enhance the regeneration and repair of diverse tissues to fight against age-associated tissue damage.

Summary

Aging leads to pathological changes in tissues. With technological advances, generation of biological substitutes to combat those changes has now been made possible. In this chapter, we have presented the recent progress on the development of scaffolds for applications in regenerative medicine. Accompanying the advent of computer-aided technologies, the versatility of interventions for tissue engineering is expected to be escalated. For instance, techniques in 3D reconstruction enable more accurate anatomical modeling for the design of artificial tissue substitutes [86]. The use of computer modeling can also increase our understanding of the relative positions of critical vascular, neural, and other anatomical structures adjacent to the site of tissue transplantation [86,87]. Despite this, aging is a systemic process. Surgically replacing damaged tissues all over the body is impossible. This is one of the problems that have to be solved before tissue engineering can be translated into antiaging interventions.

Directions for intervention development

Tissue engineering provides a route to generate tissue substitutes to restore, replace, and improve the function of a tissue damaged by aging. To apply tissue engineering to the development of an antiaging intervention, the following steps can be taken:

1. Consider the physiological properties of the damaged tissue.
2. List the properties needed to be possessed by a scaffold for the regeneration and repair of the damaged tissue.
3. Select a material showing the properties listed in Step 2.
4. Select an appropriate method of fabricating a scaffold from the selected material.
5. Choose an appropriate type of cells to be seeded on the scaffold.
6. Seed the cells on the scaffold.

(Continued)

(Continued)

7. Examine the performance of the scaffold for supporting the adhesion, proliferation, migration, differentiation, and maturation of the seeded cells.
8. Modify the properties of the scaffold, if necessary, to enhance the performance.
9. Apply the cell-laden scaffold to a biological body to mediate tissue regeneration and repair.

References

[1] F. Mottaghitalab, A. Rastegari, M. Farokhi, R. Dinarvand, H. Hosseinkhani, K.L. Ou, et al., Prospects of siRNA applications in regenerative medicine, Int. J. Pharm. 524 (2017) 312–329.

[2] Y. Liu, N.O. Nambu, M. Taya, Cell-laden microgel prepared using a biocompatible aqueous two-phase strategy, Biomed. Microdevices 19 (2017) 55.

[3] J. Jung, K. Kim, S.C. Choi, J. Oh, Microfluidics-assisted rapid generation of tubular cell-laden microgel inside glass capillaries, Biotechnol. Lett. 36 (2014) 1549–1554.

[4] L. Shapiro, S. Cohen, Novel alginate sponges for cell culture and transplantation, Biomaterials 18 (1997) 583–590.

[5] J.Y. Lai, D.H.K. Ma, M.H. Lai, Y.T. Li, R.J. Chang, L.M. Chen, Characterization of cross-linked porous gelatin carriers and their interaction with corneal endothelium: biopolymer concentration effect, PLoS One. 8 (2013) e54058.

[6] H.W. Kang, Y. Tabata, Y. Ikada, Fabrication of porous gelatin scaffolds for tissue engineering, Biomaterials 20 (1999) 1339–1344.

[7] Z. Peng, F. Chen, Hydroxyethyl cellulose-based hydrogels with various pore sizes prepared by freeze-drying, J. Macromol. Sci. B 50 (2010) 340–349.

[8] T. Miyata, T. Sode, A.L. Rubin, K.H. Stenzel, Effects of ultraviolet irradiation on native and telopeptide-poor collagen, Biochim. Biophys. Acta 229 (1971) 672–680.

[9] H.W. Kang, Y. Tabata, Y. Ikada, Effect of porous structure on the degradation of freeze-dried gelatin hydrogels, J. Bioact. Compat. Polym. 14 (1999) 331–343.

[10] M.C. Ford, J.P. Bertram, S.R. Hynes, M. Michaud, Q. Li, M. Young, et al., A macroporous hydrogel for the coculture of neural progenitor and endothelial cells to form functional vascular networks in vivo, Proc. Natl. Acad. Sci. U.S.A. 103 (2006) 2512–2517.

[11] J.S. Park, D.G. Woo, B.K. Sun, H.M. Chung, S.J. Im, Y.M. Choi, et al., In vitro and in vivo test of PEG/PCL-based hydrogel scaffold for cell delivery application, J. Control Release 124 (2007) 51–59.

[12] E. Sachlos, J.T. Czernuszka, Making tissue engineering scaffolds work. Review on the application of solid freeform fabrication technology to the production of tissue engineering scaffolds, Eur. Cells Mater. 5 (2003) 29–40.

[13] A. Zellander, R. Gemeinhart, A. Djalilian, M. Makhsous, S. Sun, M. Cho, Designing a gas foamed scaffold for keratoprosthesis, Mater. Sci. Eng. C Mater. Biol. Appl. 33 (2013) 3396–3403.

[14] R.A. Quirk, R.M. France, K.M. Shakesheff, S.M. Howdle, Supercritical fluid technologies and tissue engineering scaffolds, Curr. Opin. Solid State Mater. Sci. 8 (2004) 313–321.

[15] L.D. Harris, B.S. Kim, D.J. Mooney, Open pore biodegradable matrices formed with gas foaming, J. Biomed. Mater. Res. 42 (1998) 396–402.

[16] H. Kim, I. Park, J. Kim, C. Cho, M. Kim, Gas foaming fabrication of porous biphasic calcium phosphate for bone regeneration, Tissue Eng. Regen. Med. 9 (2012) 63–68.

[17] X. Cui, T. Boland, Human microvasculature fabrication using thermal inkjet printing technology, Biomaterials 30 (2009) 6221–6227.

[18] J.A. Lewis, Direct ink writing of 3D functional materials, Adv. Funct. Mater. 16 (2006) 2193–2204.

[19] J.N. Hanson Shepherd, S.T. Parker, R.F. Shepherd, M.U. Gillette, J.A. Lewis, R.G. Nuzzo, 3D microperiodic hydrogel scaffolds for robust neuronal cultures, Adv. Funct. Mater. 21 (2011) 47–54.

[20] R.A. Barry, R.F. Shepherd, J.N. Hanson, R.G. Nuzzo, P. Wiltzius, J.A. Lewis, Direct-write assembly of 3D hydrogel scaffolds for guided cell growth, Adv. Mater. 21 (2009) 2407–2410.

[21] J. Sun, J. Tang, J. Ding, Cell orientation on a stripe-micropatterned surface, Chin. Sci. Bull. 54 (2009) 3154–3159.

[22] S.H. Lee, J.J. Moon, J.L. West, Three-dimensional micropatterning of bioactive hydrogels via two-photon laser scanning photolithography for guided 3D cell migration, Biomaterials 29 (2008) 2962–2968.

[23] V. Liu Tsang, A.A. Chen, L.M. Cho, K.D. Jadin, R.L. Sah, S. DeLong, et al., Fabrication of 3D hepatic tissues by additive photopatterning of cellular hydrogels, FASEB J. 21 (2007) 790–801.

[24] H. Aubin, J.W. Nichol, C.B. Hutson, H. Bae, A.L. Sieminski, D. M. Cropek, et al., Directed 3D cell alignment and elongation in microengineered hydrogels, Biomaterials 31 (2010) 6941–6951.

[25] P. Zorlutuna, J.H. Jeong, H. Kong, R. Bashir, Stereolithography-based hydrogel microenvironments to examine cellular interactions, Adv. Funct. Mater. 21 (2011) 3642–3651.

[26] D.D. Deligianni, N.D. Katsala, P.G. Koutsoukos, Y.F. Missirlis, Effect of surface roughness of hydroxyapatite on human bone marrow cell adhesion, proliferation, differentiation and detachment strength, Biomaterials 22 (2001) 87–96.

[27] D.P. Dowling, I.S. Miller, M. Ardhaoui, W.M. Gallagher, Effect of surface wettability and topography on the adhesion of osteosarcoma cells on plasma-modified polystyrene, J. Biomater. Appl. 26 (2011) 327–347.

[28] A.M. Ross, Z. Jiang, M. Bastmeyer, J. Lahann, Physical aspects of cell culture substrates: topography, roughness, and elasticity, Small 8 (2012) 336–355.

[29] D.E. Discher, P. Janmey, Y.L. Wang, Tissue cells feel and respond to the stiffness of their substrate, Science 310 (2005) 1139–1143.

[30] P. Bajaj, X. Tang, T.A. Saif, R. Bashir, Stiffness of the substrate influences the phenotype of embryonic chicken cardiac myocytes, J. Biomed. Mater. Res. A. 95 (2010) 1261–1269.

[31] P. Bajaj, B. Reddy, L. Millet, C. Wei, P. Zorlutuna, G. Bao, et al., Patterning the differentiation of C2C12 skeletal myoblasts, Integr. Biol. 3 (2011) 897–909.

[32] R. McBeath, D.M. Pirone, C.M. Nelson, K. Bhadriraju, C.S. Chen, Cell shape, cytoskeletal tension, and RhoA regulate stem cell lineage commitment, Dev. Cell. 6 (2004) 483–495.

[33] D.H. Kim, P.P. Provenzano, C.L. Smith, A. Levchenko, Matrix nanotopography as a regulator of cell function, J. Cell. Biol. 197 (2012) 351–360.

[34] W. Chen, L.G. Villa-Diaz, Y. Sun, S. Weng, J.K. Kim, R.H. Lam, et al., Nanotopography influences adhesion, spreading, and self-renewal of human embryonic stem cells, ACS Nano 6 (2012) 4094–4103.
[35] M. Ghibaudo, L. Trichet, J. Le Digabel, A. Richert, P. Hersen, B. Ladoux, Substrate topography induces a crossover from 2D to 3D behavior in fibroblast migration, Biophys. J. 97 (2009) 357–368.
[36] S. Khetan, J.A. Burdick, Patterning network structure to spatially control cellular remodeling and stem cell fate within 3-dimensional hydrogels, Biomaterials 31 (2010) 8228–8234.
[37] T.M. Hammoudi, H. Lu, J.S. Temenoff, Long-term spatially defined coculture within three-dimensional photopatterned hydrogels, Tissue Eng. Part C Methods 16 (2010) 1621–1628.
[38] V. Chan, K. Park, M.B. Collens, H. Kong, T.A. Saif, R. Bashir, Development of miniaturized walking biological machines, Sci. Rep. 2 (2012) 857.
[39] C.G. Williams, A.N. Malik, T.K. Kim, P.N. Manson, J.H. Elisseeff, Variable cytocompatibility of six cell lines with photoinitiators used for polymerizing hydrogels and cell encapsulation, Biomaterials 26 (2005) 1211–1218.
[40] S.J. Bryant, C.R. Nuttelman, K.S. Anseth, Cytocompatibility of UV and visible light photoinitiating systems on cultured NIH/3T3 fibroblasts in vitro, J. Biomater. Sci. Polym. Ed. 11 (2000) 439–457.
[41] D.F. Williams, On the mechanisms of biocompatibility, Biomaterials 29 (2008) 2941–2953.
[42] J.M. Anderson, A. Rodriguez, D.T. Chang, Foreign body reaction to biomaterials, Semin. Immunol. 20 (2008) 86–100.
[43] A.J. Vegas, O. Veiseh, J.C. Doloff, M. Ma, H.H. Tam, K. Bratlie, et al., Combinatorial hydrogel library enables identification of materials that mitigate the foreign body response in primates, Nat. Biotechnol. 34 (2016) 345–352.
[44] O. Veiseh, J.C. Doloff, M. Ma, A.J. Vegas, H.H. Tam, A.R. Bader, et al., Size- and shape-dependent foreign body immune response to materials implanted in rodents and non-human primates, Nat. Mater. 14 (2015) 643–651.
[45] A. Khademhosseini, R. Langer, A decade of progress in tissue engineering, Nat. Protoc. 11 (2016) 1775–1781.
[46] B. Tian, J. Liu, T. Dvir, L. Jin, J.H. Tsui, Q. Qing, et al., Macroporous nanowire nanoelectronic scaffolds for synthetic tissues, Nat. Mater. 11 (2012) 986–994.
[47] S.R. Shin, S.M. Jung, M. Zalabany, K. Kim, P. Zorlutuna, S.B. Kim, et al., Carbon-nanotube-embedded hydrogel sheets for engineering cardiac constructs and bioactuators, ACS Nano 7 (2013) 2369–2380.
[48] WO Patent, 2010072417, Rapid preparation and use of engineered tissue and scaffolds as individual implants, 2010.
[49] US Patent, 6905678, Gene delivery vectors with cell type specificity for mesenchymal stem cells, 2005.
[50] W.F. Lai, G.P. Tang, X. Wang, G. Li, H. Yao, Z. Shen, et al., Cyclodextrin-PEI-Tat polymer as a vector for plasmid DNA delivery to placenta mesenchymal stem cells, BioNanoScience 1 (2011) 89–96.
[51] D.W. Green, W.F. Lai, H.S. Jung, Evolving marine biomimetics for regenerative dentistry, Mar. Drugs 12 (2014) 2877–2912.
[52] J.W.C. Dunlop, R. Weinkamer, P. Fratzl, Artful interfaces within biological materials, Mater. Today 14 (2011) 70–78.
[53] J. Jernvall, I. Thesleff, Tooth shape formation and tooth renewal: evolving with the same signals, Development 139 (2012) 3487–3497.
[54] C.M. Chuong, N. Patel, J. Lin, H.S. Jung, R.B. Widelitz, Sonic hedgehog signaling pathway in vertebrate epithelial appendage morphogenesis: perspectives in development and evolution, Cell. Mol. Life Sci. 57 (2000) 1672–1681.
[55] I. Thesleff, Epithelial-mesenchymal signalling regulating tooth morphogenesis, J. Cell. Sci. 116 (2003) 1647–1648.
[56] K. Ishida, M. Murofushi, K. Nakao, R. Morita, M. Ogawa, T. Tsuji, The regulation of tooth morphogenesis is associated with epithelial cell proliferation and the expression of sonic hedgehog through epithelial-mesenchymal interactions, Biochem. Biophys. Res. Commun. 405 (2011) 455–461.
[57] T. Ohara, T. Itaya, K. Usami, Y. Ando, H. Sakurai, M.J. Honda, et al., Evaluation of scaffold materials for tooth tissue engineering, J. Biomed. Mater. Res. A. 94 (2010) 800–805.
[58] J.S. Song, S.O. Kim, S.H. Kim, H.J. Choi, H.K. Son, H.S. Jung, et al., In vitro and in vivo characteristics of stem cells derived from the periodontal ligament of human deciduous and permanent teeth, Tissue Eng. Part. A 18 (2012) 2040–2051.
[59] Y. Inuyama, C. Kitamura, T. Nishihara, T. Morotomi, M. Nagayoshi, Y. Tabata, et al., Effects of hyaluronic acid sponge as a scaffold on odontoblastic cell line and amputated dental pulp, J. Biomed. Mater. Res. B 92 (2010) 120–128.
[60] S.E. Kim, D.H. Suh, Y.P. Yun, J.Y. Lee, K. Park, J.Y. Chung, et al., Local delivery of alendronate eluting chitosan scaffold can effectively increase osteoblast functions and inhibit osteoclast differentiation, J. Mater. Sci. Mater. Med. 23 (2012) 2739–2749.
[61] T. Matsunaga, K. Yanagiguchi, S. Yamada, N. Ohara, T. Ikeda, Y. Hayashi, Chitosan monomer promotes tissue regeneration on dental pulp wounds, J. Biomed. Mater. Res. A 76 (2006) 711–720.
[62] R. d'Aquino, A. De Rosa, V. Lanza, V. Tirino, L. Laino, A. Graziano, et al., Human mandible bone defect repair by the grafting of dental pulp stem/progenitor cells and collagen sponge biocomplexes, Eur. Cells Mater. 18 (2009) 75–83.
[63] E. Mahapoka, P. Arirachakaran, A. Watthanaphanit, R. Rujiravanit, S. Poolthong, Chitosan whiskers from shrimp shells incorporated into dimethacrylate-based dental resin sealant, Dent. Mater. J. 31 (2012) 273–279.
[64] K.C. Yang, C.H. Wang, H.H. Chang, W.P. Chan, C.H. Chi, T.F. Kuo, Fibrin glue mixed with platelet-rich fibrin as a scaffold seeded with dental bud cells for tooth regeneration, J. Tissue. Eng. Regen. Med. 6 (2012) 777–785.
[65] K.M. Galler, A.C. Cavender, U. Koeklue, L.J. Suggs, G. Schmalz, R.N. D'Souza, Bioengineering of dental stem cells in a PEGylated fibrin gel. Regenerative medicine, Regen. Med. 6 (2011) 191–200.
[66] N.R. Kim, D.H. Lee, P.H. Chung, H.C. Yang, Distinct differentiation properties of human dental pulp cells on collagen, gelatin, and chitosan scaffolds, Oral Surg. Oral. Med. Oral Pathol. Oral Radiol. Endod. 108 (2009) e94–e100.
[67] D. Nadeem, M. Kiamehr, X. Yang, B. Su, Fabrication and in vitro evaluation of a sponge-like bioactive-glass/gelatin composite scaffold for bone tissue engineering, Mater. Sci. Eng. C Mater. Biol. Appl 33 (2013) 2669–2678.

[68] R. Cornelini, L. Artese, C. Rubini, M. Fioroni, G. Ferrero, A. Santinelli, et al., Vascular endothelial growth factor and microvessel density around healthy and failing dental implants, Int. J. Oral Maxillofac. Implants 16 (2001) 389–393.
[69] S. Lin, A. Roguin, Z. Metzger, L. Levin, Vascular endothelial growth factor (VEGF) response to dental trauma: a preliminary study in rats, Dent. Traumatol. 24 (2008) 435–438.
[70] L. Megan, J. Jessica, H. Warren, B. Joel, Effects of VEGF-loaded chitosan coatings, J. Biomed. Mater. Res. A 102 (2014) 752–759.
[71] M. Anthonsen, O. Smidsrod, Hydrogen ion titration of chitosans with varying degrees of N-acetylation by monitoring induced 1H-NMR chemical shifts, Carbohydr. Polym. 26 (1995) 303–305.
[72] G. Berth, H. Dautzenberg, M.G. Peter, Physico-chemical characterization of chitosans varying in degree of acetylation, Carbohydr. Polym. 36 (1998) 205–216.
[73] R. Hejazi, M. Amiji, Chitosan-based gastrointestinal delivery systems, J. Control Release 89 (2003) 151–165.
[74] N. Fang, V. Chan, H.Q. Mao, K.W. Leong, Interactions of phospholipid bilayer with chitosan: effect of molecular weight and pH, Biomacromolecules 2 (2001) 1161–1168.
[75] M. Thanou, J.C. Verhoef, H.E. Junginger, Chitosan and its derivatives as intestinal absorption enhancers, Adv. Drug Deliv. Rev. 50 (Suppl 1) (2001) S91–S101.
[76] K. Subramani, M.A. Birch, Fabrication of poly(ethylene glycol) hydrogel micropatterns with osteoinductive growth factors and evaluation of the effects on osteoblast activity and function, Biomed. Mater. 1 (2006) 144–154.
[77] N.J. Castro, C.M. O'Brien, L.G. Zhang, Biomimetic biphasic 3-D nanocomposite scaffold for osteochondral regeneration, AIChE J. 60 (2014) 432–442.
[78] E.S. Place, N.D. Evans, M.M. Stevens, Complexity in biomaterials for tissue engineering, Nat. Mater. 8 (2009) 457–470.
[79] H. Cui, L. Cui, P. Zhang, Y. Huang, Y. Wei, X. Chen, In situ electroactive and antioxidant supramolecular hydrogel based on cyclodextrin/copolymer inclusion for tissue engineering repair, Macromol. Biosci. 14 (2014) 440–450.
[80] H. Cui, J. Shao, Y. Wang, P. Zhang, X. Chen, Y. Wei, PLA-PEG-PLA and its electroactive tetraaniline copolymer as multi-interactive injectable hydrogels for tissue engineering, Biomacromolecules 14 (2013) 1904–1912.
[81] H. Cui, X. Zhuang, C. He, Y. Wei, X. Chen, High performance and reversible ionic polypeptide hydrogel based on charge-driven assembly for biomedical applications, Acta Biomater. 11 (2015) 183–190.
[82] W. Zhu, F. Masood, J. O'Brien, L.G. Zhang, Highly aligned nanocomposite scaffolds by electrospinning and electrospraying for neural tissue regeneration, Nanomedicine. 11 (2015) 693–704.
[83] W. Zhu, J.K. George, V.J. Sorger, L. Grace Zhang, 3D printing scaffold coupled with low level light therapy for neural tissue regeneration, Biofabrication 9 (2017) 025002.
[84] M. Sartori, S. Pagani, A. Ferrari, V. Costa, V. Carina, E. Figallo, et al., A new bi-layered scaffold for osteochondral tissue regeneration: in vitro and in vivo preclinical investigations, Mater. Sci. Eng. C Mater. Biol. Appl. 70 (2017) 101–111.
[85] Y. Yang, T. Xia, W. Zhi, L. Wei, J. Weng, C. Zhang, et al., Promotion of skin regeneration in diabetic rats by electrospun core-sheath fibers loaded with basic fibroblast growth factor, Biomaterials 32 (2011) 4243–4254.
[86] W. Sun, P. Lal, Recent development on computer aided tissue engineering - a review, Comput. Meth. Prog. Bio. 67 (2002) 85–103.
[87] E. Keeve, S. Girod, R. Kikinis, B. Girod, Deformable modeling of facial tissue for craniofacial surgery simulation, Comput. Aided. Surg. 3 (1998) 228–238.

Chapter 13

Use of delivery technologies to mediate herbal interventions

Introduction

Herbal medicine is an integral component of oriental medicine. Active ingredients of medicinal herbs also have a long history of use in drug discovery (Table 13.1) [1]. One example of drugs discovered from herbs is Kanglaite. It is an investigational anticancer drug with its active substance extracted from *Semen coicis*. The injectable form of this drug has been approved in China for treating cancer [2]. Another example is ephedrine, which is asympathomimetic amine isolated from *Ephedra vulgaris*. It has been widely adopted as a hyperglycemic agent, a bronchodilator, and a cardiac stimulant [3]. From vision enhancement with *Lycium barbarum* fruits to cancer treatment [4–7], herbal medicine has assimilated into the life of Asian people as both foods and remedies for millennia (Table 13.2) [1]. Along with the rising global recognition of the value of oriental medicine, herbal medicine has encouraging potential in clinical use.

Medicinal herbs contain different types of active ingredients (including flavonoids, saponins, polysaccharides, and alkaloids) that show antiaging properties. These ingredients can act on different mechanisms of the aging process. One example of these mechanisms is telomere shortening. Over the years, various herbs have been found to modulate the activity of telomerase. For instance, C21 steroidal glycoside from the root of *Cynanchum auriculatum* can promote the antioxidative and antistress capacity of D-galactose-induced aging mice partly by antagonizing free radical injury, increasing SOD activity, and promoting the activity of telomerase in serum and heart tissues [8]. Ginsenoside Rg1 from *Panax ginseng* C.A. Meyer has also been reported to activate the activity of telomerase to combat *tert*-butyl hydroperoxide (t-BHP)-induced cell senescence [9]. Other examples of botanical ingredients that show modulatory effects on telomerase include the polysaccharides from *Cistanche deserticola* [10], flavonoids from *Epimedium brevicornu* Maxim. [11], acteoside from *Cistanche tubulosa* Schenk Wight [12], astragaloside from *Astragalus membranaeus* Fisch. Bge. [13], and pine pollen from *Pinus massoniana* Lamb. [14].

Apart from increasing the activity of telomerase, some ingredients and herbal extracts can regulate the production of sirtuins (which belong to a group of NAD^+-dependent deacetylases that involve in gene repair, cell cycle regulation, and other cellular processes). An example of these is resveratrol from *Polygonum cuspidatum*. This ingredient has been found to reverse the senescence of human umbilical vein endothelial cells by increasing the expression of *SIRT1* [15]. By stimulating the expression of *PGC-1α* and *SIRT1* to enhance mitochondrial biogenesis, quercetin from *Herba hyperici* has also successfully increased the maximal endurance capacity and voluntary wheel-running activity of aged mice. Other examples that may modulate the expression of *SIRT1* include butein from *Butea monosperma* Lam. Kuntze [16], and the extracts of *Ginkgo biloba* Linn [17].

Finally, the nutrient/energy-sensing pathway and the TOR pathway can be modulated by herbal ingredients. One example is the extract of *Coptis chinensis* Franch., which can improve insulin sensitivity by activating the activity of AMPK in muscle tissues [18]. Another example is 6-gingerol extracted from ginger. It can inhibit the mTOR/p70S6K pathway to attenuate senescence in vascular smooth muscle cells (VSMCs) [19]. More recently, ginsenoside Rb1, which is a protopanaxdiol extracted from the root of *Panax ginseng*, has been found to retard brain aging by acting on the mTOR/p70S6K pathway [20]. In addition, by activating signaling pathways downstream of AMPK and nuclear factor erythroid 2-related factor 2 (Nrf2), and by suppressing inflammatory processes mediated by NF-kB signaling, curcumin from *Curcuma longa* Linn. has been found to decrease the overall amyloid content and plaque burden, and to suppress indices of inflammation and oxidative damage in the brain, in a mouse model of Alzheimer's disease [21,22]. All these have supported the possibility of using herbal ingredients to combat aging at the molecular level.

Delivery of Therapeutics for Biogerontological Interventions. DOI: https://doi.org/10.1016/B978-0-12-816485-3.00013-1

TABLE 13.1 Examples of active compounds isolated from medicinal herbs as drugs [1].

Use	Active compound	Botanical origin
Muscle relaxation	Cissampelin methiodide	*Cissampelos pareira* L.
	Anabasine hydrochloride	*Alangium chinensis* (Lour.) Harms
	Tetrandrin dimethiodide	*Stephania tetrandra* S. Moore
Treatment of coronary heart diseases	Cyclovirobuxine D	*Buxus microphylla* Sieb. et
		Zucc. var. *sinica* Rehd. et Wils.
	Tanshionone II-A	*Salvia miltiorrhiza* Bunge
	Tetramethylpyrazine	*Liguistrum chuanxiong* Hort.
	Sodium ferulate	*Liguistrum chuanxiong* Hort.
Treatment of hepatitis	Sarmentosine	*Sedum sarmentosum* Bge.
	Schisantherin A	*Schizandra chinensis* (Turcz.) Baill., and *Schisandra sphenanthera* Rehd. et Wils.
Treatment of cancer	Berbamine	*Berberis poiretil* Schneid
	Harringtonine	*Cephalotaxus* hainanensis Li
	Indirubin	*Isatis tinctoria, Baphicacanthus cusia*
	Irisquinone	*Iris pallasii* Fisch. var. *chinensis* Fisch.
	Monocrotaline	*Crotalaria sessiliflora* L.
	Homoharringtonine	*Cephalotaxus hainanensis* Li
	Taxol	*Taxus brevifolia*
	Oridonin	*Rabdosia Rubescens*

TABLE 13.2 Examples of medicinal herbs used for food purposes [1].

Consumable part	Examples
Whole plant	*Taraxacum sinicum* Kitag., *Elsholtzia splendens* Nakai ex F. Maekawa, *Pogostemon cablin* (Blanco) Benth., *Houttuynia cordata* Thunberg, *Portulaca oleracea* L., *Hordeum vulgare* L.
Leaf	*Raphanus sativus* L., *Lophatherum gracile* Brongniart, *Nelumbo nucifera* Gaertn, *Perilla frutescens* (L.) Britt.
Flower	*Lonicera dasystyla* Rehd., *Citrus aurantium* L., *Nelumbo nucifera* Gaertn, *Dendranthema morifolium* (Ramat.) Tzvel., *Sophora japonica* L., *Chrysanthemum morifolium*
Rhizome and root	*Glycyrrhiza glabra* L., *Glycyrrhiza uralensis* Fisch., *Raphanus sativus* L., *Panax ginseng, Nelumbo nucifera* Gaertn, *Polygonatum cyrtonema* Hua, *Polygonatum odoratum* (Mill.) Druce, *Dioscorea opposita* Thunb., *Glycyrrhiza glabra, Dioscorea alata*
Stem and cortex	*Lophatherum gracile* Brongniart, *Glycyrrhiza uralensis* Fisch., *Glycyrrhiza glabra* L.
Seed	*Canavalia gladiata* (Jacq.) DC., *Ginkgo biloba* L., *Raphanus sativus* L., *Torreya grandis* Fort., *Cerasus japonica* (Thunb.) Lois, *Euryale ferox* Salisb., *Nelumbo nucifera, Phaseolus calcaratus, Dolichos lablab*
Fruit	*Lycium barbarum, Foeniculum vulgare* Mill., *Crataegus pinnatifida* Bunge, *Dimocarpus longan* Lour., *Cannabis sativa* L., *Citrus medica* L., *Gardenia jasminoides* Ellis, *Crataegus pinnatifida, Ziziphus jujube*

Hydrogels as delivery systems in herbal medicine

Despite the possible use of medicinal herbs in antiaging medicine, the efficiency in absorption and cellular internalization of different ingredients, which show variations in structures and properties (e.g., surface charge and molecular weight), is poor sometimes. To enhance the efficiency, a carrier may be adopted to improve the pharmacodynamics of herbal ingredients. Over the years, various materials (e.g., liposomes, phytosomes, emulsions, and hydrogels) have been developed and engineered as drug carriers [23–32]. As discussed in Chapter 7, hydrogels can absorb a substantial amount of fluids. Because most of the herbal formulations are aqueous in nature, the hydrophilic property of hydrogels favors drug loading [33]. Moreover, with advances in hydrogel design and fabrication, various stimuli-responsive hydrogels have been developed for controlled and prolonged drug release [25,34]. Along with their high biocompatibility, hydrogels are one of the most favorable systems for mediating herbal interventions.

The possibility of using hydrogels to enhance the effectiveness of herbal interventions has been demonstrated by an earlier study [35], in which a hydrogel patch containing herbal extracts has been used to treat atopic dermatitis. The study has adopted the extracts from two medicinal herbs. One is *Houttuynia cordata* Thunb. The extract of which has been used widely for treating herpes simplex [36], chronic sinusitis [37], and nasal polyps [37]. Another herb is *Ulmus davidiana* var. *japonica*, which is a deciduous broad-leaved tree commonly found in oriental countries. Its root barks and stem have therapeutic effects on rheumatoid arthritis, cancer, mastitis, edema, and inflammation [38]. After mixing the herbal extracts with polyvinyl alcohol (PVA) and propylene glycol (PG), the mixture has been subjected to the freeze-thaw process and ^{60}Co γ-ray irradiation to form a hydrogel patch [35]. Due to the moisturizing effects of the hydrogel and the activity of the herbal extracts on atopic wounds, mice treated with the patch have recovered from edema caused by contact dermatitis (Fig. 13.1) [35], and have suffered less from itchiness caused by atopic dermatitis [35]. Along with its capacity of being easily attached to or detached from the skin [35], the patch may warrant further development for clinical applications.

Example protocols for experiment planning

The method below is an example protocol for preparing an Alg-based hydrogel loaded with the extract of *H. cordata* Thunb.

1. Put 2 g of dried *H. cordata* Thunb. in a container.
2. Immerse the dried herb in 70 mL of distilled water.
3. Seal the container.
4. Incubate the container at 80°C for 16 hours.
5. Obtain the filtrate as the extract of the herb.
6. Use the filtrate to dissolve Alg to reach a concentration of 4% (w/v).

(Continued)

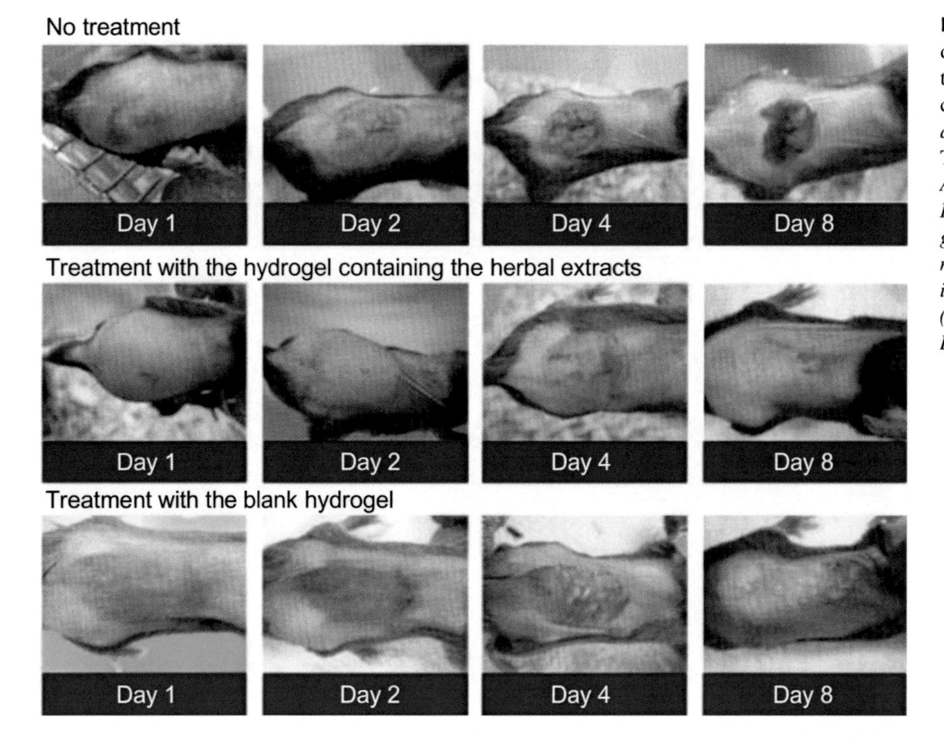

FIGURE 13.1 Photographs taken at different stages of the contact dermatitis treatment mediated by a hydrogel patch containing the extracts from *Ulmus davidiana* var. *japonica* and *Houttuynia cordata* Thunb. *Reproduced from Y.-M. Lim, S.-J. An, H.-K. Kim, Y.-H. Kim, M.-H. Youn, H.-J. Gwon, et al., Preparation of hydrogels for atopic dermatitis containing natural herbal extracts by gamma-ray irradiation, Radiat. Phys. Chem. 78 (2009) 441–444 with permission from Elsevier B.V., [35].*

(Continued)

7. Add the solution prepared in Step 6 into 500 mL of a 3% (w/v) $CaCl_2$ solution.
8. Wait for 20 minutes.
9. Retrieve the generated hydrogel by centrifugation.

The possibility of using hydrogels to deliver botanical ingredients has been further supported by the case of Astragali Radix (i.e., the root of *Astragalus membranaceus*), which is an herb that has been adopted to enhance the repair and recovery of organs and tissues (including neurons [39], heart [40], and lung [41]). In a recent study, a solid-lipid-nanoparticle-enriched hydrogel (SLN-gel) has been fabricated for topical delivery of astragaloside IV, which is an active ingredient of Astragali Radix, for wound care. In this formulation, the SLNs are used as carriers to mediate the delivery of astragaloside IV [42,43]; whereas the hydrogel helps improve the consistency of the final formulation and promotes the long-term stability of the astragaloside IV-loaded nanoparticles. The formulation has been found to enable sustained release of astragaloside IV, and to enhance the migration and proliferation of keratinocytes in vitro [44]. In a rat full-skin excision model, the formulation has increased the wound closure rate, and has facilitated angiogenesis and collagen deposition [44]. This allows the formulation to be further exploited for wound treatment.

As a matter of fact, the possible use of hydrogels as carriers in herbal medicine has already been demonstrated in clinical trials. For instance, in an earlier randomized, double-blind, placebo-controlled clinical trial conducted with ambulatory patients [45], the extract of the *Mimosa tenuiflora* cortex, which is an herb utilized in Mexico to treat skin lesions, has been incorporated into a hydrogel (which has been prepared from a mixture of PEG, Carbopol 940, and triethanol amine) for treating venous leg ulceration disease. The size of the ulcer in patients treated with the extract-loaded hydrogel has been found to be remarkably reduced; whereas the condition in those treated with the hydrogel alone has shown no observable improvement (Fig. 13.2). This has revealed the clinical feasibility of using hydrogel-based carriers in oriental medicine.

Preparation of hydrogels for herbal interventions

A lot of first-generation hydrogels are principally chemical hydrogels [46]. Many of them are prepared by cross-linking of synthetic polymers. Examples include PVA and PEG hydrogels. Chemical hydrogels can also be generated from monomers (particularly vinyl monomers) that undergo chain-addition reactions in the presence of a multifunctional cross-linker. One example of hydrogels in this category is the poly (acrylamide) hydrogel, which has been adopted to encapsulate cells and bioactive agents [47,48]. Using chemical hydrogels to deliver herbal formulations, however, might not be favorable. This is because the side reactions between the cross-linking agent and the large variety of functional groups in botanical ingredients may reduce the effectiveness of the herbal intervention.

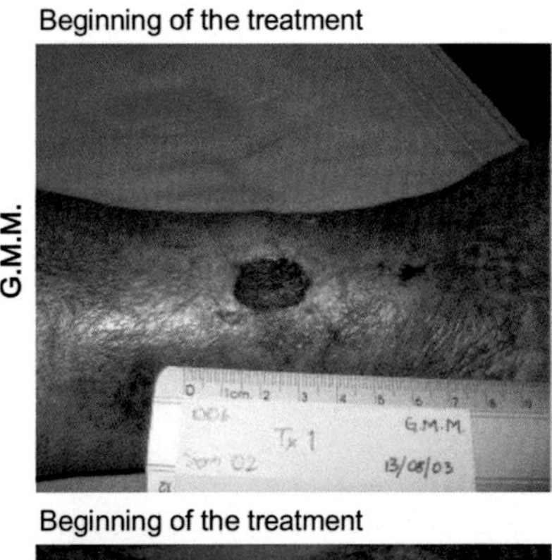

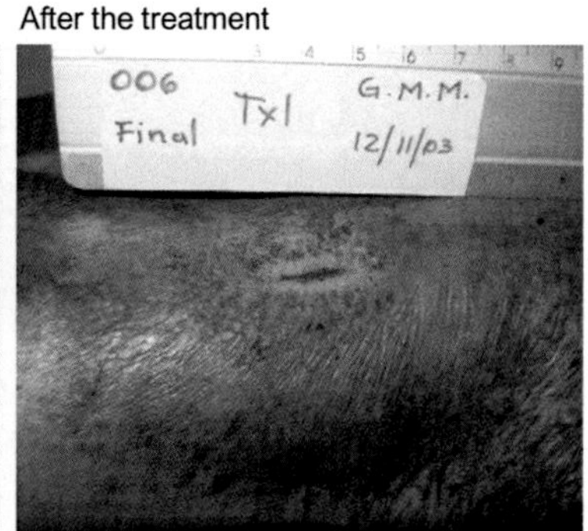

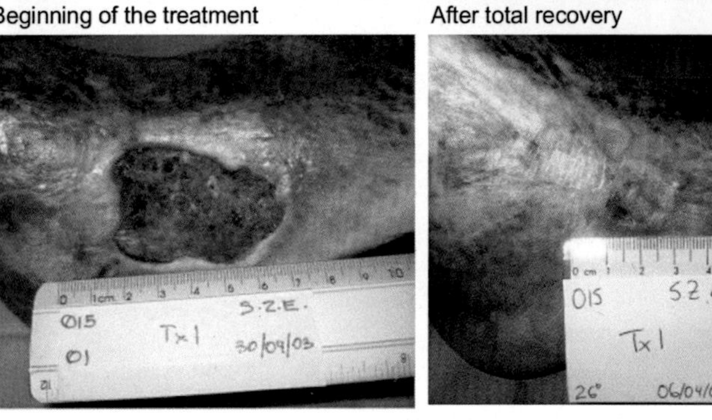

FIGURE 13.2 Skin lesions in two patients (G. M. M. and S. Z. E.), who have suffered from venous leg ulceration disease, at different stages of the treatment mediated by the hydrogel containing the extract of the *Mimosa tenuiflora* cortex. *Reproduced from E. Rivera-Arce, M.A. Chavez-Soto, A. Herrera-Arellano, S. Arzate, J. Aguero, I.A. Feria-Romero, et al., Therapeutic effectiveness of a Mimosa tenuiflora cortex extract in venous leg ulceration treatment, J. Ethnopharm. 109 (2007) 523–528 with permission from Elsevier B.V., [45].*

An alternative to chemical hydrogels is physical hydrogels. Although these hydrogels generally have lower mechanical strength and long-term stability, they possess a number of favorable properties [49], including lower toxicity, higher drug loading efficiency, and less interactions with components in herbal formulations. The use of physical hydrogels as carriers in herbal medicine has been documented in an earlier study, which has loaded the crude leaf extract of *Hemigraphis alternata*, which is a perennial herb from the family Apocyanacea, into a CS-based hydrogel (which has been prepared by first mixing an acetic acid solution of CS with the extract, followed by neutralization of the mixture with NaOH). The extract has shown antiinflammatory activity and has facilitated wound contraction and epithelialization in the mouse model of carrageenan-induced paw edema [50]. After loading into the hydrogel, the dressing generated has demonstrated hemostatic and antibacterial properties. Along with its capacity of enhancing platelet activation, dermal fibroblast attachment and blood clotting, the dressing has good potential for applications in wound treatment [51].

To produce physical hydrogels, over the years diverse methods have been adopted. One method is ionic gelation, in which oppositely charged ions are employed as cross-linkers to electrostatically interact with the polymer constituents of a hydrogel. Examples of hydrogels prepared by using this method are Ca^{2+}crosslinked Alg hydrogels and CS-TPP hydrogels. Apart from ionic gelation, physical hydrogels can be fabricated using stereocomplexation. By using this approach, a hydrogel self-assembled by stereocomplexation of enantiomeric lactic acid oligomer-grafted dextran (dex-lactate) copolymers has been generated [52]. In recent years, stimuli-responsive hydrogels have been successfully adopted to increase the effectiveness of herbal formulations. This has been shown by the case of the thermosensitive hydrogel generated from poly(ε-caprolactone-*co*-1,4,8-trioxa[4.6]spiro-9-undecanone)−PEG−poly(ε-caprolactone-*co*-1,4,8-trioxa[4.6]spiro-9-undecanone) (PECT). The hydrogel has enabled sustained release of embelin (which is a botanical ingredient whose in vivo applications have been impeded by its poor aqueous solubility), and has enhanced the antitumor effect of embelin both in vitro and in vivo [53]. This study has illustrated the prospect of using hydrogels to enhance the activity of herbal ingredients.

Example protocols for experiment planning

The method below is an example protocol for evaluating the effectiveness of a hydrogel generated from ionic gelation in encapsulating an herbal ingredient

1. Dissolve a known amount of an herbal ingredient in a 2% (w/v) Alg solution.
2. Add the solution into 500 mL of a $CaCl_2$ solution.

(Continued)

(Continued)

3. Wait for 20 minutes.
4. Retrieve the generated hydrogel by centrifugation.
5. Lyophilize the hydrogel generated.
6. Add an appropriate amount of PBS (or a simulated body fluid) to a tube containing the hydrogel.
7. Retrieve the hydrogel by centrifugation.
8. Determine the total concentration of the unloaded herbal ingredient in the supernatant using an appropriate method of quantification.
9. Calculate the loading efficiency and encapsulation efficiency using the following equations, where m_l is the mass of the herbal ingredient loaded into the hydrogel, m_t is the total mass of the ingredient-loaded hydrogel examined, and m_T is the total mass of the herbal ingredient added.

$$\text{Loading efficiency } (\%) = \frac{m_l}{m_t} \times 100\% \qquad (13.1)$$

$$\text{Encapsulation efficiency } (\%) = \frac{m_l}{m_T} \times 100\% \qquad (13.2)$$

Modulation of material properties for herbal interventions

When hydrogels are used as carriers, one important parameter to be optimized is the loading efficiency, which is partly determined by the affinity of the hydrogel matrix for the delivered agent. Based on its therapeutic role, an herb can function as "Monarch," "Minister," "Assistant," or "Guide" in an herbal formulation. This implies that, in order to have the formulation functional, multiple active ingredients have to be administered simultaneously. Many of the reported delivery systems in literature, however, are designed for single-drug therapy. Hydrogel systems that can indiscriminately codeliver multiple ingredients are lacking. In addition, the hydrogel matrix, in general, is hydrophilic. If the loaded molecules are hydrophobic, they may have a low affinity for the hydrogel, thereby diffusing out of the matrix before the loading process is complete. This reduces the loading yield obtained. To solve this problem, one method is to first load the hydrophobic components into another carrier that shows a higher affinity for the hydrogel matrix before those components are loaded into hydrogels. This method has been adopted to enhance the efficiency in loading a hydrogel with curcumin, which is an active component of turmeric (the powdered rhizome of *C. longa* L.). Curcumin is commonly used in India, China, and other Southeast Asia countries as a spice and also as a remedy for diverse inflammatory conditions and chronic diseases [54]. Upon topical or oral administration, it has been found to facilitate wound repair in vivo [55,56]; however, its therapeutic efficiency and bioavailability have still been largely limited by the first-pass metabolism [57]. To alleviate this problem, a hydrogel may be adopted to encapsulate curcumin for controlled drug release. Unfortunately, due to the high hydrophobicity of curcumin [58], the encapsulation efficiency is far from satisfactory. For this, an earlier study has first loaded curcumin into polymeric micelles [59], followed by the incorporation of the micelles (Cur–M) into a thermosensitive PEG–PCL–PEG hydrogel to form a wound dressing (Cur–M–H) [59]. The dressing behaves as a free-flowing sol at ambient conditions but is converted into a nonflowing gel at the body temperature. Wounds treated with Cur–M–H have shown marked dryness [59], and have exhibited negligible signs of inflammation and infection [59]. Histopathologic analysis has revealed that wounds treated with Cur–M–H have a higher degree of fibroblastic deposition and re-epithelialization, as compared to those treated with normal saline, blank M–H, and Cur–M [59]. This study has illustrated the viability of using carriers to modify the hydrophilicity of hydrophobic botanical ingredients to improve the loading efficiency in hydrogels.

Apart from modifying the ingredients, the loading efficiency can be enhanced by directly modulating the hydrogel properties. The effect of hydrogel properties on the loading efficiency has been documented in an earlier study, in which an Alg-based hydrogel has been adopted to carry the extract of *Piper sarmentosum* [60]. The encapsulation efficiency has been found to be affected by the mannuronic acid/guluronic acid ratio (M/G ratio) of Alg, with the hydrogel beads having a higher M/G ratio being more efficient in absorbing the extract [60]. Engineering the polymer composition, therefore, is a possible method of enhancing the loading efficiency of a hydrogel. This has been hinted at by a previous study [61], which has fabricated hydrogels with multiple functional monomers and template molecules and has successfully enhanced the efficiency of the hydrogels in encapsulating molecules with varying degrees of hydrophilicity. More recently, a study has also copolymerized CS with hypromellose using carbonyldiimidazole chemistry to generate hypromellose-*g*-CS (HC) [25], which has displayed higher aqueous solubility at physiological pH as compared to native CS [25]. Due to its positive charge, HC has complexed with CMC to form a hydrogel for drug encapsulation. By using agents (mometasone furoate, methylene blue, tetracycline hydrochloride, and metronidazole) with different degrees of hydrophobicity as drug models, the encapsulation efficiency of the hydrogel has been found to be over 90%. This has suggested the viability of manipulating the hydrogel composition to improve the efficiency of the hydrogel in encapsulating different ingredients of herbal formulations.

Apart from the loading efficiency, another important parameter to be optimized is the release sustainability of the hydrogel. Modulation of this can be achieved by altering the numbers and ratios of different reactive oligomers or polymer precursors. This enables the generation of hydrogels with different swelling properties and degradation kinetic profiles. The use of this method has been demonstrated by the case of a composite hydrogel that has been generated from poloxamer 407 (P407) and CMC for sustained release of the Cortex Moutan extract [62]. P407 is a typical thermosensitive polymer. Its hydrogel has a high moisture content that helps moisturize the skin of patients suffering from atopic dermatitis [62]. Upon the addition of CMC, modulation of the gelation transition temperature, porous structure, and rheological property of the hydrogel has resulted. Changes in the concentrations of P407 and CMC have been found to alter the bulk viscosity of the hydrogel, leading to a change in the rate of release of the extract [62]. Moreover, the presence of CMC in the hydrogel has been found to enhance the skin permeability of the extract in vivo [62].

Example protocols for experiment planning

The method below is an example protocol for evaluating the swelling capacity of a hydrogel.

1. Preweigh 0.5 g of a lyophilized hydrogel.
2. Immerse the lyophilized hydrogel in 500 mL of PBS (pH 7.4).
3. Retrieve the hydrogel from PBS at a pre-set time interval.
4. Weigh the swollen hydrogel.
5. Calculate the swelling ratio, water content, and water absorption ratio (WAR) of the hydrogel using the following formulae:

$$\text{Swelling ratio} = \frac{m_s}{m_d} \tag{13.3}$$

$$\text{Water content } (\%) = \frac{m_s - m_d}{m_s} \times 100\% \tag{13.4}$$

$$\text{WAR} = \frac{m_s - m_d}{m_d} \tag{13.5}$$

where m_s and m_d represent the masses of the swollen and dried hydrogel, respectively.

Codelivery of herbal formulations and chemical drugs

There is a proverb in China: "the experience-based oriental medicine treats humans whereas the evidence-based Western medicine treats diseases." Oriental medicine views the body as a microcosm comprising both external and internal conditions. It emphasizes the equipoise of "energies" in the five *zang*-viscera (the heart, liver, spleen, lung, and kidney) [63]. On the other hand, evidence-based medicine is more attentive to the symptoms of a disease. It helps alleviate the symptoms, but is perceived to be less attentive to the restoration of the internal balance in a body than oriental medicine [64]. In the light of this, using herbal formulations from oriental medicine and synthetic drugs from Western medicine together may become a possible direction of future medicine. The feasibility of this has been suggested by an earlier study on aged people suffering from idiopathic nephrotic syndrome [65]. Patients who have been treated simultaneously with an herbal soup and synthetic drugs (i.e., cyclophosphamide, and prednisone) have shown a lower adverse reaction rate, a higher remission rate, and a longer remission period, as compared to those treated with synthetic drugs alone. Similar promising effects of combining oriental medicine with Western medicine has been reported by Yao and colleagues [66], who have adopted both herbal remedies (including dandelion, herba patriniae, barbed stullcap, and giant knotweed) and synthetic drugs (e.g., rapamycin, tacrolimus, cyclosporine A, azathioprine, mycophenolate, and prednisone) to treat severe postkidney-transplant lung infection. Those herbal remedies have promoted patients' immunity [66], thereby improving the treatment outcome in 15 out of the 18 patients who have participated in the study.

Despite the potential mentioned above, the possible incompatibility of synthetic drugs with herbal remedies is one of the challenges to be solved, or the effectiveness of the combined therapy will be jeopardized. The validity of this concern has been supported by the capacity of *Hypericum perforatum* to lower the serum concentration of synthetic drugs (including cyclosporine, indinavir, irinotecan, and nevirapine) by inducing the expression of genes for cytochrome P450 enzymes and P-glycoprotein [67]. In addition, *Caulis Tripterygium* wilfordii has been found to reduce the effectiveness of azathioprine in vivo in treating pulmonary fibrosis [68]. These have indicated the importance of taking herb–drug interactions into account when a combined therapy is administered. Intriguingly, possible solutions to the incompatibility problem have been illuminated by recent advances in microfabrication technologies. For instance, multicompartment microgel beads have been prepared by using microfluidic electrospray for codelivery of incompatible agents [23]. Upon the incorporation of the polymer blending technique, the drug release sustainability of different compartments of the bead has been tuned to control the release profiles of the codelivered agents [23] (Fig. 13.3). The versatility provided by these beads may bring new opportunities to the development of interventions that combine herbal formulations with synthetic drugs in the future.

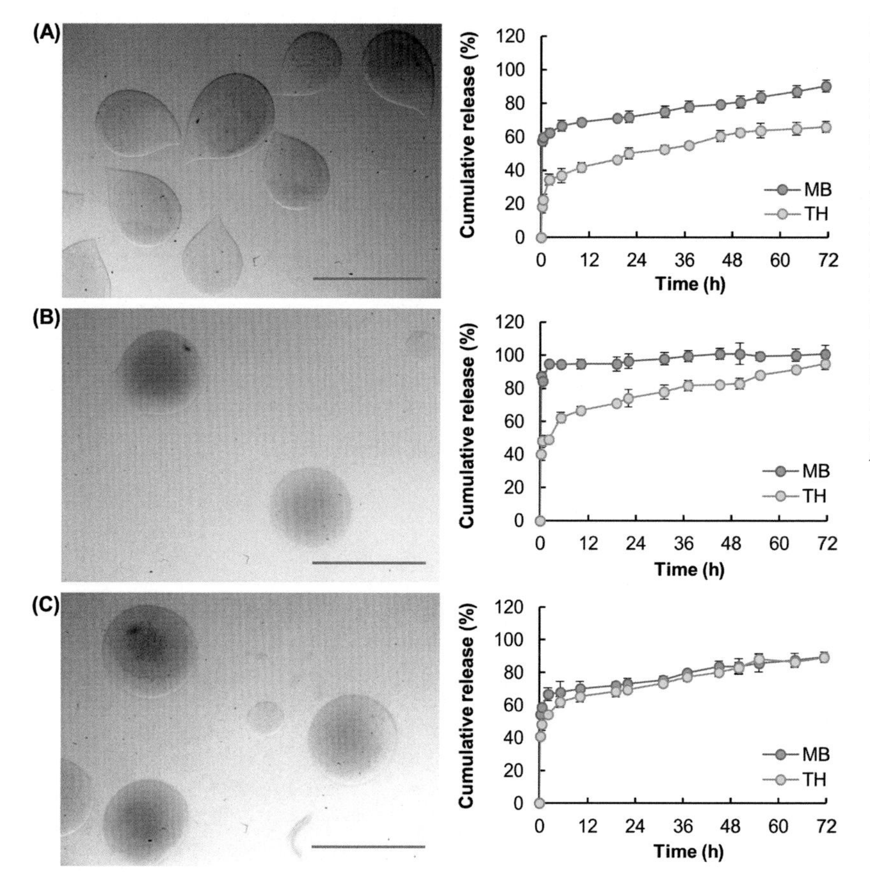

FIGURE 13.3 Photos and drug release profiles of the multicompartment microgel beads generated from an Alg/CMC blend. The microgels have been loaded with both methylene blue (MB) and tetracycline hydrochloride (TH). The blend has contained different weight percentages of Alg: (A) 100 wt.% for both MB-loaded and TH-loaded compartments; (B) 25 wt.% for both MB-loaded and TH-loaded compartments; and (C) 100 wt.% for the MB-loaded compartment and 25 wt.% for the TH-loaded compartment. Scale bar = 500 μm. *Reproduced from W.F. Lai, A.S. Susha, A.L. Rogach, Multicompartment microgel beads for co-delivery of multiple drugs at individual release rates, ACS Appl. Mater. Interfaces 8 (2016) 871–880 with permission from the American Chemical Society [23].*

Summary

With the rapid research progress in drug delivery, the possibility of using delivery systems to improve the effectiveness of herbal remedies has been evidenced in both clinical and pre-clinical studies. Before the routine execution of herbal interventions can be facilitated by the use of delivery technologies, much optimization, however, is required. This is partly shown by a randomized, double-blind clinical trial on the use of a hydrogel containing tepescohuite, which is an extract of the bark of the *Mimosa tenuiflora* tree, in treating venous leg ulcers (VLUs) [69]. The study has recruited 41 patients with venous ulcers. The recruited patients have been instructed to cleanse the ulcer, and to apply the hydrogel and compression to the ulcer daily. Histological analysis has found no difference in the clinical outcome between the patients treated with the extract-loaded hydrogel and those treated with the hydrogel alone. The effectiveness of using delivery systems to enhance the performance of an herbal intervention, therefore, is not guaranteed, and is affected by multiple factors including the properties of the carrier and the choice of herbal remedies. Nevertheless, cumulative works in literature have already laid a theoretical foundation for the development of delivery technologies to mediate herbal treatment. What remains is to translate the theories into practicable interventions.

Directions for intervention development

Herbal medicine has a long history of applications in disease prevention and treatment, with various herbs containing active ingredients that can modulate the aging network. To develop an antiaging intervention based on herbal remedies or ingredients, the following steps can be taken:

1. Select an herb that has the potential to modulate the aging network.
2. Choose an appropriate usage form of the herb. The herb can be used as an extract, as a formulation in which multiple remedies are adopted simultaneously, or simply as a source of an active compound.
3. Examine the toxicity and biological properties of the proposed usage form in vitro and in vivo.
4. Design a carrier to facilitate the execution of the proposed herbal intervention.
5. Evaluate the performance of the carrier.
6. Tune the structure and properties of the carrier to optimize the performance.

References

[1] W.F. Lai, Z.C.Y. Chan, Chinese herbal medicine in health care: what can be learnt from the context of Hong Kong? in: Z.C.Y. Chan (Ed.), Health Issues in Chinese Contexts, Nova Publishers, New York, 2009, pp. 173–190.

[2] P. Basu, Trading on traditional medicines, Nat. Biotechnol. 22 (2004) 263–265.

[3] A. Cruz, M. Juarez-Juarez, Heterocyclic compounds derived from ephedrines, Curr. Org. Chem. 8 (2004) 671–693.

[4] C.T. Ting, W.C. Li, C.Y. Chen, T.H. Tsai, Preventive and therapeutic role of traditional Chinese herbal medicine in hepatocellular carcinoma, J. Chin. Med. Assoc. 78 (2015) 139–144.

[5] K.Y. Tan, C.B. Liu, A.H. Chen, Y.J. Ding, H.Y. Jin, F. Seow-Choen, The role of traditional Chinese medicine in colorectal cancer treatment, Tech. Coloproctol. 12 (2008) 1–6.

[6] H. Rui, Recent progress of traditional Chinese medicine and herbal medicine for the treatment and prevention of cancer, Gan. To. Kagaku. Ryoho. 29 (Suppl 1) (2002) 67–75.

[7] I. Cohen, M. Tagliaferri, D. Tripathy, Traditional Chinese medicine in the treatment of breast cancer, Semi. Oncol. 29 (2002) 563–574.

[8] S.X. Zhang, X. Li, J.L. Yin, L.L. Chen, H.Q. Zhang, Effect of C21 steroidal glycoside from root of Cynanchum auriculatum on D-galactose induced aging model mice, Zhongguo. Zhong. Yao. Za. Zhi. 32 (2007) 2511–2514.

[9] Y. Zhou, R. Jiang, B. Yang, X. Yao, P. Wang, D. Liu, et al., Changes of telomere and telomerase in effect of ginsenoside Rg1 to delay hematopoietic stem cell senescence, Zhongguo. Zhong. Yao. Za. Zhi. 36 (2011) 3172–3175.

[10] H.Q. Zhang, Y. Li, Y.Y. Song, Effect of polysaccharides of Cistanche deserticola on immune cells and telomerase activity in aging mice, Chin. Pharm. J. 14 (2011) 1081–1083.

[11] Z.W. Hu, Z.Y. Shen, J.H. Huang, Experimental study on effect of Epimedium brevicornu flavonoids in protecting telomere length of senescence cells, Chin. J. Integr. Tradit. West Med. 12 (2004) 1094–1097.

[12] H.Q. Zhang, X.J. Weng, L.L. Chen, X. Li, Effect of Cistanche tubulosa (Scheuk) Whight acteoside on telomerase activity and mimunity of aging mice, Chin. J. Pharmacol. Toxicol. 4 (2008) 270–273.

[13] L. Guo, X.D. Wei, Q. Ou, S. Wang, G.M. Zhu, Effect of astragaloside on the expression of telomerase activity and klotho gene in aged HELF cells, Chin. J. Gerontol. 13 (2010) 1819–1822.

[14] L.X. Zhao, L. Yu, Pine pollen delays cell senescence and its effects on telomerase activity, Sichuan J. Tradit. Chin. Med. 4 (2004) 11–13.

[15] Z.H. Jiang, S. Ma, J.W. Chen, T. Guo, X.J. Li, M.M. Fan, et al., Resveratrol protects endothelial cells from senescence via inhibition of apoptosis, Chin. Heart J. 6 (2016) 638–641.

[16] K.T. Howitz, K.J. Bitterman, H.Y. Cohen, D.W. Lamming, S. Lavu, J.G. Wood, et al., Small molecule activators of sirtuins extend Saccharomyces cerevisiae lifespan, Nature 425 (2003) 191–196.

[17] L. Hao, X.H. Ren, Y. Zhao, C.Y. Xia, C.X. Guo, Y.C. Wang, et al., The effects of Ginkgo biloba extract on anti oxidantive DNA damage and delaying telomere shortening in prefrontal cortex of natural aging rats and its mechanisms, Pharmacol. Clin. Chin. Mater. Med. 6 (2013) 38–42.

[18] L.L. Qiao, F. Huang, X.G. Yan, H. Gong, Y. Li, Effect of Rhizoma Coptidis apozem on expression of AMP-activated protein kinase in skeletal muscle of metabolic syndrome rats, Chin. J. Tradit. Chin. Med. Pharm. 1 (2010) 145–148.

[19] Y.F. Zhou, G.H. Zhang, H. Wang, A study on 6-gingerol attenuate vascular smooth muscle cells senescence through inhibition of

mTOR pathway molecular, Chongqing Med. 14 (2014) 1687–1689.

[20] P. Peng, Z.M. Song, Y. Liu, B.S. Hao, S.J. Yu, B. Zhou, et al., Effect of ginsenoside Rb1 on the brain aging of mouse and mTOR/p70S6K pathway, J. Sun Yat-sen Univ. (Med. Sci.) 36 (2015) 176–180.

[21] S. Salvioli, E. Sikora, E.L. Cooper, C. Franceschi, Curcumin in cell death processes: a challenge for CAM of age-related pathologies, Evid. Based Complement. Alternat. Med. 4 (2007) 181–190.

[22] G.P. Lim, T. Chu, F. Yang, W. Beech, S.A. Frautschy, G.M. Cole, The curry spice curcumin reduces oxidative damage and amyloid pathology in an Alzheimer transgenic mouse, J. Neurosci. 21 (2001) 8370–8377.

[23] W.F. Lai, A.S. Susha, A.L. Rogach, Multicompartment microgel beads for co-delivery of multiple drugs at individual release rates, ACS Appl. Mater. Interfaces 8 (2016) 871–880.

[24] W.F. Lai, H.C. Shum, A stimuli-responsive nanoparticulate system using poly(ethylenimine)-graft-polysorbate for controlled protein release, Nanoscale 8 (2016) 517–528.

[25] W.F. Lai, H.C. Shum, Hypromellose-graft-chitosan and Its polyelectrolyte complex as novel systems for sustained drug delivery, ACS Appl. Mater. Interfaces 7 (2015) 10501–10510.

[26] W.F. Lai, Z.D. He, Design and fabrication of hydrogel-based nanoparticulate systems for in vivo drug delivery, J. Control Release 243 (2016) 269–282.

[27] G.M. Gelfuso, M.S. Cunha-Filho, T. Gratieri, Nanostructured lipid carriers for targeting drug delivery to the epidermal layer, Ther. Deliv. 7 (2016) 735–737.

[28] P. Kidd, K. Head, A review of the bioavailability and clinical efficacy of milk thistle phytosome: a silybin-phosphatidylcholine complex (Siliphos), Altern. Med. Rev. 10 (2005) 193–203.

[29] B. Hazra, B. Kumar, S. Biswas, B.N. Pandey, K.P. Mishra, Enhancement of the tumour inhibitory activity, in vivo, of diospyrin, a plant-derived quinonoid, through liposomal encapsulation, Toxicol. Lett. 157 (2005) 109–117.

[30] A. Kumari, S.K. Yadav, Y.B. Pakade, V. Kumar, B. Singh, A. Chaudhary, et al., Nanoencapsulation and characterization of Albizia chinensis isolated antioxidant quercitrin on PLA nanoparticles, Colloids Surf. B Biointerfaces 82 (2011) 224–232.

[31] F.T. Vicentini, T.R. Simi, J.O. Del Ciampo, N.O. Wolga, D.L. Pitol, M.M. Iyomasa, et al., Quercetin in w/o microemulsion: in vitro and in vivo skin penetration and efficacy against UVB-induced skin damages evaluated in vivo, Eur. J. Pharm. Biopharm. 69 (2008) 948–957.

[32] C.H. Zhang, C.Y. Jia, W. Li, F.P. Zhao, Q.N. Yu, J.Y. Yu, et al., Preparation of Sanqi Hydrogel Patch used for setting a bone and study on its transdermal permeability in vitro, Chin. Tradit. Herb. Drugs 46 (2015) 654–664.

[33] K. Raemdonck, J. Demeester, S. De Smedt, Advanced nanogel engineering for drug delivery, Soft Matter 5 (2009) 707–715.

[34] Z. Lin, W. Gao, H. Hu, K. Ma, B. He, W. Dai, et al., Novel thermo-sensitive hydrogel system with paclitaxel nanocrystals: high drug-loading, sustained drug release and extended local retention guaranteeing better efficacy and lower toxicity, J. Control Release 174 (2014) 161–170.

[35] Y.-M. Lim, S.-J. An, H.-K. Kim, Y.-H. Kim, M.-H. Youn, H.-J. Gwon, et al., Preparation of hydrogels for atopic dermatitis containing natural herbal extracts by gamma-ray irradiation, Radiat. Phys. Chem. 78 (2009) 441–444.

[36] L.C. Chiang, J.S. Chang, C.C. Chen, L.T. Ng, C.C. Lin, Anti-Herpes simplex virus activity of Bidens pilosa and Houttuynia cordata, Am. J. Chin. Med. 31 (2003) 355–362.

[37] C. Li, Y. Zhao, C. Liang, H. An, Observations of the curative effect with various liquid for post operative irrigation of ESS of treating chronic sinusitis and nasal polyps, Lin Chuang Er Bi Yan Hou Ke Za Zhi 15 (2001) 53–54.

[38] S.J. Lee, Korean Folk Medicine, Monographs Series No. 3, Seoul National University Press, Korea, 1966.

[39] W.K. Fang, F.Y. Ko, H.L. Wang, C.H. Kuo, L.M. Chen, F.J. Tsai, et al., The proliferation and migration effects of huangqi on RSC96 Schwann cells, The Am. J. Chin. Med 37 (2009) 945–959.

[40] L. Fengqin, W. Yulin, Z. Xiaoxin, J. Youpeng, C. Yan, W. Qingqing, et al., The heart-protective mechanism of Qishaowuwei formula on murine viral myocarditis induced by CVB3, J. Ethnopharm. 127 (2010) 221–228.

[41] X.R. Dong, J.N. Wang, L. Liu, X. Chen, M.S. Chen, J. Chen, et al., Modulation of radiation-induced tumour necrosis factor-alpha and transforming growth factor beta1 expression in the lung tissue by Shengqi Fuzheng injection, Mol. Med. Rep. 3 (2010) 621–627.

[42] H. Chen, X. Chang, D. Du, W. Liu, J. Liu, T. Weng, et al., Podophyllotoxin-loaded solid lipid nanoparticles for epidermal targeting, J. Control Release 110 (2006) 296–306.

[43] M. Uner, G. Yener, Importance of solid lipid nanoparticles (SLN) in various administration routes and future perspectives, Int. J. Nanomed. 2 (2007) 289–300.

[44] X. Chen, L.H. Peng, Y.H. Shan, N. Li, W. Wei, L. Yu, et al., Astragaloside IV-loaded nanoparticle-enriched hydrogel induces wound healing and anti-scar activity through topical delivery, Int. J. Pharm. 447 (2013) 171–181.

[45] E. Rivera-Arce, M.A. Chavez-Soto, A. Herrera-Arellano, S. Arzate, J. Aguero, I.A. Feria-Romero, et al., Therapeutic effectiveness of a Mimosa tenuiflora cortex extract in venous leg ulceration treatment, J. Ethnopharm. 109 (2007) 523–528.

[46] S.J. Buwalda, K.W. Boere, P.J. Dijkstra, J. Feijen, T. Vermonden, W.E. Hennink, Hydrogels in a historical perspective: from simple networks to smart materials, J. Control Release 190 (2014) 254–273.

[47] A. Freeman, Y. Aharonowitz, Immobilization of microbial-cells in crosslinked, pre-polymerized, linear polyacrylamide gels-antibiotic production by immobilized streptomyces-clavuligerus cells, Biotechnol. Bioeng. 23 (1981) 2747–2759.

[48] G.P. Hicks, S.J. Updike, The preparation and characterization of lyophilized polyacrylamide enzyme gels for chemical analysis, Anal. Chem. 38 (1966) 726–730.

[49] Y. Tang, C.L. Heaysman, S. Willis, A.L. Lewis, Physical hydrogels with self-assembled nanostructures as drug delivery systems, Expert. Opin. Drug. Deliv. 8 (2011) 1141–1159.

[50] A. Subramoniam, D.A. Evans, S. Rajasekharan, G. Sreekandan Nair, Effect of Hemigraphis colorata (Blume) H. G Hallier on wound healing and inflammation in mice, Indian J. Pharmacol. 33 (2001) 283–285.

[51] M. Annapoorna, P.T. Sudheesh Kumar, L.R. Lakshman, V.K. Lakshmanan, S.V. Nair, R. Jayakumar, Biochemical properties of Hemigraphis alternata incorporated chitosan hydrogel scaffold, Carbohydr. Polym. 92 (2013) 1561–1565.

[52] S.J. de Jong, B. van Eerdenbrugh, C.F. van Nostrum, J.J. Kettenes-vande Bosch, W.E. Hennink, Physically crosslinked

dextran hydrogels by stereocomplex formation of lactic acid oligomers: degradation and protein release behavior, J. Control Release 71 (2001) 261–275.

[53] M. Peng, S. Xu, Y. Zhang, L. Zhang, B. Huang, S. Fu, et al., Thermosensitive injectable hydrogel enhances the antitumor effect of embelin in mouse hepatocellular carcinoma, J. Pharm. Sci. 103 (2014) 965–973.

[54] G.C. Jagetia, B.B. Aggarwal, "Spicing up" of the immune system by curcumin, J. Clin. Immunol. 27 (2007) 19–35.

[55] D. Gopinath, M.R. Ahmed, K. Gomathi, K. Chitra, P.K. Sehgal, R. Jayakumar, Dermal wound healing processes with curcumin incorporated collagen films, Biomaterials 25 (2004) 1911–1917.

[56] G.C. Jagetia, G.K. Rajanikant, Role of curcumin, a naturally occurring phenolic compound of turmeric in accelerating the repair of excision wound, in mice whole-body exposed to various doses of gamma-radiation, J. Surg. Res. 120 (2004) 127–138.

[57] P. Anand, A.B. Kunnumakkara, R.A. Newman, B.B. Aggarwal, Bioavailability of curcumin: problems and promises, Mol. Pharm. 4 (2007) 807–818.

[58] A. Safavy, K.P. Raisch, S. Mantena, L.L. Sanford, S.W. Sham, N. R. Krishna, et al., Design and development of water-soluble curcumin conjugates as potential anticancer agents, J. Med. Chem. 50 (2007) 6284–6288.

[59] C. Gong, Q. Wu, Y. Wang, D. Zhang, F. Luo, X. Zhao, et al., A biodegradable hydrogel system containing curcumin encapsulated in micelles for cutaneous wound healing, Biomaterials 34 (2013) 6377–6387.

[60] E.-S. Chan, Z.-H. Yim, S.-H. Phan, R.F. Mansa, P. Ravindra, Encapsulation of herbal aqueous extract through absorption with ca-alginate hydrogel beads, Food Bioprod. Process 88 (2010) 195–201.

[61] S. Venkatesh, S.P. Sizemore, M.E. Byrne, Biomimetic hydrogels for enhanced loading and extended release of ocular therapeutics, Biomaterials 28 (2007) 717–724.

[62] W. Wang, P.C. Hui, E. Wat, F.S. Ng, C.W. Kan, X. Wang, et al., In vitro drug release and percutaneous behavior of poloxamer-based hydrogel formulation containing traditional Chinese medicine, Colloids Surf. B Biointerfaces 148 (2016) 526–532.

[63] Y. Liu, L. Dong, Basic Theories of Traditional Chinese Medicine, Academic Press, Beijing, 1998.

[64] T.P. Lam, Strengths and weaknesses of traditional Chinese medicine and Western medicine in the eyes of some Hong Kong Chinese, J. Epidemiol. Community Health 55 (2001) 762–765.

[65] L. Wei, R. Ye, X. Chen, Clinical observation of elderly idiopathic nephrotic syndrome treated with integrated traditional Chinese and Western medicine, Zhongguo. Zhong. Yao. Za. Zhi. 20 (2000) 99–101.

[66] Q. Yao, S.W. Zhang, H. Wang, A.M. Ren, A. Li, B.E. Wang, Treatment of severe post-kidney-transplant lung infection by integrative Chinese and Western medicine, Chin. J. Integr. Med. 12 (2006) 55–58.

[67] A.A. Izzo, Herb-drug interactions: an overview of the clinical evidence, Fundam. Clin. Pharm. 19 (2005) 1–16.

[68] L.J. Dai, J. Hou, H.R. Cai, Experimental study on treatment of pulmonary fibrosis by Chinese drugs and integrative Chinese and Western medicine, Zhongguo. Zhong. Xi. Yi. Jie. He. Za. Zhi. 24 (2004) 130–132.

[69] L. Lammoglia-Ordiales, M.E. Vega-Memije, A. Herrera-Arellano, E. Rivera-Arce, J. Aguero, F. Vargas-Martinez, et al., A randomised comparative trial on the use of a hydrogel with tepescohuite extract (Mimosa tenuiflora cortex extract-2G) in the treatment of venous leg ulcers, Int. Wound. J. 9 (2012) 412–418.

Part IV

From Interventions to Practice

Chapter 14

Biological barriers to cellular interventions

Introduction

Therapeutics delivery at the cellular level is a multistage process [1]. Multiple obstacles have to be overcome before the delivered agent can reach an appropriate site for action. The exact location of the "appropriate site" is decided by the nature of the agent to be delivered. If small-molecule compounds (especially those initiating immunotherapeutic effects) are involved, they may simply interact with cell surface receptors to elicit action. The therapeutic effect of these compounds, therefore, will still be observed even if the carrier releases the delivered agent outside target cells. If nucleic acids are involved, that will, however, be a different story.

Nucleic acids have to get into target cells, or the manipulation of the target gene will hardly be achieved [2–4]. For RNA molecules, the cytosol is the site of action. For plasmids, successful nuclear translocation is necessary. With a special focus on the case of polymer-mediated gene delivery, this chapter will introduce different obstacles a carrier will encounter during intervention administration at the cellular level (Fig. 14.1). Thorough understanding of these cellular impediments is pivotal to the subsequent optimization of the performance of a developed carrier. Apart from presenting the cellular obstacles that have been limiting the efficiency of existing carriers in transfection, some of the strategies to enhance the performance of nonviral carriers will be highlighted for future intervention development to combat aging.

Strategies to enhance cellular internalization and endolysosomal escape

In general, lipoplexes and polyplexes enter cells through endocytosis that can occur via two pathways: clathrin-dependent pathway and clathrin-independent pathway. Lipoplexes usually enter cells via the clathrin-dependent pathway; however, when they are coated with serum albumin, they tend to be internalized via the caveolar pathway. The internalization of polyplexes, on the contrary, is determined by the cell type and the nature of the polymer [5]. In HepG2 cells, the cellular uptake of PEI polyplexes is mediated by the clathrin-dependent pathway [6], but in HeLa cells, those polyplexes tend to be internalized via the caveolar pathway. In addition, the size of the polyplexes partially decides which endocytotic mechanism the polyplexes undertake. Polyplexes with an average diameter of 200 nm tend to follow the clathrin-dependent pathway; whereas those with a diameter larger than 500 nm are more likely to be internalized via the clathrin-independent mechanism [7,8]. Despite this, there is no clear correlation between the choice of the pathway and the efficiency in transfection.

To facilitate the process of cellular uptake, modulation of the size and zeta potential of the polyplexes is a possible way. Alternatively, ligands can be incorporated into the carrier so that the carrier can bind to specific cell surface receptors to elicit receptor-mediated endocytosis. One commonly used ligand is transferrin, which is an iron-binding and iron-transport protein that can serve as a targeting moiety toward different cancer cell lines (including those of ovarian cancer, glioblastoma, and colon cancer) [9]. By conjugating transferrin to the PEG-adamantane (PEG-AD) conjugate, the product obtained has been found to be able to coat the surface, via inclusion complexation between AD and the CD moieties on the particle surface, of DNA nanoparticles generated from the linear imidazole-conjugated β-CD-based polymer. In K562 leukemia cells, the transfection efficiency of the coated nanoparticles has been shown to be fourfold higher than that of the uncoated ones [10]. This has demonstrated the feasibility of using ligand conjugation to enhance the performance of polymeric vectors in gene transfer. Other ligands adopted for cellular targeting, particularly targeting to tumor cells, include PSMA-specific monoclonal antibody J591, anti-CD3 antibody, OV-TL16, HER-2 antibody, GRP-78 targeting peptide, RGD peptide, lactose, mannose, and folic acid. Despite this, it is worth noting that receptors usually transfer their cargoes to lysosomes. The capability of the delivered agent to escape from lysosomes is crucial to transfection.

Low efficiency in endolysosomal escape is common in nonviral vectors. To facilitate endolysosomal escape, enhancing the proton sponge capacity of the vector can

Delivery of Therapeutics for Biogerontological Interventions. DOI: https://doi.org/10.1016/B978-0-12-816485-3.00014-3

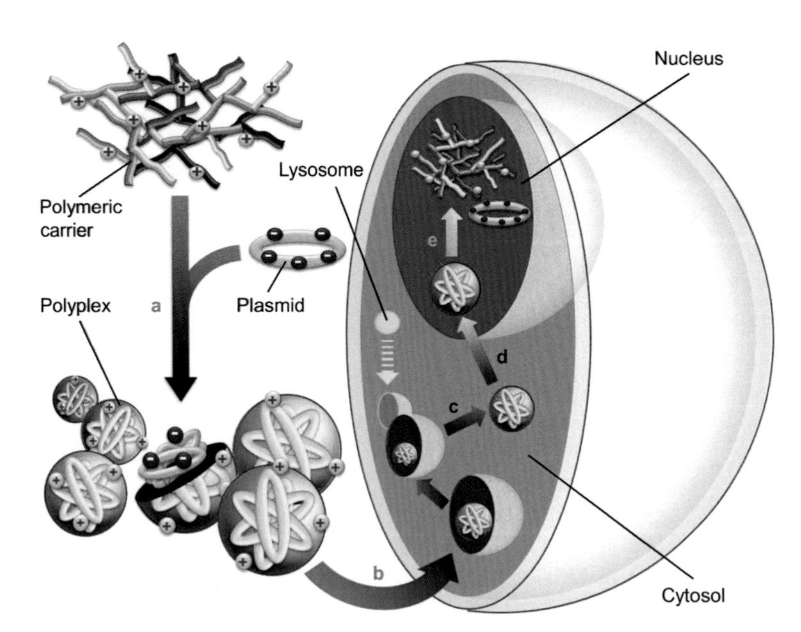

FIGURE 14.1 Overview of major processes involved in the intracellular delivery of nucleic acids mediated by a polymeric gene carrier: (a) electrostatic interactions, (b) cellular internalization, (c) endolysosomal escape, (d) nuclear translocation, and (e) polyplex dissociation. *Reproduced from W.F. Lai, W.T. Wong, Design of polymeric gene carriers for effective intracellular delivery, Trends Biotechnol. 36 (2018) 713–728 with permission from Elsevier B.V., [1]*

help. This can be achieved by incorporating imidazole groups or unprotonated amine groups into the vector. These groups can be protonated in endosomes, leading to endosomal swelling and rupture [11]. Apart from these groups, pH-sensitive fusogenic peptides may be incorporated into the carrier. Examples of these peptides include ppTG20, ppTG1, penetratin, transportan, trans-activating transcriptional activator peptide, and KALA. Many of these peptides can undergo conformational changes under mildly acidic conditions, and can improve the transfection efficiency of nonviral vectors [12–14]. Recently, melittin, a lytic peptide extracted from honey bee venom, has emerged as a favorable ligand for vector modification because of its ability to induce endosomal membrane dissociation. Cells, however, can hardly withstand the damage caused by melittin to cytoplasmic membranes. To address this problem, dimethylmaleic anhydride has been adopted to cover the cationic charge of melittin so that the lytic activity can be hindered at physiological pH but can be maintained at pH 5 [15].

Outstanding questions for clinical translation

1. How can the stability of a peptide-conjugated carrier be enhanced? Peptides are one of the important classes of ligands that can enhance the target specificity of a carrier. Because of the susceptibility of peptides to degradation, the stability of the carrier is often reduced after peptide conjugation. Methods of enhancing or maintaining the stability of the peptide are important when a peptide is adopted for carrier modification.
2. How can the endolysosomal membrane be disrupted in a way that does not increase cytotoxicity? Membrane disruption and cytotoxicity are closely related to each other. This is because the mechanism used to disrupt the endolysosomal membrane may cause damage to other important membranes (e.g., plasma membrane and nuclear membrane). If the damage is too severe, cell lysis occurs. Methods of eliciting the membrane-disrupting activity in a temporal-spatial manner may help solve the cytotoxicity problem.

(Continued)

(Continued)

Challenges in cytosolic and nuclear transport

Apart from endolysosomal escape, cytosolic transport, in which cytoskeletal components play an important role, is another barrier to be overcome. If the delivered agent is RNA, reaching the cytosol may be sufficient; however, effective delivery to the nucleus is required if the delivered agent is a plasmid. The situation is complicated by the fact that the transfer of the plasmid to the nucleus is often compromised by the random binding of DNA with keratin, actin, and vimentin [16–18]. Because of this, DNA delivery is, in general, more technically challenging than RNA transfer, with the efficiency in nuclear import being one of the impediments to effective transfection.

As far as nuclear import is concerned, it is mediated predominately by the NPC [19–21]. Structurally, an NPC is a structural component of the nuclear envelope (Fig. 14.2) [22]. It comprises multiple copies of around 30 different nucleoporins (Nups), and its mass is approximately 15 times that of a ribosome. Each NPC is a supramolecular assembly that contains the nucleoplasmic part (i.e., the nuclear ring and nuclear baskets), the cytoplasmic part (i.e., the cytoplasmic ring and cytoplasmic filaments), and the central structure [23–25].

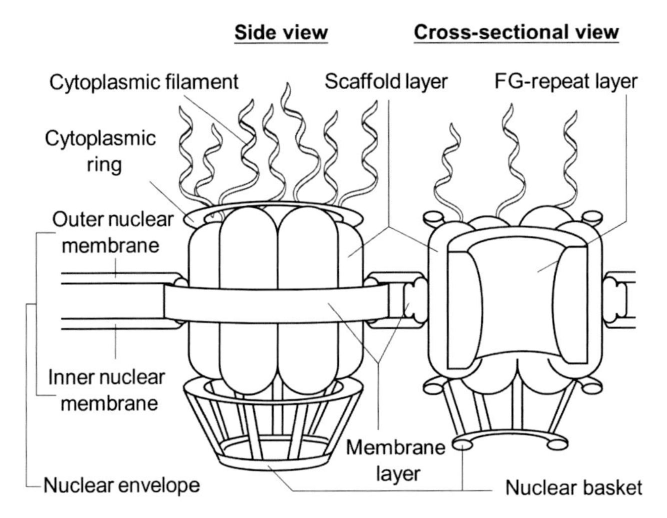

FIGURE 14.2 A schematic illustration of the NPC structure. *Reproduced from W.F. Lai, W.T. Wong, Design of polymeric gene carriers for effective intracellular delivery, Trends Biotechnol. 36 (2018) 713–728 with permission from Elsevier B.V., [1]*

The outermost part of an NPC consists of transmembrane Nups that can anchor the complex to the nuclear envelope, whereas the innermost part contains phenylalanine-glycine (FG)-repeat Nups, which also form part of the peripheral structure that extends from the channel to the nuclear and cytoplasmic environments. Because the NPC is a key player in numerous cellular processes (e.g., cell cycle regulation, basal replication, and gene expression), dysfunctions in its activity may lead to pathological manifestations [26]. For example, one mutant of the scaffold Nup, ALADIN(I482S), has been reported to compromise the karyopherin-α/β-mediated import of aprataxin and DNA ligase I, subjecting the affected cells to DNA damage [27]. Changes in the expression level of the gene for the oxalate binding protein Nup62 (an FG-Nup located at the central gated channel [28]) have also been associated with the nephrotic syndrome [29], which is a glomerular disease commonly found in the elderly population [30,31]. All these have evidenced the role played by the NPC in maintaining normal physiological processes.

In eukaryotic cells, the NPC functions as a conduit for macromolecular exchange across the nuclear envelope. Over the years, few models have been promulgated to explain its selectivity. The virtual gate model holds that the highly dense phenylalanine-glycine-repeat (FG-repeat) filaments build noncohesive, entropic bristles, which subsequently prevent nonkaryophilic molecules from passing through the NPC via Brownian motion [32]. On the other hand, according to the selective phase model, molecules passing through the NPC are sieved by size exclusion, which is caused by the presence of a selective meshwork generated by hydrophobic interactions among FG-repeat Nups [33]. Due to the presence of the meshwork, import and export receptors (collectively known as karyopherins), as well as translocation signals, have to be involved before nucleocytoplasmic transport (NCT) of macromolecules can occur [34]. Small molecules/ions, however, may still be transported across the nuclear pore via passive diffusion, as suggested by the reduction of dimensionality model [35] and the spaghetti oil model [36], if the size of those molecules/ions is below the 40–60 kDa size exclusion threshold. During NCT, the directionality of the transport has to be controlled. This can be attained by regulating the concentration gradient of the GDP- and GTP-bound forms of Ran, which is a Ras-related small guanosine triphosphatase (GTPase), between the cytoplasm and the nucleus. In the cytoplasm, the Ran GTPase activating protein (RanGAP) causes the hydrolysis of GTP, resulting in conformational changes of Ran to generate RanGDP, which is subsequently converted back to RanGTP in the nucleus under the action of the Ran guanine nucleotide exchange factor (RanGEF). The action of RanGAP and RanGEF leads to the dissociation of export and import complexes, resulting in the release of the cargo in the cytoplasm and the nucleus, respectively. Because the dynamics of NCT has been reviewed elsewhere, readers are referred to those articles for details [25,37].

To facilitate nuclear import, one approach is to incorporate the cell-penetrating peptide TAT, which can facilitate the transport of nanoparticles across the nuclear membrane, into a carrier. The feasibility of using this approach has been shown by Wu et al. [38], who have incorporated large-pore ultrasmall mesoporous organosilica nanoparticles with TAT for more effective nuclear translocation. Besides TAT, nuclear localization signal (NLS) peptides can be used. For example, an earlier study has incorporated an NLS peptide derived from the N-terminal tail of histone H3 into polyplexes to enhance interactions between the polyplexes and histone effectors [39]. The modified polyplexes have been found to preferentially accumulate in the endoplasmic reticulum (ER), and have re-distributed into the nucleus after cell division [40]. The success in using NLS peptides has also been shown in a recent study [41] in which the synthesis of a hydroxyl-terminal PAMAM dendrimer derivative (namely PAMS) has been reported. The derivative contains a β-thiopropionate bond that can degrade under acidic conditions. Upon interactions with DNA that has precomplexed with the NLS peptide of the SV40T antigen (CGGGPKKKRKVED, $M_w = 1401.7$), polyplexes targeting RanGAP1 have been obtained [41]. In RanGAP1-overexpressing cells, both the nuclear import and the transfection efficiency of the polyplexes have been shown to be enhanced [41]. A similar phenomenon has been noted in vivo, in which the transfection efficiency of the polyplexes has been found to be remarkably increased upon overexpression of *RANGAP1* in mice

[41]. Apart from targeting RanGAP1, inhibition of histone deacetylase activity may lead to stabilization of acetylate microtubules [16–18], which are cytosolic peptides needed for nuclear translocation, thereby enhancing nuclear import of polyplexes.

Finally, nuclear translocation of the delivered agent can be facilitated by structural engineering of the carrier per se. This has been documented in an earlier study [42], which has cross-linked *o*-nitrobenzyl urethane (NBU) linker molecules with low-molecular-weight PEI to generate a UV-degradable carrier. Upon UV irradiation, the carrier can produce anionic carbamic acid groups and nitrosobenzyl aldehyde groups to facilitate the polyplexes to interact with nuclear transcription proteins for enhanced transgene expression [42]. More recently, alveolar epithelial type I (ATI) cell–specific nuclear import of a plasmid has been achieved after the plasmid has been incorporated with a sequence that can interact with ATI-enriched transcription factors [43]. This has suggested the possibility to restrict transgene expression to particular cell types by engineering the sequence of a plasmid.

Outstanding questions for clinical translation

1. How can the time of polyplex dissociation be controlled? If polyplexes dissociate in the cytosol, manipulating the carrier per se might not be an effective way to enhance the efficiency of the delivered agent in nuclear import. In this case, modification should be performed on the nucleic acid to be delivered. Conversely, if polyplexes undergo nuclear import before they dissociate, manipulating the structure of the carrier may be helpful. Controlling the time of polyplex dissociation, therefore, is important.
2. How can the efficiency of a carrier be predicted by using the SAR? At this moment, carriers are usually developed by trial and error. Their performance is known only after experimental evaluation. If the performance of a carrier can be estimated in silico based on the SAR, with candidates that are less likely to succeed being able to be excluded from further experimental studies, carriers will be developed more effectively.

Reprogramming an aged cell for rejuvenation

Rejuvenation of an aged cell can be possibly achieved by nuclear transfer, which replaces a damaged nucleus with a healthy one. The success of this method, however, relies on the source of the donor cell. This has been shown in the case of somatic cell nuclear transfer (SCNT), which has been adopted for reproductive and therapeutic cloning (Fig. 14.3) [44]. SCNT involves three stages: enucleation, injection/fusion, and activation. After the removal of the nucleus from an oocyte, an exogenous nucleus is introduced. The M-phase-promoting factors (MPF) present in the ooplasm trigger the occurrence of premature chromosome condensation (PCC) [45]. This leads to the breakdown of the nuclear membrane of the donor nucleus, and causes the formation of condensed metaphase-like chromosomes. After PCC, the activation of the reconstructed oocyte occurs, with the genome of the activated donor cell entering the G1 phase and undergoing diverse processes (ranging from nuclear expansion and DNA replication to transcriptome reprogramming). Despite the plausibility of cloning animals by using SCNT, at the moment the rate of cloned animals that can reach term is far from satisfactory. Approximately 70% of SCNT embryos has been reported to undergo developmental arrest before entering the blastocyst stage [46], and only 1%–2% of embryos transferred to surrogate mothers have ultimately developed to term [46]. The impediment to successful SCNT reprogramming comes partly from the defects in somatic donor cells (including ectopic activation of Xist, aberrant DNA methylation, and H3K27me3-mediated genomic imprinting defects) [44]. The source of the donor nucleus, therefore, has to be carefully selected as it may affect the functioning of the reconstructed cell.

Apart from the selection of the cell source, right now nuclear transfer is performed manually and physically. Chemical methods of transferring an exogenous nucleus into a recipient cell are absent. Owing to the fact that there are approximately 4×10^{13} cells in a human body [47], transferring an exogenous nucleus to each of these cells using the prevalent physical method is not possible to succeed. In addition, after the introduction of an exogenous nucleus into an aged cell, the original nucleus has to be removed. In cloning, the commonly used method of removing a nucleus from a cell is the "blind" mechanical removal of metaphase II chromosomes [48]. This method, as well as few alternative methods that have been reported recently (e.g., centrifugation of zona-free mature bovine oocytes in a Percoll density gradient [49], or cell treatment with antimitotic agents such as demecolcine and nocodazole [50]), can be executed only in vitro. Applications in the in vivo context are almost impossible. Development of strategies to avoid the coexistence of multiple nuclei in one cell is required before rejuvenation of body cells can be achieved by nuclear transfer.

Finally, considering the fact that SCNT-reconstructed oocytes show totipotent properties but encounter epigenetic barriers to successful reprogramming [44], it is likely that the outcome of nuclear transfer is determined by the interactions between the exogenous nucleus and the cytoplast. This has been partially evidenced in an earlier study [51], which has adopted porcine embryonic fibroblasts (PEF), adult porcine ear skin fibroblasts (APEF), and adipose-derived stem cells (ADSC) as donor cells of exogenous nuclei. The total number of blastocysts

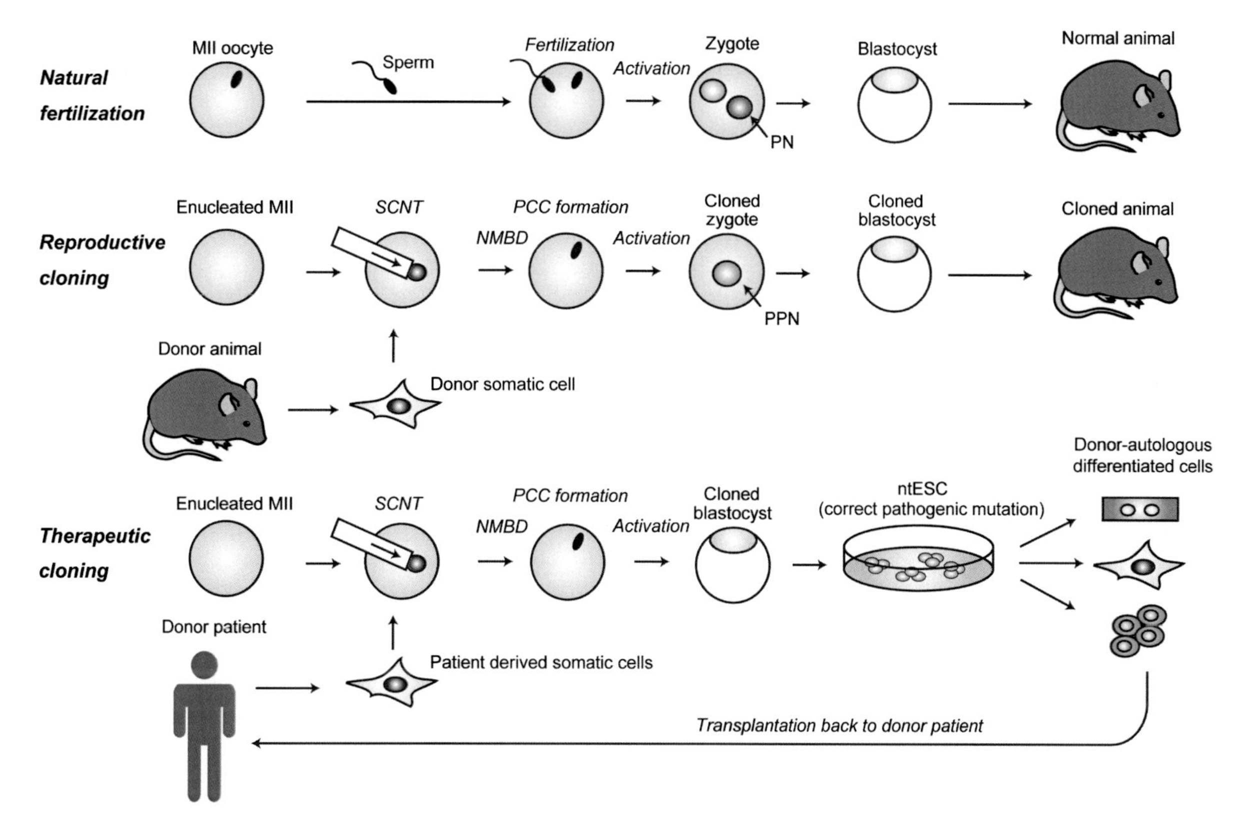

FIGURE 14.3 Procedures for natural fertilization, reproductive cloning, and therapeutic cloning. *MII*, metaphase II; *PN*, pronuclei; *NMBD*, nuclear membrane breakdown; *PCC*, premature chromosome condensation; *ntESCs*, embryonic stem cells derived from the blastocysts of nuclear transferred embryos. *Reproduced from S. Matoba, Y. Zhang, Somatic cell nuclear transfer reprogramming: mechanisms and applications, Cell Stem Cell 23 (2018) 471–485 with permission from Elsevier B.V., [44]*

generated from the reconstructed cells in the APEF group has been found to be lower than that in the PEF and ADSC groups, suggesting the impact of the donor cell type on the success of nuclear transfer. In addition, organelles (especially mitochondria) other than the nucleus can be damaged during aging. The damage may account for the malfunctioning of the reconstructed cell. Unfortunately, similar to the case of nuclear transfer, organelle transfer right now is largely performed in a manual manner [52–54] and can hardly be applied in vivo. To rejuvenate aged cells bodywide by replacing damaged nuclei and organelles with healthy ones, chemical methods that can be effectively used in a living body have to be first developed. More studies are also needed to examine the effect of the cytoplast from an aged cell on the functioning of the reconstructed cell.

Instead of rejuvenating all cells bodywide, one may consider rejuvenating only stem cells in aged tissues. The rationale for this is that, although somatic cells experience cumulative intrinsic alternations (e.g., telomere attrition, oxidative damage, and mitochondrial dysfunction) during aging, postnatal stem cells in organs, in general, maintain their regenerative capacity throughout life [55]. The regenerative capacity of aged stem cells can be possibly rescued after manipulation of biochemical cues at the cellular or molecular levels [56–61]. To achieve this goal, manipulation of the genome of stem cells can be applied; however, nonviral carriers that can effectively deliver nucleic acids to stem cells are lacking at the moment. Apart from genetic manipulation, injecting systemic factors present in young serum to an aged individual may help restore, at least partially, the functioning of aged cells. This has been documented in an earlier study [62], which has established parabiotic pairings between old and young mice. Upon heterochronic parabiosis, the regenerative capacity of satellite cells, as well as the proliferative activity of hepatocytes, in old mice has been restored [62]. The possibility of changing the aging phenotype via exposure to the microenvironment from the youth has also been shown in a recent study [63], which has cocultured young endothelial cells with aged hematopoietic stem and progenitor cells (HSPCs) and has successfully restored the repopulating activity of the aged stem cells. Despite this, various studies have suggested that stem cells from young tissues can be inhibited by the systemic milieu of an aged body [64–66]. This implies that even though an aged stem cell can be rejuvenated, the systemic milieu of an aged body may still impair the functioning of the rejuvenated cell. Detailed studies on the physiology of rejuvenated cells and the long-term effects of the rejuvenation intervention, particularly the influence on the overall life span, are required before stem cell rejuvenation can be an option for the design and development of a biogerontological intervention.

Outstanding questions for clinical translation

1. How can exogenous organelles be transferred to aged cells bodywide, followed by the removal of redundant organelles? Right now organelle transfer (particularly nuclear and mitochondrial transfer) and organelle removal rely mainly on physical means. These methods can hardly be applied in vivo.
2. How does the systemic milieu of an aged body affect the functioning of rejuvenated cells? Even though an aged cell can be rejuvenated, the functioning of the rejuvenated cell may still be impaired by the systemic milieu of the aged body. Understanding the mechanism behind can provide clues to improve the efficiency of the rejuvenation intervention.

Summary

Compared to the viral counterparts, nonviral carriers display lower immunogenicity and pathogenicity. Intracellular delivery of these carriers, however, is a multistage process, with each stage involving specific barriers to be overcome. In this chapter, we have highlighted some of the barriers to interventions at the cellular level, and have suggested solutions for carrier optimization. We have also discussed the possibility of rejuvenating aged cells by replacing damaged nuclei and organelles with exogenous ones. In the next chapter, we will continue to discuss the impediments to the execution of an antiaging intervention, but will focus on those at the organismal level. It is hoped that, with a better understanding of the barriers involved, more rational design of delivery systems can be made available to facilitate the process of bench-to-clinic translation.

Directions for intervention development

There are barriers to different stages, ranging from cellular internalization to nuclear translocation, of the delivery process at the cellular level. Knowledge of these barriers can help guide carrier design and can facilitate the process of carrier optimization. This can be done by following the steps below:

1. Determine the subcellular region that is expected to be the site of action of the delivered agent.
2. List major barriers that limit the efficiency of the delivery process.

(Continued)

(Continued)

3. Select appropriate strategies to overcome those barriers.
4. Modify the carrier, if necessary, to enhance the delivery performance.
5. Evaluate and optimize the performance of the modified carrier in vitro and in vivo.

References

[1] W.F. Lai, W.T. Wong, Design of polymeric gene carriers for effective intracellular delivery, Trends Biotechnol. 36 (2018) 713–728.
[2] W.F. Lai, Z.D. He, Design and fabrication of hydrogel-based nanoparticulate systems for in vivo drug delivery, J. Control Release 243 (2016) 269–282.
[3] W.F. Lai, Nucleic acid delivery: roles in biogerontological interventions, Ageing Res. Rev. 12 (2013) 310–315.
[4] W.F. Lai, Nucleic acid therapy for lifespan prolongation: present and future, J. Biosci. 36 (2011) 725–729.
[5] P. Midoux, G. Breuzard, J.P. Gomez, C. Pichon, Olymer-based gene delivery: a current review on the uptake and intracellular trafficking of polyplexes, Curr. Gene Ther. 8 (2008) 335–352.
[6] K. von Gersdorff, N.N. Sanders, R. Vandenbroucke, S.C. De Smedt, E. Wagner, M. Ogris, The internalization route resulting in successful gene expression depends on polyethylenimine both cell line and polyplex type, Mol. Ther. 14 (2006) 745–753.
[7] S. Grosse, Y. Aron, G. Thevenot, D. Francois, M. Monsigny, I. Fajac, Potocytosis and cellular exit of complexes as cellular pathways for gene delivery by polycations, J. Gene Med. 7 (2005) 1275–1286.
[8] J. Rejman, V. Oberle, I.S. Zuhorn, D. Hoekstra, Size-dependent internalization of particles via the pathways of clathrin-and caveolae-mediated endocytosis, Biochem. J. 377 (2004) 159–169.
[9] A. Calzolari, I. Oliviero, S. Deaglio, G. Mariani, M. Biffoni, N.M. Sposi, et al., Transferrin receptor 2 is frequently expressed in human cancer cell lines, Blood Cells Mol. Dis. 39 (2007) 82–91.
[10] N.C. Bellocq, S.H. Pun, G.S. Jensen, M.E. Davis, Transferrin-containing, cyclodextrin polymer-based particles for tumor-targeted gene delivery, Bioconj. Chem. 14 (2003) 1122–1132.
[11] W.F. Lai, In vivo nucleic acid delivery with PEI and its derivatives: current status and perspectives, Expert Rev. Med. Devices 8 (2011) 173–185.
[12] P. Midoux, C. Pichon, J.J. Yaouanc, P.A. Jaffres, Chemical vectors for gene delivery: a current review on polymers, peptides and lipids containing histidine or imidazole as nucleic acids carriers, Brir. J. Pharmacol. 157 (2009) 166–178.
[13] E. Wagner, Application of membrane-active peptides for nonviral gene delivery, Adv. Drug Deliv. Rev. 38 (1999) 279–289.
[14] T. Kakudo, S. Chaki, S. Futaki, I. Nakase, K. Akaji, T. Kawakami, et al., Transferrin-modified liposomes equipped with a pH-sensitive fusogenic peptide: an artificial viral-like delivery system, Biochemistry 43 (2004) 5618–5628.
[15] M. Meyer, A. Zintchenko, M. Ogris, E. Wagner, A dimethylmaleic acid-melittin-polylysine conjugate with reduced toxicity, pH-triggered endosomolytic activity and enhanced gene transfer potential, J. Gene Med. 9 (2007) 797–805.

[16] S. Grosse, Y. Aron, G. Thevenot, M. Monsigny, I. Fajac, Cytoskeletal involvement in the cellular trafficking of plasmid/PEI derivative complexes, J. Control Release 122 (2007) 111–117.
[17] E.E. Vaughan, R.C. Geiger, A.M. Miller, P.L. Loh-Marley, T. Suzuki, N. Miyata, et al., Microtubule acetylation through HDAC6 inhibition results in increased transfection efficiency, Mol. Ther. 16 (2008) 1841–1847.
[18] E.E. Vaughan, D.A. Dean, Intracellular trafficking of plasmids during transfection is mediated by microtubules, Mol. Ther. 13 (2006) 422–428.
[19] G. Kabachinski, T.U. Schwartz, The nuclear pore complex – structure and function at a glance, J. Cell Sci. 128 (2015) 423–429.
[20] M. Iwamoto, Y. Hiraoka, T. Haraguchi, The nuclear pore complex acts as a master switch for nuclear and cell differentiation, Commun. Integr. Biol. 8 (2015) e1056950.
[21] M. Eibauer, M. Pellanda, Y. Turgay, A. Dubrovsky, A. Wild, O. Medalia, Structure and gating of the nuclear pore complex, Nat. Commun. 6 (2015) 7532.
[22] C. Li, A. Goryaynov, W. Yang, The selective permeability barrier in the nuclear pore complex, Nucleus 7 (2016) 430–446.
[23] C.P. Lusk, G. Blobel, M.C. King, Highway to the inner nuclear membrane: rules for the road, Nat. Rev. Mol. Cell Biol. 8 (2007) 414–420.
[24] S. Walde, R.H. Kehlenbach, The part and the whole: functions of nucleoporins in nucleocytoplasmic transport, Trends Cell Biol. 20 (2010) 461–469.
[25] T. Jamali, Y. Jamali, M. Mehrbod, M.R. Mofrad, Nuclear pore complex: biochemistry and biophysics of nucleocytoplasmic transport in health and disease, Int. Rev. Cell Mol. Biol. 287 (2011) 233–286.
[26] C.M. Doucet, M.W. Hetzer, Nuclear pore biogenesis into an intact nuclear envelope, Chromosoma 119 (2010) 469–477.
[27] M. Hirano, Y. Furiya, H. Asai, A. Yasui, S. Ueno, ALADINI482S causes selective failure of nuclear protein import and hypersensitivity to oxidative stress in triple A syndrome, Proc. Natl Acad. Sci. U.S.A. 103 (2006) 2298–2303.
[28] T. Guan, S. Muller, G. Klier, N. Pante, J.M. Blevitt, M. Haner, et al., Structural analysis of the p62 complex, an assembly of O-linked glycoproteins that localizes near the central gated channel of the nuclear pore complex, Mol. Biol. Cell 6 (1995) 1591–1603.
[29] P. Sivakamasundari, P. Kalaiselvi, R. Sakthivel, R. Selvam, P. Varalakshmi, Nuclear pore complex oxalate binding protein p62: expression in different kidney disorders, Clin. Chim. Acta 347 (2004) 111–119.
[30] H.E. Yoon, M.J. Shin, Y.S. Kim, B.S. Choi, B.S. Kim, Y.J. Choi, et al., Clinical impact of renal biopsy on outcomes in elderly patients with nephrotic syndrome, Nephron Clin. Pract. 117 (2011) c20–c27.
[31] J. Prakash, A.K. Singh, R.K. Saxena, Usha, Glomerular diseases in the elderly in India, Int. Urol. Nephrol. 35 (2003) 283–288.
[32] M.P. Rout, J.D. Aitchison, M.O. Magnasco, B.T. Chait, Virtual gating and nuclear transport: the hole picture, Trends Cell Biol. 13 (2003) 622–628.
[33] K. Ribbeck, D. Gorlich, Kinetic analysis of translocation through nuclear pore complexes, EMBO J. 20 (2001) 1320–1330.
[34] E.J. Tran, S.R. Wente, Dynamic nuclear pore complexes: life on the edge, Cell 125 (2006) 1041–1053.
[35] R. Peters, Translocation through the nuclear pore complex: selectivity and speed by reduction-of-dimensionality, Traffic 6 (2005) 421–427.
[36] I.G. Macara, Transport into and out of the nucleus, Microbiol. Mol. Biol. Rev. 65 (2001) 570–594.
[37] A. Hoelz, E.W. Debler, G. Blobel, The structure of the nuclear pore complex, Annu. Rev. Biochem. 80 (2011) 613–643.
[38] M. Wu, Q. Meng, Y. Chen, Y. Du, L. Zhang, Y. Li, et al., Large-pore ultrasmall mesoporous organosilica nanoparticles: micelle/precursor co-templating assembly and nuclear-targeted gene delivery, Adv. Mater. 27 (2015) 215–222.
[39] N.L. Ross, M.O. Sullivan, Importin-4 regulates gene delivery by enhancing nuclear retention and chromatin deposition by polyplexes, Mol. Pharm. 12 (2015) 4488–4497.
[40] N.L. Ross, E.V. Munsell, C. Sabanayagam, M.O. Sullivan, Histone-targeted polyplexes avoid endosomal escape and enter the nucleus during postmitotic redistribution of ER membranes, Mol. Ther. Nucleic Acids 4 (2015) e226.
[41] K. Chen, L. Guo, J. Zhang, Q. Chen, K. Wang, C. Li, et al., A gene delivery system containing nuclear localization signal: increased nucleus import and transfection efficiency with the assistance of RanGAP1, Acta Biomater. 48 (2017) 215–226.
[42] H. Lee, Y. Kim, P.G. Schweickert, S.F. Konieczny, Y.Y. Won, A photo-degradable gene delivery system for enhanced nuclear gene transcription, Biomaterials 35 (2014) 1040–1049.
[43] L. Gottfried, X. Lin, M. Barravecchia, D.A. Dean, Identification of an alveolar type I epithelial cell-specific DNA nuclear import sequence for gene delivery, Gene Ther. 23 (2016) 734–742.
[44] S. Matoba, Y. Zhang, Somatic cell nuclear transfer reprogramming: mechanisms and applications, Cell Stem Cell 23 (2018) 471–485.
[45] K.H. Campbell, P. Loi, P.J. Otaegui, I. Wilmut, Cell cycle co-ordination in embryo cloning by nuclear transfer, Rev. Reprod. 1 (1996) 40–46.
[46] A. Ogura, K. Inoue, T. Wakayama, Recent advancements in cloning by somatic cell nuclear transfer, Philos. Trans. R. Soc. Lond. B. Biol. Sci. 368 (2013) 20110329.
[47] E. Bianconi, A. Piovesan, F. Facchin, A. Beraudi, R. Casadei, F. Frabetti, et al., An estimation of the number of cells in the human body, Ann. Hum. Biol. 40 (2013) 463–471.
[48] J. Fulka Jr., P. Loi, H. Fulka, G. Ptak, T. Nagai, Nucleus transfer in mammals: noninvasive approaches for the preparation of cytoplasts, Trends Biotechnol. 22 (2004) 279–283.
[49] B.G. Tatham, A.T. Dowsing, A.O. Trounson, Enucleation by centrifugation of in vitro-matured bovine oocytes for use in nuclear transfer, Biol. Reprod. 53 (1995) 1088–1094.
[50] N. Costa-Borges, M.T. Paramio, J. Santalo, E. Ibanez, Demecolcine- and nocodazole-induced enucleation in mouse and goat oocytes for the preparation of recipient cytoplasts in somatic cell nuclear transfer procedures, Theriogenology 75 (2011) 527–541.
[51] X. Li, P. Zhang, S. Jiang, B. Ding, X. Zuo, Y. Li, et al., Aging adult porcine fibroblasts can support nuclear transfer and transcription factor-mediated reprogramming, Anim. Sci. J. 89 (2018) 289–297.
[52] S.A. Johnston, P.Q. Anziano, K. Shark, J.C. Sanford, R.A. Butow, Mitochondrial transformation in yeast by bombardment with microprojectiles, Science 240 (1988) 1538–1541.

[53] R.A. Butow, T.D. Fox, Organelle transformation: shoot first, ask questions later, Trends Biochem. Sci. 15 (1990) 465–468.
[54] M.J. Kim, J.W. Hwang, C.K. Yun, Y. Lee, Y.S. Choi, Delivery of exogenous mitochondria via centrifugation enhances cellular metabolic function, Sci. Rep. 8 (2018) 3330.
[55] T.A. Rando, T. Wyss-Coray, Stem cells as vehicles for youthful regeneration of aged tissues, J. Gerontol. A. Biol. Sci. Med. Sci. 69 (2014) S39–S42.
[56] M. Wahlestedt, C.J. Pronk, D. Bryder, Concise review: hematopoietic stem cell aging and the prospects for rejuvenation, Stem Cells Transl. Med. 4 (2015) 186–194.
[57] P. Sousa-Victor, L. Garcia-Prat, A.L. Serrano, E. Perdiguero, P. Munoz-Canoves, Muscle stem cell aging: regulation and rejuvenation, Trends Endocrinol. Metab. 26 (2015) 287–296.
[58] F. Zarei, A. Abbaszadeh, Stem cell and skin rejuvenation, J. Cosmet. Laser Ther. 20 (2018) 193–197.
[59] S. Bi, H. Wang, W. Kuang, Stem cell rejuvenation and the role of autophagy in age retardation by caloric restriction: an update, Mech. Ageing Dev. 175 (2018) 46–54.
[60] P. Paliwal, N. Pishesha, D. Wijaya, I.M. Conboy, Age dependent increase in the levels of osteopontin inhibits skeletal muscle regeneration, Aging 4 (2012) 553–566.
[61] M.E. Carlson, C. Suetta, M.J. Conboy, P. Aagaard, A. Mackey, M. Kjaer, et al., Molecular aging and rejuvenation of human muscle stem cells, EMBO Mol. Med. 1 (2009) 381–391.
[62] I.M. Conboy, M.J. Conboy, A.J. Wagers, E.R. Girma, I.L. Weissman, T.A. Rando, Rejuvenation of aged progenitor cells by exposure to a young systemic environment, Nature 433 (2005) 760–764.
[63] M.G. Poulos, P. Ramalingam, M.C. Gutkin, P. Llanos, K. Gilleran, S.Y. Rabbany, et al., Endothelial transplantation rejuvenates aged hematopoietic stem cell function, J. Clin. Invest. 127 (2017) 4163–4178.
[64] M. Sinha, Y.C. Jang, J. Oh, D. Khong, E.Y. Wu, R. Manohar, et al., Restoring systemic GDF11 levels reverses age-related dysfunction in mouse skeletal muscle, Science 344 (2014) 649–652.
[65] I.M. Conboy, T.A. Rando, Aging, stem cells and tissue regeneration: lessons from muscle, Cell Cycle 4 (2005) 407–410.
[66] A.S. Brack, M.J. Conboy, S. Roy, M. Lee, C.J. Kuo, C. Keller, et al., Increased Wnt signaling during aging alters muscle stem cell fate and increases fibrosis, Science 317 (2007) 807–810.

Chapter 15

Technical barriers to systemic interventions

Introduction

With the advent of computational technologies, a search for age-related genes has been facilitated in recent years, with the genome-wide expression profiles of diverse tissues (e.g., brain [1], skeletal muscles [2], and kidney [3]) on individuals of various ages documented in literature. More recently, genome-wide analysis has been performed on low-dose irradiated *D. melanogaster* [4], and a set of genes [e.g., cytochrome-related genes (*Cyp4p3*, *Cyp6a9*, *Cyp1*, *Cyp4d21*, *Cyp6g1*, and *Cyp318a1*), genes associated with the regulation of oxidative stress (e.g., *Trxr-1*, *Trxr-2*, *mmd*, *GstS1*, *Jon65Ai*, *Jon65Aiv*, *Jon66Ci*, *Jon99Ci*, and *Jon99Cii*), and genes associated with protein turnover and ubiquitination (e.g., *Roc1b*, *Ubc84D*, *CG2924*, *CG7220*, *crl*, *neur*, and *Ubc-E2H*)] have been shown to contribute to ionizing radiation-induced life span extension. The discoveries made from gene expression profiling have offered a platform for potential targets to be identified during intervention development.

The identification of potential intervention targets has been further enhanced by the development of SNP genotyping technologies, which have made genome-wide association studies (GWAS) feasible. By using GWAS, hundreds of thousands of SNPs across the entire genome can be examined (Fig. 15.1) [5]. Genetic signatures associated with the phenotype of interest can also be identified [6]. In addition, the implementation of GWAS requires no prior assumption on gene functions. Although the poor reproducibility of the collected data is one of the problems that has to be tackled, GWAS provide an effective way to identify new genes whose mechanistic roles are poorly elucidated. By examining the GWAS data of around 400 nonagenarians from long-lived sibships and over 1600 younger population controls, *POT1* has been found to involve in determining the association between longevity and telomere maintenance in humans [7]. Moreover, the role of insulin/IGF-1 signaling in life span determination has been shown to be partially mediated by *AKT3*, *AKT1*, *FOXO4*, *IGF2*, *INS*, *PIK3CA*, *SGK*, *SGK2*, *YWHAG*, and *POT1* [7]. Moreover, GWAS have been adopted to study the genetic contributions to diverse age-related diseases (including cancer, diabetes mellitus, and atherosclerosis) [6]. Along with the recognition of different age-associated pathways (including the TOR pathway [8] and the insulin/IGF-1 signaling pathway [9]), manipulation of the aging network at the molecular, subcellular, or even tissue levels has emerged as a feasible direction for the development of biogerontological interventions.

Gaps in intervention execution

In spite of the progress in basic biogerontological research and the technological advances in genetic manipulation, the success in intervening in biological processes in vivo cannot be equated with the success of intervention development. This is because many techniques applied in the laboratory context can hardly be translated into practicable interventions. For instance, by crossing *UAS-D-GADD45* females with *GAL4-1407* males, overexpression of the *D-GADD45* gene has been attained, resulting in life span extension in fruit flies [10]; however, this mating technique cannot possibly be used to prolong life span in mature adults. Moreover, genetic manipulation is always adopted to establish mammalian knock-out models. For example, by deleting the *p66Shc* longevity gene in mice, an earlier study has generated the long-lived p66(Shc − / −) mice that are protected from vascular cell apoptosis and early atherogenesis [11]. Yet, such manipulation can hardly be executed in mature adults.

The gap between laboratory research and intervention development is a major barrier that has to be tackled before an executable biogerontological intervention can be developed. For example, to prevent tumorigenesis, one

Delivery of Therapeutics for Biogerontological Interventions. DOI: https://doi.org/10.1016/B978-0-12-816485-3.00015-5

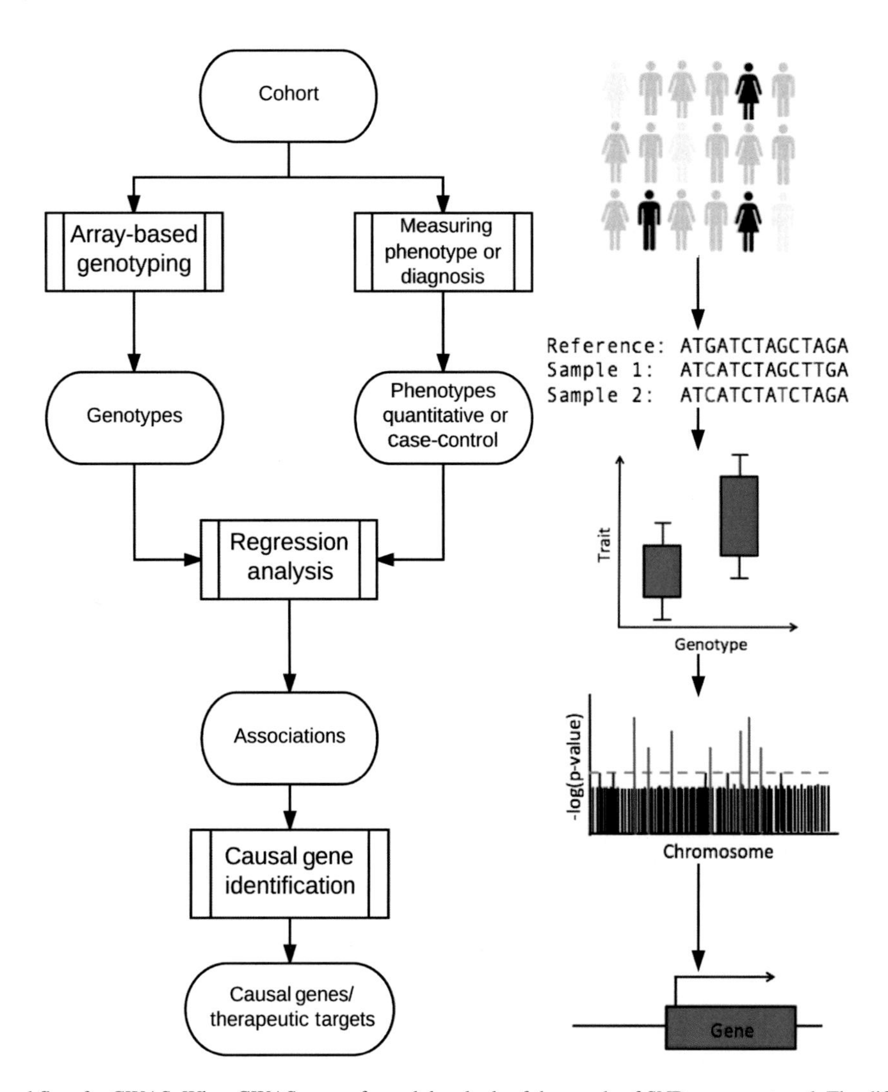

FIGURE 15.1 The workflow for GWAS. When GWAS are performed, hundreds of thousands of SNPs are genotyped. The differences in phenotypes as a function of genotypes for a large number of genetic variants are then identified by using regression analysis. After the identification of genetic loci that are associated with either changes in the quantitative trait or the diseased phenotype, the genes within those associated regions will be further examined to look for causal genes that are related to the phenotype. *Reproduced from O.L. Sabik, C.R. Farber, Using GWAS to identify novel therapeutic targets for osteoporosis, Transl. Res. 181 (2017) 15–26 with permission from Elsevier B.V. [5].*

may delete the telomere subunit from all cells other than germ lines and stem cell pools (in which telomere maintenance is needed for proper functioning) [12]. With the use of genetic manipulation technologies, modulation of gene expression is no longer a technical problem in vitro; however, how the therapeutic nucleic acids (e.g., siRNA molecules and plasmids) can be delivered to somatic cells bodywide is a challenge. Furthermore, to solve the problem caused by the age-associated accumulation of lysosomal aggregates and the subsequent lysosomal storage diseases, substrate reduction therapy (SRT) [13], enzyme replacement therapy (ERT) [14], and ex vivo gene therapy [15] can be adopted. Unfortunately, SRT and ERT are mainly palliative in nature; whereas ex vivo gene therapy can only be applied to tissues accessible for cell transplantation. They are not true solutions to the problem of lysosomal aggregate accumulation. With the advent of genetic engineering techniques, plasmids encoding the enzyme desired can now be easily synthesized. For this, genetic manipulation can be achieved by using an expression vector to produce a persistent endogenous reservoir for the enzyme desired in target cells to combat the accumulation of lysosomal aggregates [15]. Those aggregates, however, are present in cells throughout the body. How the expression vector can be delivered to all body cells for action is the problem that has yet to be solved.

Limitations to systemic delivery

To enhance the development of an executable biogerontological intervention, the availability of effective technologies for therapeutics delivery is required. As delineated in

Part II of this book, over the years, advances have already been achieved in the design of delivery technologies. For example, after aerosol delivery of PEI-RNAi complexes to silence the expression of *WT1* in mice with lung metastasis, the growth of blood vessels in tumors has been suppressed, with apoptosis induced in the lung tumor foci [16]. More recently, polycationic liposomes have also been used to deliver a plasmid encoding extracellular SOD to mice to protect the animals from acute liver injury [17]. Advances in therapeutics delivery have been further enhanced by the availability of techniques for site-specific genomic integration. One of these techniques is the use of the φC31 integrase system, which is a phage-derived system that enables the integration of a plasmid bearing an *attB* site into the pseudo-*attP* site in the genome [18,19]. DNA damage and chromosomal rearrangements induced by the φC31 integrase system in mammalian cells, however, are some of the problems that have to be addressed prior to bench-to-clinic translation [20,21]. Site-specific genomic integration can also be achieved by first incorporating recombinase target sites into the genome using the *piggyBac* transposase, followed by the integration of plasmids into those sites or into multiple transposons via the action of an engineered zinc-finger recombinase (ZFR) [22]. These techniques, along with the possibility of confining transgene expression spatially and temporally as depicted in Chapter 6 [23–25], have streamlined the process of intervention development.

Despite this, technologies for effective systemic delivery are lacking right now. Although viral vectors show high transduction efficiency and may enable gene delivery to both dividing and nondividing cells, pathogenicity and immunogenicity are some of the problems that have caused safety concerns [26]. These problems may be addressable if the virus-mediated intervention is administered to a highly localized area (e.g., a primary tumor) with a low viral dose. Aging, however, is a systemic process. The intervention has to act on the whole body [12]. Because of this, a high viral dose may be needed, and the risks involved could be tremendous. The practical potential of executing a biogerontological intervention by using viruses is further limited by the low cargo capacity of the viral vector, which has restricted the size of the nucleic acid to be delivered. An alternative to viral vectors is nonviral carriers. For instance, with the use of CS-coated polyisohexylcyanoacrylate nanoparticles to carry anti-RhoA siRNA for i.v. administration, the growth of xenografted aggressive breast tumors in mice has been reported to be inhibited [27]. By using a dendrimer having poly(L-glutamic acid)-grafted PEI as the surrounding multiple arms and poly(amidoamine) (G 4.0) as the inner core for gene delivery, cells have been successfully transfected in the serum-containing medium [28]. In spite of these successful cases, the efficiency of nonviral methods in gene delivery is always inferior to that of viral vectors. Along with the transient nature of transgene expression mediated by nonviral systems, most of the nonviral technologies have not yet been mature enough for applications in the clinical context.

To achieve systemic administration, proper selection of an administration method is needed (Table 15.1) [29]. Injection via the i.v. route is commonly used for systemic delivery. As the administered agent can be passively deposited into a number of tissues (e.g., liver, bone marrow, and spleen) [30], this administration route is suitable for interventions that tackle localized age-associated diseases. If the intervention is designed to address the systemic degenerative process and hence requires the administered agent to be delivered to cells bodywide, the passive targeting phenomenon, together with the removal of the administered agent by defense systems of the body, will substantially reduce the effective dose of the therapeutic agent in target cells and hence the overall efficiency of the intervention. To address this problem, one strategy is to increase the dose administered to compensate for the dose loss. But there are tens of trillions of cells in a body; the dose will be too high to be practical if we just rely on increasing the dose to enhance the delivery efficiency.

Outstanding questions for clinical translation

1. How can an agent be delivered effectively bodywide? Aging occurs in all tissues and organs in a body. Technologies that can effectively deliver an agent to the whole body are needed for the execution of an antiaging intervention. Such technologies, however, are lacking at the moment. Overcoming the barrier to systemic delivery is a challenge that is well worth undertaking.
2. Which carrier shall we select for further development and optimization? Over the years, different delivery systems (ranging from liposomes to nanoparticles) have been developed, but their performance has rarely been compared. If candidate systems showing the potential to succeed can be identified, among a plethora of reported systems in literature, for further development, the chance of obtaining an effective carrier for systemic delivery will be enhanced.

Biomarkers for outcome assessment

To accurately and validly assess the performance of a biogerontological intervention, biomarkers are often needed to be used. Clinically, a number of standard biomarkers are available for studies on age-associated diseases. For instance, the fasting serum glucose level can be adopted to diagnose diabetes. The risk of cardiovascular diseases can be assessed based on the circulating level of dehydroepiandosterone (DHEA) [31]. The levels of aggrecan,

TABLE 15.1 Some commonly used methods of administering therapeutic agents [29].

Positives	Negatives
Intradermic injection	
• High vascularization of the injection site and hence rapid absorption • Less discomfort due to the relatively small amount of nerve endings present in the injection site	• Risks of hitting a large blood vessel or nerve
Subcutaneous injection	
• Slow and constant absorption • Less discomfort due to the relatively small amount of nerve endings present in the injection site	• Inflammation, ulcers, and/or abscesses at the injection site
Intravenous injection	
• A straightforward method of injecting the therapeutics into the systemic circulation • Rapid absorption	• Electrolyte imbalance • Hyperthermia • Phlebitis • Risks of embolism and extravasation • Pulmonary edema, hypertension, and heart failure caused by fluid overload • Bleeding, infection, and inflammation at the injection site
Intraperitoneal injection	
• An alternative method to i.v. injection if the recipient has low blood pressure • Ease of operation (compared to other parenteral methods)	• Laceration of vessels or organs • Bacterial peritonitis/ileus • Chemical peritonitis/ileus
Oral administration	
• Painlessness • Higher safety • Ease of dose regulation • Convenience	• The efficiency of therapeutics delivery can be hindered by the protection mechanisms in the GI tract
Nasal administration	
• A possible method of delivering the therapeutics to the brain • Painlessness • Higher safety • Convenience	• Loss of the delivered agent due to the filtering function of the nasal vestibular area and the metabolizing action of the olfactory epithelium

proteoglycans, and glycosaminoglycans in discs can be employed to diagnose age-associated intervertebral disc degeneration [32]. The plasma levels of von Willebrand factors VIII and IX, triglycerides, homocysteine, glucose, creatinine, cystatin C, fibrinogen, IL-6, selenium, magnesium, testosterone, adiponectin, and carotenoids have been found to change with age [33–35], and can be used as biomarkers to assess the fragility of an individual. Notwithstanding the availability of these biomarkers, if the intervention to be assessed is designed to tackle the systemic aging process, the selection of biomarkers is no longer so straightforward. At the moment, one method of evaluating the performance of a systemic antiaging intervention is to determine the extent of life span extension. This can be done by scoring the death of *C. elegans* based on the nematode's touch-provoked movement and pumping of the pharynx [36]. Survival distributions can also be estimated by using the Kaplan–Meyer method, and then be compared by employing the log-rank test [36]. Life span prolongation, however, as discussed later in Chapter 17, is not the only outcome intended to be accomplished by the biogerontological intervention. Restoration of other signs of aging, and the health span of an individual, should be taken into consideration.

In reality, different mammalian tissues and organs age at different rates, and may contribute to aging differently. Because of this, some efforts in literature have been paid to target specific organs to combat aging. For instance, the reproductive system has been used as a target in regenerative medicine, leading to the popularity of chimpanzee testicular transplantation among the old and wealthy in the past [37]. Even though rejuvenation operations on the reproductive system have already been denounced [37], the concept of progeroid organs persists. This concept may explain why, compared with parameters of other organs, indicators of neurological and cardiovascular health are more commonly used when the performance of an antiaging intervention is assessed. For example, in an earlier study, the antiaging effect of an herbal formulation has been evaluated in mice based solely on the endurance capacity, the learning ability, the memory performance, the neuromuscular coordination ability, and the level of monoamines in the brain [38]. The inclination to assess the improvement in neurological functioning as the outcome of an antiaging intervention has also been observed in another study [39], which has evaluated the learning and memory performance of mice to determine the antiaging effect of polypeptides from wolfberry, although changes in the SOD activity, the malondialdehyde (MDA) content, and the telomerase activity in serum and few selected organs (e.g., heart, liver, and brain) have been examined.

To assess more comprehensively the efficiency of an intervention, gene expression profiling has recently been employed. One earlier study has carried out microarray experiments to evaluate the effect of short-term CR in male mice [40]. Based on the microarray data collected from livers, changes in the expression of over 3000 genes have been found to be induced by CR. Yet, how these changes can be used to quantify the efficiency of an intervention is a problem that has to be solved. For life span extension, changes in the expression of *ATF4* can be used as a molecular biomarker [41]. The involvement of *ATF4* in aging has been suggested by the role played by *Gcn4* (which is the yeast homologue of *ATF4*) in life span prolongation in yeast cells that have undergone nutrient deprivation [41]. The importance of *ATF4* has later been observed in mice, in which an increase in the expression of *ATF4* has been found to be associated with life span extension induced by CR, methionine restriction, or other antiaging interventions [42]. Another biomarker related to life span determination is the activity of enzymes involved in xenobiotic metabolism. Upregulation of these enzymes has been detected in various tissues in mice upon life span prolongation mediated by CR, litter crowding, and rapamycin supplementation [43,44].

As the age-associated decline in the regenerative capacity (and the ability to maintain homeostasis) of tissues is partly contributed by the decline in the capacity of cell proliferation [45], markers of cell proliferation may be used to assess the efficiency of an antiaging intervention. The correlation between cell proliferation and life span determination has been shown in fibroblasts from long-lived Snell dwarf mice. Those fibroblasts have been found to be more resistant to growth arrest than those from normal mice [46]. A similar observation has been reported by Harper and coworkers [47], who have studied primary fibroblast cultures from 35 species of free-living birds. Compared to those from short-lived bird species, fibroblasts from longer-lived counterparts have been found to proliferate more rapidly, have displayed resistance not only to metabolic inhibition induced by the low-glucose medium but also to cell death induced by diverse chemicals (including methyl methanesulfonate, paraquat, hydrogen peroxide, and cadmium) [47]. These findings suggest the relationship between the proliferative capacity of cells and longevity.

Apart from the aforementioned, changes in the serum levels of various biochemical parameters may be used for outcome assessment. For example, the serum levels of some peptides (e.g., FGF19, sFlt1, and osteoprotegerin) are higher in centenarians [48], whose serum also has lower levels of pigment epithelium-derived factor (PEDF), chemerin, FGF21, and fetuin-A [48]. These peptides may be used as biomarkers in aging research. In addition, the urinary albumin/creatinine ratio (ACR), which is an indicator of kidney damage, has been observed to increase with age in mice [49]. It may be used when the effectiveness of an antiaging intervention is assessed. To evaluate the metabolic status of an individual, several biochemical agents (including follicular stimulating hormone, testosterone, triglycerides, glucose, adiponectin, cholesterol, IGFBP-1, and leptin) can be used as markers as the serum levels of these agents are known to change with age [50–52]. Finally, changes in the relative level of oxidative stress before and after the administration of an intervention can be possibly determined by quantifying the lipid hydroperoxide content, the amount of oxidative DNA lesions, the extent of lipofuscin accumulation, or the expression level of stress response genes [52]. Despite the availability of diverse biomarkers, each biomarker may indicate only one aspect of the aging process. The number of biomarkers that are needed to be used for obtaining objective and comprehensive evaluation is a question to be settled when the efficacy of an experimental antiaging intervention is determined.

Outstanding questions for clinical translation

1. How should biomarkers be selected for outcome assessment? Over the years, biomarkers of different physiological processes have been identified. Genes that express differently in different stages of the aging process have also been found. Proper selection of representative markers, among a plethora of markers reported in literature, is key to valid outcome assessment.
2. How can the outcome of an antiaging intervention be objectively assessed? The most direct method of assessing the outcome of an antiaging intervention is to examine the change in life span, but the restoration of health span should be considered, too. Health, however, is a holistic concept. It involves different parts of the body. Biomarkers have to be selected carefully so that the health status of an individual can be properly determined.

Other challenges for intervention execution

To predict the clinical performance of an experimental intervention, the use of in vivo models is required. Previously an aging model has been reported to be established by injecting D-galactose into mice [53]. Upon injection, D-galactose has generated advanced glycation end products from lipids, proteins, and nucleic acids in cells via the Maillard reaction [53], and has induced ROS production [53]. Oxidative stress, however, is only one of the many factors that contribute to aging. Together with the fact that age-associated changes in the expression of genes (or gene sets) in mice are different from those in humans, there are concerns about how well the model can predict the clinical performance of an intervention. Apart from the murine model and other commonly used aging models (including *D. melanogaster*, *C. elegans*, and *S. cerevisiae*), birds have been proposed to be used as models in aging research [54]. Birds, however, are distant from mammals evolutionarily. More efforts are needed to evaluate the validity of the data collected from birds for predicting the clinical outcome of an intervention.

Another barrier to intervention development is the effect of genetic polymorphisms on the clinical performance of an intervention. This effect has been shown in an earlier study [55], which has analyzed 17 SNPs in selected genes in cancer patients, and has identified the effect of genotypes on the responsiveness of the tumor to chemo-radiotherapy. Another study has administered oxycodone to human subjects [56], and has found that subjects with different CYP2D6 genotypes display responses in different extent (in terms of the change in the pupil size, the level of sedation, the antinociceptive effect, and the level of respiratory depression) after oxycodone administration. These findings corroborate the impact of individual differences on the outcome of an intervention.

Outstanding questions for clinical translation

1. How can we establish a model that can accurately predict the clinical outcome of an intervention? At the moment, models commonly adopted in aging research include *D. melanogaster*, *C. elegans*, and *S. cerevisiae*; however, the transferability of results from these models to humans has been questioned. In addition, aging research is often carried out on genetically homogeneous laboratory strains of in vivo models. This further reduces the transferability of results to human populations.
2. How can the performance of an intervention be predicted for a general population? The outcome of an intervention is contributed not only by the effectiveness of an intervention per se but also by genetic variations among individuals. Such variations can affect the transferability of results from a sample population to a larger population of interest, and have to be taken into account when the performance of an intervention is assessed.

Summary

The lack of technologies for effective systemic delivery, the challenge for biomarker selection, and the absence of a faithful aging model are some of the technical barriers to the execution of antiaging interventions. Despite this, the technical possibility of delaying the aging process and extending life span by genetic manipulation has been verified in vivo [57], in which life span has been extended in mice after the administration of an AAV vector for overexpression of the gene for telomerase reverse transcriptase. Such success has confirmed the feasibility of confronting the aging process, and has justified more efforts to be devoted to the development of practicable biogerontological interventions. Apart from the technical barriers presented in this chapter, there are barriers in the ethical and social dimensions. These barriers will be discussed in the next two chapters.

Directions for intervention development

Carriers for systemic delivery are required for the execution of antiaging interventions. The development of such carriers can be facilitated by following the steps below:

1. Select a carrier that warrants further development and optimization.
2. Generate a plasmid encoding a reporter gene.

(Continued)

(Continued)

3. Administer the carrier, which now carries the plasmid as mentioned above, to a mammalian model via a systemic administration route.
4. Examine the biodistribution profile of the carrier.
5. Modify the physicochemical properties of the carrier to enhance the delivery performance.
6. Repeat the steps above until the efficiency of the carrier in systemic delivery is satisfactory.

References

[1] M.G. Hong, A.J. Myers, P.K. Magnusson, J.A. Prince, Transcriptome-wide assessment of human brain and lymphocyte senescence, PLoS One 3 (2008) e3024.

[2] J.M. Zahn, R. Sonu, H. Vogel, E. Crane, K. Mazan-Mamczarz, R. Rabkin, et al., Transcriptional profiling of aging in human muscle reveals a common aging signature, PLoS Genet. 2 (2006) e115.

[3] G.E. Rodwell, R. Sonu, J.M. Zahn, J. Lund, J. Wilhelmy, L. Wang, et al., A transcriptional profile of aging in the human kidney, PLoS Biol. 2 (2004) e427.

[4] K.M. Seong, C.S. Kim, S.W. Seo, H.Y. Jeon, B.S. Lee, S.Y. Nam, et al., Genome-wide analysis of low-dose irradiated male Drosophila melanogaster with extended longevity, Biogerontology 12 (2011) 93–107.

[5] O.L. Sabik, C.R. Farber, Using GWAS to identify novel therapeutic targets for osteoporosis, Transl. Res. 181 (2017) 15–26.

[6] F. Kronenberg, Genome-wide association studies in aging-related processes such as diabetes mellitus, atherosclerosis and cancer, Exp. Gerontol. 43 (2008) 39–43.

[7] J. Deelen, H.W. Uh, R. Monajemi, D. van Heemst, P.E. Thijssen, S. Bohringer, et al., Gene set analysis of GWAS data for human longevity highlights the relevance of the insulin/IGF-1 signaling and telomere maintenance pathways, Age (Dordr.) 35 (2013) 235–249.

[8] K.L. Sheaffer, D.L. Updike, S.E. Mango, The Target of Rapamycin pathway antagonizes pha-4/FoxA to control development and aging, Curr. Biol. 18 (2008) 1355–1364.

[9] J. Papaconstantinou, Insulin/IGF-1 and ROS signaling pathway cross-talk in aging and longevity determination, Mol. Cell. Endocrinol. 299 (2009) 89–100.

[10] E.N. Plyusnina, M.V. Shaposhnikov, A.A. Moskalev, Increase of Drosophila melanogaster lifespan due to D-GADD45 overexpression in the nervous system, Biogerontology 12 (2011) 211–226.

[11] C. Napoli, I. Martin-Padura, F. de Nigris, M. Giorgio, G. Mansueto, P. Somma, et al., Deletion of the p66Shc longevity gene reduces systemic and tissue oxidative stress, vascular cell apoptosis, and early atherogenesis in mice fed a high-fat diet, Proc. Natl. Acad. Sci. U.S.A. 100 (2003) 2112–2116.

[12] A.D. de Grey, B.N. Ames, J.K. Andersen, A. Bartke, J. Campisi, C.B. Heward, et al., Time to talk SENS: critiquing the immutability of human aging, Ann. N.Y. Acad. Sci. 959 (2002) 452–462; discussion 463–455.

[13] T.M. Cox, Substrate reduction therapy for lysosomal storage diseases, Acta Paediatr. Suppl. 94 (2005) 69–75; discussion 57.

[14] R.H. Lachmann, Enzyme replacement therapy for lysosomal storage diseases, Curr. Opin. Pediatr. 23 (2011) 588–593.

[15] M.S. Sands, B.L. Davidson, Gene therapy for lysosomal storage diseases, Mol. Ther. 13 (2006) 839–849.

[16] D.E. Zamora-Avila, P. Zapata-Benavides, M.A. Franco-Molina, S. Saavedra-Alonso, L.M. Trejo-Avila, D. Resendez-Perez, et al., WT1 gene silencing by aerosol delivery of PEI-RNAi complexes inhibits B16-F10 lung metastases growth, Cancer Gene Ther. 16 (2009) 892–899.

[17] J. Wu, L. Liu, R.D. Yen, A. Catana, M.H. Nantz, M.A. Zern, Liposome-mediated extracellular superoxide dismutase gene delivery protects against acute liver injury in mice, Hepatology 40 (2004) 195–204.

[18] M. Karow, M.P. Calos, The therapeutic potential of φC31 integrase as a gene therapy system, Expert. Opin. Biol. Ther. 11 (2011) 1287–1296.

[19] B. Thyagarajan, E.C. Olivares, R.P. Hollis, D.S. Ginsburg, M.P. Calos, Site-specific genomic integration in mammalian cells mediated by phage φC31 integrase, Mol. Cell. Biol. 21 (2001) 3926–3934.

[20] J. Liu, T. Skjorringe, T. Gjetting, T.G. Jensen, PhiC31 integrase induces a DNA damage response and chromosomal rearrangements in human adult fibroblasts, BMC Biotechnol. 9 (2009) 31.

[21] A. Ehrhardt, J.A. Engler, H. Xu, A.M. Cherry, M.A. Kay, Molecular analysis of chromosomal rearrangements in mammalian cells after φC31-mediated integration, Hum. Gene Ther. 17 (2006) 1077–1094.

[22] C.A. Gersbach, T. Gaj, R.M. Gordley, A.C. Mercer, C.F. Barbas, Targeted plasmid integration into the human genome by an engineered zinc-finger recombinase, Nucleic Acids Res. 39 (2011) 7868–7878.

[23] M.K. Jayakumar, A. Bansal, B.N. Li, Y. Zhang, Mesoporous silica-coated upconversion nanocrystals for near infrared light-triggered control of gene expression in zebrafish, Nanomedicine. 10 (2015) 1051–1061.

[24] Y. Yang, F. Liu, X. Liu, B. Xing, NIR light controlled photorelease of siRNA and its targeted intracellular delivery based on upconversion nanoparticles, Nanoscale 5 (2013) 231–238.

[25] H. Guo, D. Yan, Y. Wei, S. Han, H. Qian, Y. Yang, et al., Inhibition of murine bladder cancer cell growth in vitro by photocontrollable siRNA based on upconversion fluorescent nanoparticles, PLoS One 9 (2014) e112713.

[26] W.F. Lai, Nucleic acid delivery: roles in biogerontological interventions, Ageing Res. Rev. 12 (2013) 310–315.

[27] J.Y. Pille, H. Li, E. Blot, J.R. Bertrand, L.L. Pritchard, P. Opolon, et al., Intravenous delivery of anti-RhoA small interfering RNA loaded in nanoparticles of chitosan in mice: safety and efficacy in xenografted aggressive breast cancer, Hum. Gene Ther. 17 (2006) 1019–1026.

[28] X. Zeng, S. Pan, J. Li, C. Wang, Y. Wen, H. Wu, et al., A novel dendrimer based on poly(L-glutamic acid) derivatives as an efficient and biocompatible gene delivery vector, Nanotechnology 22 (2011) 375102.

[29] W.F. Lai, M.C. Lin, Nucleic acid delivery with chitosan and its derivatives, J. Control Release 134 (2009) 158–168.

[30] U. Lungwitz, M. Breunig, T. Blunk, A. Gopferich, Polyethylenimine-based non-viral gene delivery systems, Eur. J. Pharm. Biopharm. 60 (2005) 247–266.

[31] T. Mannic, J. Viguie, M.F. Rossier, In vivo and in vitro evidences of dehydroepiandrosterone protective role on the cardiovascular system, Int. J. Endocrinol. Metab. 13 (2015) e24660.

[32] N. Vo, L.J. Niedernhofer, L.A. Nasto, L. Jacobs, P.D. Robbins, J. Kang, et al., An overview of underlying causes and animal models for the study of age-related degenerative disorders of the spine and synovial joints, J. Orthop. Res. 31 (2013) 831–837.

[33] M. Yamato, A. Ishimatsu, Y. Yamanaka, T. Mine, K. Yamada, Tempol intake improves inflammatory status in aged mice, J. Clin. Biochem. Nutr. 55 (2014) 11–14.

[34] C.C. Spaulding, R.L. Walford, R.B. Effros, Calorie restriction inhibits the age-related dysregulation of the cytokines TNF-α and IL-6 in C3B10RF1 mice, Mech. Ageing Dev. 93 (1997) 87–94.

[35] G. Lippi, F. Sanchis-Gomar, M. Montagnana, Biological markers in older people at risk of mobility limitations, Curr. Pharm. Des. 20 (2014) 3222–3244.

[36] A. Olsen, M.C. Vantipalli, G.J. Lithgow, Lifespan extension of Caenorhabditis elegans following repeated mild hormetic heat treatments, Biogerontology 7 (2006) 221–230.

[37] M.A. Kozminski, D.A. Bloom, A brief history of rejuvenation operations, J. Urol. 187 (2012) 1130–1134.

[38] H.C. Shih, K.H. Chang, F.L. Chen, C.M. Chen, S.C. Chen, Y.T. Lin, et al., Anti-aging effects of the traditional Chinese medicine bu-zhong-yi-qi-tang in mice, Am. J. Chin. Med. 28 (2000) 77–86.

[39] W.Z. Jiang, H.Q. Zhang, Anti-aging effect of polypeptides from Fructus Lycii on D-gal nduced aging model mice and the possible mechanism, J. Int. Pharm. Res. 37 (2010) 47–50.

[40] P.W. Estep 3rd, J.B. Warner, M.L. Bulyk, Short-term calorie restriction in male mice feminizes gene expression and alters key regulators of conserved aging regulatory pathways, PLoS One 4 (2009) e5242.

[41] K.K. Steffen, V.L. MacKay, E.O. Kerr, M. Tsuchiya, D. Hu, L.A. Fox, et al., Yeast life span extension by depletion of 60s ribosomal subunits is mediated by Gcn4, Cell 133 (2008) 292–302.

[42] W. Li, X. Li, R.A. Miller, ATF4 activity: a common feature shared by many kinds of slow-aging mice, Aging Cell. 13 (2014) 1012–1018.

[43] M.J. Steinbaugh, L.Y. Sun, A. Bartke, R.A. Miller, Activation of genes involved in xenobiotic metabolism is a shared signature of mouse models with extended lifespan, Am. J. Physiol. Endocrinol. Metab. 303 (2012) E488–E495.

[44] W.R. Swindell, Comparative analysis of microarray data identifies common responses to caloric restriction among mouse tissues, Mech. Ageing. Dev. 129 (2008) 138–153.

[45] N.S. Wolf, W.R. Pendergrass, The relationships of animal age and caloric intake to cellular replication in vivo and in vitro: a review, J. Gerontol. A. Biol. Sci. Med. Sci. 54 (1999) B502–B517.

[46] S.P. Maynard, R.A. Miller, Fibroblasts from long-lived Snell dwarf mice are resistant to oxygen-induced in vitro growth arrest, Aging Cell. 5 (2006) 89–96.

[47] J.M. Harper, M. Wang, A.T. Galecki, J. Ro, J.B. Williams, R.A. Miller, Fibroblasts from long-lived bird species are resistant to multiple forms of stress, J. Exp. Biol. 214 (2011) 1902–1910.

[48] F. Sanchis-Gomar, H. Pareja-Galeano, A. Santos-Lozano, N. Garatachea, C. Fiuza-Luces, L. Venturini, et al., A preliminary candidate approach identifies the combination of chemerin, fetuin-A, and fibroblast growth factors 19 and 21 as a potential biomarker panel of successful aging, Age 37 (2015) 9776.

[49] S.W. Tsaih, M.G. Pezzolesi, R. Yuan, J.H. Warram, A.S. Krolewski, R. Korstanje, Genetic analysis of albuminuria in aging mice and concordance with loci for human diabetic nephropathy found in a genome-wide association scan, Kidney Int. 77 (2010) 201–210.

[50] M. Sanchez-Rodriguez, A. Garcia-Sanchez, R. Retana-Ugalde, V. M. Mendoza-Nunez, Serum leptin levels and blood pressure in the overweight elderly, Arch. Med. Res. 31 (2000) 425–428.

[51] M. Adamczak, E. Rzepka, J. Chudek, A. Wiecek, Ageing and plasma adiponectin concentration in apparently healthy males and females, Clin. Endocrinol. (Oxf) 62 (2005) 114–118.

[52] L.J. Niedernhofer, J.L. Kirkland, W. Ladiges, Molecular pathology endpoints useful for aging studies, Ageing Res. Rev. 35 (2017) 241–249.

[53] S.C. Ho, J.H. Liu, R.Y. Wu, Establishment of the mimetic aging effect in mice caused by D-galactose, Biogerontology 4 (2003) 15–18.

[54] D.J. Holmes, M.A. Ottinger, Birds as long-lived animal models for the study of aging, Exp. Gerontol. 38 (2003) 1365–1375.

[55] M. Tanaka, M. Javle, X. Dong, C. Eng, J.L. Abbruzzese, D. Li, Gemcitabine metabolic and transporter gene polymorphisms are associated with drug toxicity and efficacy in patients with locally advanced pancreatic cancer, Cancer 116 (2010) 5325–5335.

[56] C.F. Samer, Y. Daali, M. Wagner, G. Hopfgartner, C.B. Eap, M. C. Rebsamen, et al., The effects of CYP2D6 and CYP3A activities on the pharmacokinetics of immediate release oxycodone, Br. J. Pharmacol. 160 (2010) 907–918.

[57] B. Bernardes de Jesus, E. Vera, K. Schneeberger, A.M. Tejera, E. Ayuso, F. Bosch, et al., Telomerase gene therapy in adult and old mice delays aging and increases longevity without increasing cancere, EMBO Mol. Med. 4 (2012) 691–704.

Chapter 16

Ethical barriers to intervention development

Introduction

As the development of biogerontological interventions becomes increasingly possible and begins to attract attention from the community, it is unavoidable that different opinions emerge, both positive and negative, about the impact and righteousness of related research. Some of the ethical concerns are genuine and are directly related to the practice of aging research, while some criticisms are more prospective in nature. These critics, however, should not be overlooked because they may influence the research environment (e.g., by influencing policy-making decisions) and may directly be translated into barriers to the development of biogerontological interventions.

As a matter of fact, biogerontological research can be divided into three stages: (1) basic research, (2) in vivo experimentation, and (2) bench-to-clinic translation (Fig. 16.1). In this chapter, ethical issues involved in each stage will be revisited. Issues needed to be considered during the design of clinical tests will also be highlighted. It is hoped that by examine some of the ethical concerns involved in intervention development, light can be shed on possible directions for future efforts to overcome the ethical barriers that stand in the way of having biogerontological interventions to be practicable in reality.

Righteousness of aging research

Bioethicists have been criticizing research initiatives in antiaging medicine [1–3]. Some critics have propelled the idea that efforts to prolong life span are driven merely by humans' fantasies of control over life and death, and may eventually lead to detrimental changes in the economic structure and human development. For instance, Leon Kass has raised concerns that the development of interventions to combat aging may cause social injustice, with the society divided not only into the rich and the poor but also into the mortal and the immortal [4]. Because of this, efforts to retard aging have sometimes been opposed by bioethicists such as Daniel Callahan, a cofounder of the Hastings Center, who has argued that "it would not seem wise or humanly beneficial to carry out research designed to extend the life of those who live to the age of 90, for instance, as if that were not a long enough life for most people" [5].

Criticisms of antiaging medicine have arisen partly from the unstated fear of the need of reformulation (e.g., in terms of the social structure and the society security scheme) when life span is indefinitely extended [6]; however, such fear is vague. First, taking the current status of intervention development into account, there is a long way to go before aging can be truly tackled, not to mention the possibility of reaching immortality. Most of the criticisms of antiaging medicine, therefore, are focusing more on speculative, biologically implausible possibilities rather than realistic situations. Very little of the commentators' discussions on the consequence of contemporary efforts to tackle aging are based on a firm grasp of the likely course of research and development. Second, even though immortality can be reached to cause a social impact, the need of adapting to changes should never be a logical reason to stop a change from happening. Humans are expected to have the ability to adapt to changing circumstances. If the success of antiaging medicine, at the end, really leads to an impact on the social and cultural structures, all these structures should be adjusted to acknowledge the evolving circumstances as what they have done over the history of mankind. This is because making adjustments is an expected act from humans as a manifestation of adaptation.

Delivery of Therapeutics for Biogerontological Interventions. DOI: https://doi.org/10.1016/B978-0-12-816485-3.00016-7

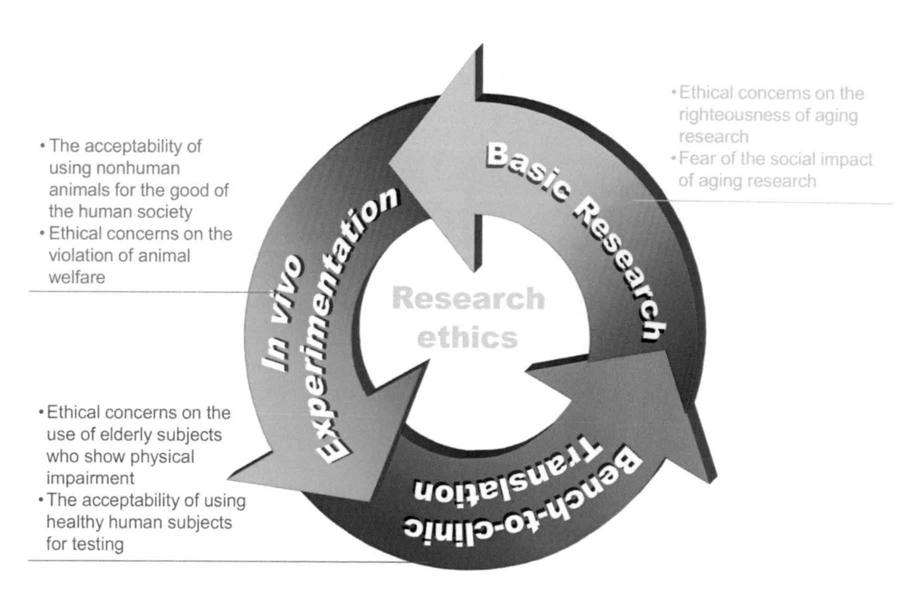

FIGURE 16.1 Overview of the ethics of biogerontological research.

> **Outstanding questions for clinical translation**
>
> 1. How can the impact of antiaging medicine on the prevalent social and cultural structures be fairly estimated? At this moment, most of the concerns about the advent of antiaging technologies are speculative rather than objective. Methods of assessing the impact objectively and fairly are needed to be developed.
> 2. How can the concerns about the perceived "damage" caused by the development of biogerontological interventions be alleviated? Education may help address the problem; however, if the bias against antiaging medicine is so deeply rooted, changing it may not be easy.

Logics of prevalent in vivo studies

After in vitro evaluation, an experimental intervention will usually proceed into preclinical trials. Animal experimentation is deemed morally "acceptable" if it finally contributes to the good of humans. This has been shown in Magalhaes-Sant'Ana et al.'s earlier paper [7], which has advocated the ethical acceptability of animal-based pain research under the sole premise of scientific advancement. In fact, the history of using nonhuman animals in scientific research can be dated back to the ancient Greek as Aristotle and Hippocrates dissected nonhuman animals in an attempt to understand the anatomical structure of the human body. In the 17th century when Cartesian philosophy developed, people performed animal experiments without having much ethical consideration [8]. At that time, nonhuman animals were regarded as a machine for humans' use. This was advocated by René Descartes, who believed that humans were distinct to all other nonhuman animals because of language and reason [9]. According to him, man has little responsibilities to other species unless the human society is negatively influenced [9]. At present, the absolute authoritarianism of humans no longer goes to that extreme. Yet, animal experimentation has still been extensively adopted in research. The number of nonhuman animals used in research, teaching, and testing reaches 75–100 million per year worldwide [10]. Most of these nonhuman animals are used in drug and cancer research, and in testing vaccines and other biological agents [8].

To secure "good welfare" for nonhuman animals in the laboratory setting, we may gain insights from the standards proposed by the Brambell Committee [11], which has stated that "good welfare" for farm animals can be attained if the animals have the following forms of freedom: (1) freedom from thirst, malnutrition, and hunger; (2) freedom from fear, and distress; (3) freedom to express normal behavior; (4) freedom from discomfort; and (5) freedom from pain, injury, and disease. Such standards, however, have not drawn much attention from researchers. Some researchers have even regarded themselves to be "morally" correct to compromise the welfare of nonhuman animals that tend to have lower evolutionary fitness [12]. To enhance the ethics of animal research, Russell and Nurch's [13] three R's approach (replacement, reduction, and refinement) has also been proposed as a nonbinding reference for the maintenance of laboratory animal welfare. But, as with most other discourses addressing the topic [8,12], it has still been grounded on the conviction that animal welfare is subordinated to human needs. The approach allows a compromise in favor of the human side to be made when animal welfare and human needs come into conflicts. It is the legitimacy of this long-held ground that should be questioned.

Power struggle in aging research

Over the years, the logic of using nonhuman animals in research has been grounded in the seemingly inherent connection between animal experimentation and scientific advances, which have been equated with human betterment. The latter has been revealed in Francis Bacon's *New Atlantis*, in which a scientist has been portrayed as a priest to absolve the misery of the human race through scientific practices [14]. Presently, anthropocentrism prevails. Because of this, humans are thought to be superior to other species and the nature. The sustained use of nonhuman animals in research for social good is deemed inexorably "rational" and "moral." But here a question comes. In what can the determination of "rationality" and "morality" be grounded? Or, putting it further, can universal "rationality" and "morality" be possible at all? Over history, rationality has been conceived to be objective, universal, and singular. As suggested by Lukes [15], the criteria of rationality simply "are". Such an objectivist perspective has been further ensured in the process of commensuration in which a common language has been created in an attempt to fit all paradigms. As stated by Hollis [16], "Western rational thought is not just one species of rational thought nor rational thought just one species of thought."

Under the mainstream value system, reality has been viewed as being immutable. Paradigms other than the dominant rational one has been marginalized or absent. The dominion of anthropocentrism has turned the motive of human betterment into a moral apologia for man's overexploitation of animal welfare in scientific research. Compromising any trifles of scientific research simply for animal welfare has often been concluded to be "irrational" in the anthropocentric sense, and this conclusion has been taken as a universal truth in society. Yet, our society is currently operated by modernity, which symbolizes the progressive economic/administrative rationalization and differentiation of the world. There is a close link between modernity and the development of the capitalist state [17]. Modernism, which has arisen in response to the premodern notions of the world, is controlled by a divine and unquestionable order [18]. It believes that the world is based on universal principles that are discoverable. Some major notions, therefore, are held. These include reason, objectivity, neutrality, and systematic inquiry that helps humans to make sense of the world. Grand narratives have been raised. Foundationalism, which has subjugated alternative discourses and forms of inquiry, has been admitted. In these grand narratives, the majority voice has become the golden standard; while other voices have become colonies only. Grand narratives have long been deemed unshakable and unquestionable.

After centuries of living with anthropocentrism and of exploiting animal rights under the belief that human good is everything, it is high time for laboratory workers to revisit the legitimacy of the prevalent research practice, particularly when the data collected from animal models fail to predict the outcome of clinical studies. It is also necessary to revisit the common belief on the righteousness of compromising animal welfare for perceived advances in the human society. This belief has been challenged by postmodernism, which holds that truth and rational acceptability are fluid, relative, contextual, and dependent on a preset paradigm [19], which in this case is the society of man. In other words, the rationality of man's superiority over animal rights is open to question. This idea may be confronted by the prevalent modern thoughts that insist on the objectivity and universality of social norms; but instead of sticking to the notion of closed ethnocentrism, if the possibility of using multiple paradigms to determine the limit of scientific research could be recognized, conflicts between animal rights and human needs would be handled in a more enlightened manner.

At present, the paramountcy of human needs forms the mainstream moral paradigm. This cornerstone of modern morality has seldom been challenged and appears to be unchallengeable. But quoting the words from Adorno [20], "There is no moral certainty. Its mere assumption would be immoral, would falsely relieve the individual of anything that might be called morality." Moral values are not inerrable, objective, singular, and cast in stone. They should neither be accepted simply because of the authority nor be fashioned by one group. Instead, they should comfort the interests of various parties, including nonhuman animals. Animal experimentation is impossible to cease in the forthcoming decades, and neither are debates over animal welfare. Yet, solving the ethical problems related to animal experimentation in aging research, in fact, is much simpler than we think. Establishing new nonanimal models that can better predict the clinical outcome of an experimental intervention is a way out, and should always be one of the research imperatives in antiaging medicine. Indeed, technological advances have already brought this so-called "fantasy" closer to reality. Owing to the advent of microengineering and microfluidic techniques, it is now possible to mimic key aspects of living organs (e.g., critical microarchitectures, spatiotemporal cell-cell interactions, and extracellular milieus) faithfully using "organ-on-a-chip" systems [21,22]. Such advances have potentially enabled biological processes to be studied without the need of animal experimentation [21], and may even one day serve as replacements for animal testing in intervention development [23,24].

Outstanding questions for clinical translation

1. How can animal rights be protected during our pursuit of biogerontological interventions? Scientific advances are important to the development of the society; however, such development should not jeopardize the rights of nonhuman animals. These animals are not obliged to sacrifice themselves for the advances in the society of man. Developing antiaging interventions while respecting animal rights is a mission we need to fulfil.
2. How can the discordance between in vivo and clinical results be solved when the performance of an intervention is evaluated? Right now in vivo experimentation is a major method of evaluating the efficiency of an intervention; however, the clinical outcome of the intervention can hardly been known unless the intervention is tested in human models. Strategies for solving the discordance between in vivo and clinical studies shall be developed as they cannot only increase the validity of preclinical studies but can also alleviate the ethical problems associated with in vivo experimentation.

Morality in bench-clinic translation

Even though the ethical concerns about in vivo experimentation may be settled by the development of nonanimal models, ethical problems involved in clinical studies can hardly be avoided. Over the years, few clinical trials on antiaging interventions have been carried out. One example is the study "Metformin in Longevity Study" (MILES) performed in the United States [25]. Its objective is to evaluate the off-label use of metformin, which has been shown to extend life span in mice, in increasing longevity in humans [26,27]. Clinical trials have also been performed on nicotinamide mononucleotide (NMN) [28], which has shown the potential to retard or reverse the aging process [29,30]. In fact, when aging research and intervention development are carried out, the final goal always points to benefiting, or applying to, humans. The involvement of human subjects during the research process is, therefore, inevitable.

As far as clinical studies are concerned, it is inevitable to involve issues of "morality." Morality is, broadly speaking, a set of social norms and value judgments with which the "good" and the "bad" are determined. In the context of modernism, moral values are homogenous, coherent, and inexorable. As denoted by Kekes [31], "there is one and only one reasonable system of values" and such values are believed to be "the same for all human beings, always, everywhere." This ideology is consistent with the Kantian view of morality, which argues that morality is grounded in reason and reason is universal [32–34]. In fact, if we subscribe to the utilitarian point of view, use of human subjects in clinical trials on experimental biogerontological interventions is unquestionably ethical. This is because, according to utilitarianism, which is an ideology of consequentialism and a modern form of the Hedonistic ethical theory [35], an act is deemed ethical as long as its end results can contribute to the utility (i.e., happiness and pleasure) of a society [36]. This notion has first been proposed by Epicurus [37,38] and has later been systematized by Bentham [39], who has argued that pleasure and pain are two constitutional factors determining the operation of a society. Any act that brings the greatest amount of happiness to a society is morally correct [40]. Irrefutably, clinical trials can enhance the development of medicine, further our understanding of different biological processes, and contribute to the utility of the society by advancing the treatment and diagnosis of a variety of diseases.

Judging from this fact, testing the safety and efficacy of an antiaging intervention on human subjects is doubtless moral in the utilitarian sense. Yet, in the postmodern framework, the existence of an unchanging, objective, and universal external moral reality is disputable. As Beck has recognized [41], "morality" is constructed culturally, and it varies over time and from one culture to another. Different groups create moral standards to suit their own needs and cultures in a self-centered manner. In another word, the paramountcy of human needs is moral, rational, and infallible just because it is appraised under a belief system in which values are created by a human society in accordance solely to the interests, motivations, and circumstances of man [41].

Practical considerations in clinical trials

To use human subjects in a clinical trial, one unavoidable issue to be considered is the age group to be recruited. As a rule of thumb, the enrolled population should represent the population that ultimately uses the intervention in the real world. This may be easy if we are planning to perform a study on an intervention that targets a disease (which is more population-specific in general). Aging, however, is a problem encountered by every single being. Setting the inclusion and exclusion criteria for subject recruitment may not be easy. This is because if the inclusion and exclusion criteria are too broad, the scale of the study will be too big to be affordable. On the other hand, using too restricted criteria may reduce the recruitment rate, causing a delay in obtaining the research results and jeopardizing the validity of the collected data.

When the performance of a biogerontological intervention is evaluated, either the younger population or the elderly population can be used. If we decide to take the younger population as the subject group, the goal of the research will be to evaluate the efficiency of the intervention for preventing aging. In this case, objective and valid study outcomes should first be established. After that, the outcome and duration of the study should be carefully set. In general, if the incidence of the outcome is very low, a longer study period may be adopted, though an increase in the level of discomfort to the subjects may result. Alternatively, one may consider performing interim measures of the intervention outcome, or even using some surrogate markers (e.g., the serum level of low-density lipoprotein-cholesterol as an indicator of cardiovascular risk [42]) to evaluate the performance of the intervention. The validity of such markers, in this case, have to be supported by the literature.

If we would like to test the efficiency of the intervention in reversing the aging process, the elderly subjects can be used. One of the ethical concerns about this, however, is the vulnerability of the elderly people. Issues such as cognitive impairment and hearing problems may impair the autonomy of the subject participating in a clinical trial. This is because, if the subject cannot hear and read properly, his/her right to make an informed decision about research participation will be jeopardized (owing to his/her failure to understand the risk imposed to him/her). Apart from this, elderly people sometimes have chronic comorbidities, and take multiple medicines on a daily basis. This may turn out to be a confounding factor for the correct interpretation and translation of the results obtained in clinical trials. Setting an exclusion criteria to filter out those having chronic comorbidities may help solve the problem, but if the occurrence of those comorbidities is associated with aging, excluding those subjects will exclude some of the aging phenotypes from being investigated.

Outstanding questions for clinical translation

1. How can the cost and benefit be estimated when a clinical trial on a biogerontological intervention is performed? The success in developing a biogerontological intervention is beneficial to humans, but the subjects may have to bear the risks imposed by the study. Being able to assess the cost and benefit objectively is pivotal.
2. How should we handle elderly subjects who are physically and cognitively frail? Due to physical and cognitive impairment, some elderly people may have difficulties in understanding the risks before they agree to participate in a clinical trial. Recruiting them as subjects may cause ethical concerns, but excluding them may make the sample population too skewed to draw any robust conclusions.

Summary

Intervention development requires different barriers to be overcome, including those at the cellular and tissue levels. Attainment of technologies to rejuvenate or repair aged tissues are challenging to succeed and requires extensive efforts from the field. On top of these technical barriers, a host of ethical issues have formed an obstacle to intervention development, making the already-difficult task more unreachable. In this chapter, we have discussed the ethical issues involved in different stages of biogerontological research. It is worth noting that ethical and philosophical discussions are subjective in nature. Intellectual debates about ethical and moral issues should remain open [43]. Finally, the availability of social support and the distribution of public funding are some of the factors that may influence the success of intervention development. They should not be overlooked during our fight against aging. This will be discussed in the next chapter.

Directions for intervention development

Ethical barriers to the development of biogerontological interventions are difficult to be overcome because they involve social conventions and values. Despite this, some of the following efforts can still be paid by an individual:

1. Understand the ethical concerns, if any, raised by the public on related research.
2. Be familiar with the policies governing the use of in vivo and human models.
3. Carry out a study in a way that is well-accredited by the scientific community.
4. Adjust the sample size or research plan, if feasible, to minimize criticisms and ethical concerns.

References

[1] L. Partridge, D. Gems, Ageing: a lethal side-effect, Nature 418 (2002) 921.

[2] S.L. Helfand, S.K. Inouye, Rejuvenating views of the ageing process, Nat. Rev. Genet. 3 (2002) 149–153.

[3] Y.J. Lin, L. Seroude, S. Benzer, Extended life-span and stress resistance in the Drosophila mutant methuselah, Science 282 (1998) 943–946.

[4] L. Kass, L'Chaim and its limits: why not immortality? First. Things 113 (2001) 17–24.

[5] D. Callahan, Death and the research imperative, N. Engl. J. Med. 342 (2000) 654–656.

[6] L. Hayflick, The future of ageing, Nature 408 (2000) 267–269.

[7] M. Magalhaes-Sant'Ana, P. Sandoe, I.A.S. Olsson, Painful dilemmas: the ethics of animal-based pain research, Anim. Welf. 18 (2009) 49–63.

[8] V. Baumans, Science-based assessment of animal welfare: laboratory animals, Rev. Sci. Tech. 24 (2005) 503–514.

[9] R. Descartes, Animals are machines, in: S.J. Armstrong, R.G. Botzler (Eds.), Environmental Ethics: Divergence and Convergence, McGraw-Hill, New York, 1993, pp. 281–285.
[10] L.F.M. Van Zutphen, History of animal use, in: L.F.M. Van Zutphen, V. Baumans, A.C. Beynen (Eds.), Principles of Laboratory Animal Science, Elsevier, Amsterdam, 2001, pp. 2–5.
[11] Brambell Committee, Report of the Technical Committee to enquire into the welfare of animals kept under intensive livestock husbandry systems, Her Majesty's Stationery Office, London, 1965.
[12] C.J. Barnard, J.L. Hurst, Welfare by design: the natural selection of welfare criteria, Anim. Welf. 5 (1996) 405–434.
[13] W.N.S. Russell, R.L. Burch, The Principles of Humane Experimental Techniques, Methuen, London, 1959.
[14] C. Merchant, The Death of Nature: Women, Ecology and the Scientific Revolution, Harper San Francisco, San Francisco, 1989.
[15] S. Lukes, Some problems about rationality, in: B.R. Wilson (Ed.), Rationality, Basil Blackwell, Oxford, 1970, pp. 194–213.
[16] M. Hollis, The limits of irrationality, in: B.R. Wilson (Ed.), Rationality, Harper Torchbooks, New York, 1970, pp. 214–220.
[17] M. Sarup, An Introductory Guide to Post-Structuralism and Postmodernism, University of Georgia Press, Atlanta, 1993.
[18] D. Howe, Modernity, postmodernity and social work, Br. J. Soc. Work 24 (1994) 513–532.
[19] H. Putnam, Representation and Reality, MTT Press, Cambridge, 1988.
[20] T. Adorno, Negative Dialectics, The Continuum Publishing Company, New York, 1973.
[21] F. Zheng, F. Fu, Y. Cheng, C. Wang, Y. Zhao, Z. Gu, Organ-on-a-chip systems: microengineering to biomimic living systems, Small 12 (2016) 2253–2282.
[22] F. An, Y. Qu, X. Liu, R. Zhong, Y. Luo, Organ-on-a-chip: new platform for biological analysis, Anal. Chem. Insights 10 (2015) 39–45.
[23] J.B. Lee, J.H. Sung, Organ-on-a-chip technology and microfluidic whole-body models for pharmacokinetic drug toxicity screening, Biotechnol. J. 8 (2013) 1258–1266.
[24] J. Ribas, H. Sadeghi, A. Manbachi, J. Leijten, K. Brinegar, Y.S. Zhang, et al., Cardiovascular organ-on-a-chip platforms for drug discovery and development, Appl. In Vitro Toxicol. 2 (2016) 82–96.
[25] ClinicalTrials.gov, Metformin in Longevity Study (MILES). <https://clinicaltrials.gov/ct2/show/NCT02432287> (accessed 26.10.18).
[26] A. Martin-Montalvo, E.M. Mercken, S.J. Mitchell, H.H. Palacios, P.L. Mote, M. Scheibye-Knudsen, et al., Metformin improves healthspan and lifespan in mice, Nat. Commun. 4 (2013) 2192.
[27] V.N. Anisimov, L.M. Berstein, P.A. Egormin, T.S. Piskunova, I. G. Popovich, M.A. Zabezhinski, et al., Metformin slows down aging and extends life span of female SHR mice, Cell Cycle 7 (2008) 2769–2773.
[28] S.K. Poddar, A.E. Sifat, S. Haque, N.A. Nahid, S. Chowdhury, I. Mehedi, Nicotinamide mononucleotide: exploration of diverse therapeutic applications of a potential molecule, Biomolecules 9 (2019) 34.
[29] J. Yoshino, K.F. Mills, M.J. Yoon, S. Imai, Nicotinamide mononucleotide, a key NAD^+ intermediate, treats the pathophysiology of diet- and age-induced diabetes in mice, Cell. Metab. 14 (2011) 528–536.
[30] K.F. Mills, S. Yoshida, L.R. Stein, A. Grozio, S. Kubota, Y. Sasaki, et al., Long-term administration of nicotinamide mononucleotide mitigates age-associated physiological decline in mice, Cell. Metab. 24 (2016) 795–806.
[31] J. Kekes, The Morality of Pluralism, Princeton University Press, Princeton, 1993.
[32] I. Kant, Groundwork of the Metaphysic of Morals, Hutchinson, London, 1785.
[33] I. Kant, Critique of Practical Reason, Bobbs-Merrill, Indianapolis, 1788.
[34] I. Kant, Metaphysics of Morals, Cambridge University Press, Cambridge, 1991.
[35] Z.C.Y. Chan, W.F. Lai, Revisiting the melamine contamination event in China: implications for ethics in food technology, Trends Food Sci. Technol. 20 (2009) 366–373.
[36] R.E. Goodin, Utilitarianism as a Public Philosophy, Cambridge University Press, Cambridge, 1995.
[37] N.W. DeWitt, Epicurus and His Philosophy, University of Minnesota Press, Minneapolis, 1954.
[38] P. Mitsis, Epicurus' Rthical Theory: The Pleasures of Invulnerability, Cornell University Press, New York, 1988.
[39] J. Bentham, An Introduction to the Principles of Morals and Legislation, Methuen, London, 1982.
[40] F. Rosen, Classical Utilitarianism From Hume to Mill, Routledge, New York, 2003.
[41] C. Beck, Postmodernism, ethics, and moral education, in: W. Kohli (Ed.), Critical Conversations in Philosophy of Education, Routledge, New York, 1995, pp. 127–136.
[42] T.J. Colatsky, Reassessing the validity of surrogate markers of drug efficacy in the treatment of coronary artery disease, Curr. Opin. Investig. Drugs 10 (2009) 239–244.
[43] J. Derrida, Passions: an oblique offering, in: T. Dutoit (Ed.), On the Name, Stanford University Press, Stanford, 1995, pp. 15–17.

Chapter 17

Social barriers to intervention success

Introduction

With increasing understanding of the aging process and the continuous development of biotechnological techniques, diagnosis and treatment of a variety of age-associated diseases have been enhanced. This has also made life span prolongation more possible. Life span prolongation, therefore, should be much more than merely postponing the time of death. It should also be accompanied by the extension of one's healthy and quality life. The health span is largely influenced by a variety of nonbiological factors in the social context. This has been documented in an earlier study, which, based on the data collected from a cross-sectional survey of a sample of 1106 elderly respondents, has demonstrated that family social support is a potential mediator of the relationship between income sources and depression [1]. Lately, Phillips et al. have interviewed more than 500 persons (224 males, 294 females) aged 60 and over, and have revealed that subjective measures of informal support (e.g., satisfaction with support received from family members) are predictors of psychological well-being among the elderly [2].

As a matter of fact, extension of physical life without extending one's healthy period would only cause medical burdens on society at last. Yet, the need for the nonbiological factors as mentioned above could hardly be fulfilled merely by using laboratory studies. Serious discussions and exploration with a sociological perspective are a sine qua non. The objective of this chapter is to highlight the importance of social support as one of the indispensable yet intangible measures coordinating with the research efforts, as depicted in other chapters in this book, to attain technological advances in the development and execution of biogerontological interventions.

Life span extension across cultures

The term "longevity" encompasses multidimensional socio-cultural meanings, yet it is in general defined as the state of having a "long life" (p. 918) [3]. According to Roman philosophers, humans' mistaken horror of death has led to the desire of longevity. Such interpretation of prolongevism, however, has later been objected by Roy Perret, who believed that "Most people consider their lives worth living: they do not contemplate suicide, but rather attempt to prolong their lives" (p. 222) [4]. In other words, humans pursue longevity solely because they treasure the state of living [5]. To extend the life span, different tactics have been adopted over the years. In the Middle Ages, alchemists attempted to mix and experiment with different chemicals to create life. Similar attempts of fabricating elixirs occurred in ancient China. For example, in the Qin Dynasty, memorial temples were built for people to worship the Star of Longevity [6]. Extensive efforts on preserving "body fluids" were also made by Taoism followers to prolong the life span. Even though the generation of elixirs has not succeeded, the efforts have made important contributions to the development of chemistry and contemporary medicine.

From the perspectives of traditional Chinese medicine (TCM), different mutually interrelated components work together in a human body to attain an internal balance among life-supporting substances, such as "vital energy," "essence," "blood," "body fluids," and "spirit" [7]. As suggested by *Yellow Emperor's Internal Classic*, aging is a consequence of gradual breakdown of the internal balance. Such breakdown may disturb the replenishment and nourishment of the body and vital organs, resulting in physiological deterioration. According to TCM theories, man has to fill "blood" to nourish the internal organ systems (i.e., "liver, "kidney," "heart," "stomach, and "spleen") to retard the aging process and to achieve longevity [7]. This ideology is very similar to the idea of holistic care in modern medicine.

Taking the evidence above into account, longevity appears to be a common goal of the human race across cultures. For decades, advances have been made to improve diagnosis and treatment of a variety of diseases. The life span of humans has been extended considerably, and is expected to be further extended in the future. Yet, are the objectives of life span extension merely a numerical increase in the age humans can attain? Our goal of

Delivery of Therapeutics for Biogerontological Interventions. DOI: https://doi.org/10.1016/B978-0-12-816485-3.00017-9

life span extension, indeed, should go beyond that. We need to have a quality and fruitful life instead of having extended years of being fragile in bed. This is the reason why the concept of health span engineering should be emphasized during our fight against aging.

Importance of health span maintenance

The objective of health span engineering is to use various behavioral, biomechanical, pharmacological, or regenerative approaches to maintain or restore the functional reserve of critical tissues/organs to shorten the period of morbidity at increasing chronological age [8]. In humans, individuals grow to mature and reproduce, with maximum vitality generally achieved at the age of 20s. After that, a gradual decline in physiological functions occurs, followed by an increase in the morbidity/mortality rate. Such a life cycle seems to be determined by natural selection, which aims to maximize vitality during the years of reproduction [8]. Until now, the longest documented life span of humans is 122 years and 164 days; whereas that of whales is over 200 years [9]. The latter is reported by George and co-workers, who have examined the extent of aspartic acid racemization in the lens nuclei of the eye globes collected from bowhead whales [9]. Because the resting heart rate has been reported by an earlier study to reflect autonomic nervous system activity and to be inversely related to the life span of homeothermic mammals [10], researchers have proposed that, by engineering the heartbeats via deliberate cardiac slowing (e.g., by pharmacological interventions and lifestyle modifications), the biological clock of aging might be slowed down, with the maximum life span extended [10]. In addition, with the development of effective drugs, some diseases that were life-threatening can now be cured. Examples include some cancers, chicken pox, diphtheria, measles, and tetanus.

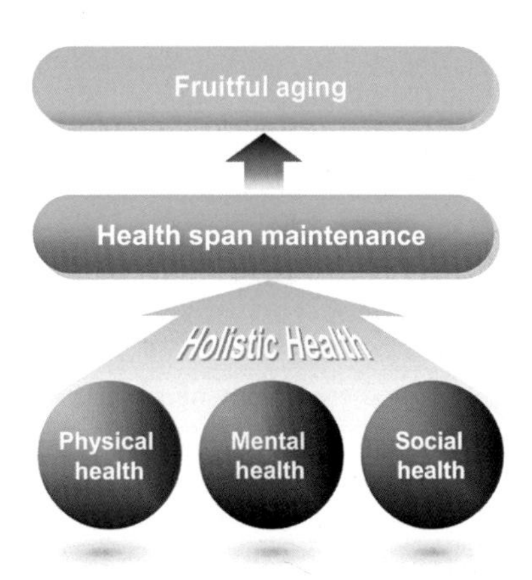

FIGURE 17.1 The relevance of holistic health to fruitful aging.

In reality, in comparison with interventions that target aging as a whole, the efficiency of interventions to restore the physiological reserve can be easier to be evaluated. For instance, to assess the pulmonary function, measuring the forced expiratory volume in 1s (FEV1) can serve the purpose [11,12]. To evaluate the hearing capacity, the Speech, Spatial, and Qualities of Hearing Scale (SSQ) can be adopted [13]. The availability of reliable standards and markers to evaluate the health status of tissues and organs can facilitate the development and optimization of practices to maintain the health span. Despite this, as far as the health span is concerned, the absence of physical illnesses has often been regarded as an indicator of being healthy. Such an over-simplistic concept of "health span" has become a barrier to the holistic maintenance of health during our pursuit of life span extension.

To engineer the health span, elimination of physical diseases is not sufficient to fully maintain the functional reserve in tissues and organs. In fact, the term "health" stems from the Sanskrit term "Hal," which means totality and completeness. In TCM, "health" is fulfilled through the harmony that exists between the Ying and the Yang in the body, as well as among individuals, communities, and the cosmos. There is a similar emphasis on holism as the essence of health in the WHO's Constitution, which holds that "health" is a state of complete well-being in the physical, mental, and social aspects of an individual, instead of merely the absence of disease or infirmity (Fig. 17.1) [14]. Because of this, while advances in medicine could partially fulfill the goal of health span engineering, sociological factors, such as social relations experienced by the elderly, should also be considered. Unfortunately, ignorance of the effects of individuals' mental and social health to the outcome of an antiaging intervention prevails, impeding the establishment of a suitable environment to embrace the emergence of holistic antiaging interventions.

Outstanding questions for clinical translation

1. How can we promote awareness of the importance of health span maintenance in prolonging life span? As far as longevity is concerned, right now most of the efforts focus on the extension of life span per se. The importance of keeping an individual in a healthy state has seldom been emphasized. This concept is needed for successful prolongation of longevity.
2. How can the concept of holistic health be promoted among the scientific sector? "Health" should not be confined to the absence of ailments, but should involve the maintenance of the state of well-being in the mental and social aspects of an individual. This holistic concept of health is the essence of health span engineering.

Social impacts on the health span

Social relations are important to the mental health and physical health of aged people [15–17], who may adapt to the surroundings more effectively via their promoted sense of satisfactory functioning of their social networks. With intangible social support, aged people can also obtain subtle yet essential information, which is needed for them to have proper and positive self-evaluation. The latter has been documented in Caplan's study [18], in which elderly people have been reported to feel being respected (and being held in high esteem by others) after they have received support and assistance from their family members and friends. Based on this, support from social relations is important to the aged people because it can replenish their feelings of perceived control. It can also provide them concrete help for dealing with personal problems. The subtle relationship between social support and problem-solving performance has been confirmed by a recent study, which has demonstrated that a higher level of social support is associated with a more positive problem orientation, helping individuals to interpret stressful aspects of daily life problems in a more benign manner [19].

Social relations cannot only enable elderly people to evaluate the surrounding situation more positively, but can also neutralize the negative impact of life strains experienced by those people, thereby restoring them a sense of morale, helping them to avoid distress symptoms, and enhancing their self-perception. In a study on a community sample of 1106 Chinese people aged 60 or above [20], depressive symptomatology and all dimensions of support (e.g., social network size, network composition, social contact frequency, satisfaction with social support, and the availability of instrumental/emotional support) from social relations have been found to display significant bivariate relationships. Among different factors, "satisfaction with social support" has been suggested to be one of the most important ones determining the depression level among the senior population, has been shown to be related to individuals' perception of health-related quality of life (HRQoL) [21,22]. This illustrates the relationship between mental health and social relations, and demonstrates the role played by social support in the prevention of mental disorders. In fact, depression is one of the most prevalent functional mental disorders among elderly people [23]. In Hong Kong, around 20%–40% of the elderly people face the problem of depression [24]. A recent study has also demonstrated that Chinese older people with depression in general have a poorer perception of HRQoL [21]. Nevertheless, with the positive effect of social support on mental health [20–22], promotion of intangible supportive networks would be one of the possible methods of facilitating mental health in the aged community when the human life span is extended by biogerontological advances.

Apart from mental health, social relations can influence physical health [16]. As suggested by social control and social identity theorists, social relations promote healthier behavior (e.g., exercising and adapting healthy diets) directly and indirectly [16], leading to an improvement in physical health. In addition, intangible support from social relations can increase life satisfaction and can boost the capacity of an individual to face stress It can, therefore, function as a mental health promoter. In fact, social relations can modulate an individual's immune, neuroendocrine, and cardiovascular functions. For instance, by using electron beam tomography, coronary artery calcification has been illustrated to be associated with social network indices (e.g., the number of people in the household, and the marital status) [25]. This is consistent with the finding of Wang et al.'s study [26], in which substantially greater coronary atherosclerosis progression has been observed in women experiencing improper interpersonal social relations, social isolation and inadequate emotional support [26]. This phenomenon has been found to be independent of conventional clinical and lifestyle factors (e.g., smoking history, age, body mass index, menopausal status, and diagnosis of the index event of acute myocardial infarction). Immunologically, social support can modulate immune functions, causing changes in NK cell counts and Th1/Th2 balance [27]. It can also influence the immunity of an individual [27]. This partly explains the phenomenon that limited social support often leads to adverse health outcomes [27]. Taking all these into consideration, research efforts devoted to aging retardation should take biological, socio-demographic, environmental, and social aspects of aging into account. Furnishing the aged population with good social relations is one of the preconditions for promoting holistic health among the elderly.

Despite this, integrating the social aspect of aging into works on aging retardation is challenging. This is because of the gaps between disciplines, and also of the confusion caused by the high diversity of proposed mechanisms explaining the protective effects of social support. For example, based on the stress and coping perspective [28,29], social relations influence health by enhancing one's coping performance and protecting one from strains; however, according to the social constructionist perspective [30], they contribute to health by promoting one's self-esteem and self-regulation, regardless of the presence of stress in life. The diversity of theoretical perspectives has led to a large variety of support measures used in research; however, at present the validity of support measures adopted in research is still under some sort of debate. For instance, though self-report of the quality and quantity of social support is a commonly adopted measure of supportive actions, some scholars have raised concerns on its validity [28,31]. At the moment, no support measures have been totally accredited by scholars in the field. Under this situation, communication among research findings is difficult. This impedes the use of social support in both social and health practices for the elderly.

Outstanding questions for clinical translation

1. How can the efficiency of different social support activities be objectively and accurately evaluated? Research methods, such as questionnaires and interviews, are some of the tools that might be used; however, the design of the questions and the setup of the study can affect the validity of the research findings. Careful evaluation of the validity of a study is, therefore, required before the obtained results can be generalized to other contexts.
2. How can we ensure that individuals with different personalities can obtain the same level of social support? People with different personalities may experience differently the level of support from the same social support activity. This should always be considered when social support activities are designed and implemented.

Social constraints on research efforts

Another social barrier to intervention success is the current perception of aging, which is sometimes regarded as a normal stage of life rather than a disease. This understanding of aging has been limiting the amount of public funding to be spent on aging research and on the development of biogerontological interventions. In addition, people sometimes hold that aging is intractable, thinking that efforts devoted to retarding the aging process (or to developing interventions to manipulate it) will turn out to be quixotic [32]. This is not a big problem if it prevails only in the general community; however, it is devastating when it becomes a norm in the academic society as it does right now. The reason is simple: science is expensive. This is particularly true for longitudinal aging research, which may take years to complete. Keeping the project running requires public funding.

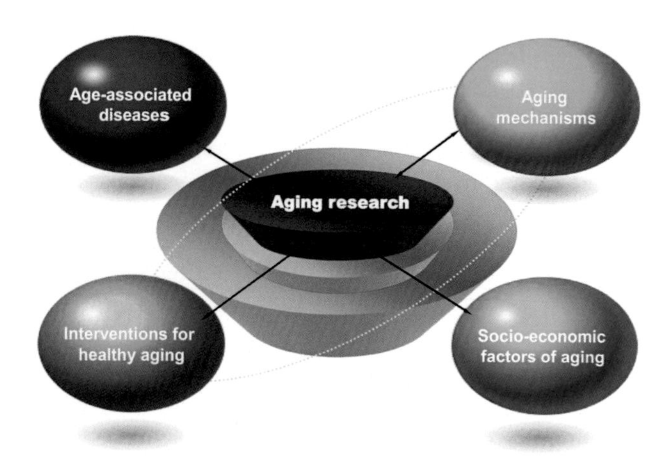

FIGURE 17.2 Major areas of aging research.

In fact, funding for biological research is becoming tighter than before [33]. For instance, the National Institute of Health in the United States currently funds only around 10% of the submitted applications [34]. Similar situations of tight funding also occur in other industrialized countries. The net outcome is that researchers spend less and less time on performing research, but are overwhelmed by writing and submitting applications to maximize their chances in the grant lottery. When getting funding on the development of antiaging therapies is becoming difficult, young and emerging researchers may be driven to fields other than aging. This further retards the pace of the emergence of antiaging interventions. In addition, as far as aging research is concerned, it involves different areas (ranging from basic research on the mechanisms of aging to the development of interventions for healthy aging) (Fig. 17.2). All these areas are interrelated, contributing to the establishment of comprehensive understanding of the aging process. Unfortunately, the amount of resources provided to research on different areas has not been equal, with a substantial amount of public funding devoted only to research initiatives associated with age-associated diseases [35]. This causes efforts to understand and tackle aging unstructured and poorly coordinated. To address this situation, funding has to be secured to achieve an integrated, multidisciplinary means of approaching aging research. Investments in aging research actually have the potential to lead to immense gains to society, particularly in terms of the self-sufficiency and productivity of the elderly population. Further efforts to raise the awareness of the importance of understanding and tackling aging are also needed so that aging research could be recognized as a societal and governmental priority.

Outstanding questions for clinical translation

1. How can we help the society to recognize the importance of antiaging medicine? Over the years, aging has often been regarded as only an irreversible process that belongs to part of the life cycle. Efforts to combat aging, therefore, have gained a high level of pessimism over the possible success. Although recent advances in technologies have shed light on the viability of tackling aging, before the success of the development of biogerontological interventions can be made possible, overcoming the social barrier to intervention success is necessary.

(Continued)

(Continued)

2. How can we ensure that different areas in aging research can be funded in a more balanced manner? Aging research encompasses different areas, ranging from the elucidation of the mechanisms of aging (and age-associated diseases) to the development of antiaging therapies; however, right now most of the funding has been directed to research on age-associated diseases. Works relating to aging per se have gained less recognition. This problem should be resolved to ensure that research on aging can be carried out in a more comprehensive way.

Summary

When life span is extended, prolongation of the health span becomes important because this directly affects the quality of the extended life. Health span of this generation can be possibly increased by eliminating life-shortening infectious diseases, enhancing the quality of medical care, maintaining a healthy lifestyle, and developing new drugs; however, the importance of support from social relations (as well as tangible social support) to health maintenance should not be overlooked. In addition, proper recognition of the importance of aging research in society is vital as it may change the societal and governmental priorities when research initiatives are funded. Achieving this, however, is expected to be challenging because of the prevalent concept of aging, which has often been regarded as a normal stage of life rather than a problem that has to be urgently solved. We need a change in our way to think about aging. Once this could be attained, an evolutionary leap in the development of antiaging medicine would be conceivable.

Directions for intervention development

When we develop biogerontological interventions to postpone or ward off aging, we hope that the extended life is healthy, fruitful, and supportive. This goal can be achieved by following the steps below. Fulfilling these steps, however, can hardly be accomplished by researchers alone. Collaborative efforts from different parties (ranging from policymakers to community members) are required.

1. Estimate the social impact brought by the development of biogerontological interventions.
2. Predict the social needs caused by the impact.
3. Propose appropriate policies or measures to address those needs.
4. Review the efficiency of those policies or measures.
5. Amend those policies or measures regularly to accommodate the unmet needs.

References

[1] K.L. Chou, I. Chi, N.W. Chow, Sources of income and depression in elderly Hong Kong Chinese: mediating and moderating effects of social support and financial strain, Aging Ment. Health 8 (2004) 212–221.

[2] D.R. Phillips, O.L. Siu, A.G. Yeh, K.H. Cheng, Informal social support and older persons' psychological well-being in Hong Kong, J. Cross. Cult. Gerontol. 23 (2008) 39–55.

[3] M. Makins (Ed.), Collins English Dictionary, Harper Collins, Glasgow, 1991.

[4] R. Perrett, Regarding immortality, Religious Stud. 22 (1986) 219–233.

[5] T. Nagel, Death., in: J. Rachels (Ed.), Moral Problems: A Collection of Philosophical Essays, Harper & Row, New York, 1975, pp. 401–409.

[6] Q. Sima, Shi Ji, Yue Lu Shu She, Chang Sha, 1988.

[7] C.C. Zhou, The Yellow Emperor's Medicine Classic: Treatise on Health and Long Life, ASIAPAC, Singapore, 1996.

[8] J.W. Larrick, A. Mendelsohn, Applied healthspan engineering, Rejuvenat. Res. 13 (2010) 265–280.

[9] J.C. George, J. Bada, J. Zeh, L. Scott, S.E. Brown, T. O'Hara, et al., Age and growth estimates of bowhead whales (Balaena mysticetus) via aspartic acid racemization, Can. J. Zool. 77 (1999) 571–580.

[10] G.Q. Zhang, W. Zhang, Heart rate, lifespan, and mortality risk, Ageing Res. Rev. 8 (2009) 52–60.

[11] M.L. North, M. Ahmed, S. Salehi, K. Jayasinghe, M. Tilak, J. Wu, et al., Exposomics-based analysis of environmental factors associated with forced expiratory volume in 1 second at 6 months post lung transplantation, Ann. Am. Thorac. Soc. 15 (2018) S122.

[12] Y. Matsuda, S. Eba, F. Hoshi, H. Oishi, T. Sado, M. Noda, et al., Sustained-release tacrolimus stabilizes decline of forced expiratory volume in 1 second through decreasing fluctuation of its trough blood level, Transplant. Proc. 50 (2018) 2768–2770.

[13] K. Demeester, V. Topsakal, J.J. Hendrickx, E. Fransen, L. van Laer, G. Van Camp, et al., Hearing disability measured by the speech, spatial, and qualities of hearing scale in clinically normal-hearing and hearing-impaired middle-aged persons, and disability screening by means of a reduced SSQ (the SSQ5), Ear. Hear. 33 (2012) 615–616.

[14] World Health Organization, Constitution of the World Health Organization, in: T.L. Beauchamp, J.F. Childress (Eds.), Principles of Biomedical Ethics, Oxford University Press, New York, 1979, pp. 284–285.

[15] N. Krause, Negative interaction and satisfaction with social support among older adults, J. Gerontol. B Psychol. Sci. Soc. Sci. 50 (1995) P59–P73.

[16] B.N. Uchino, Social support and health: a review of physiological processes potentially underlying links to disease outcomes, J. Behav. Med. 29 (2006) 377–387.

[17] A. Vaux, D. Harrison, Support network characteristics associated with support satisfaction and perceived support, Am. J. Commun. Psychol. 13 (1985) 245–268.

[18] G. Caplan, Mastery of stress: psychosocial aspects, Am. J. Psychiatry 138 (1981) 413–420.

[19] J.S. Grant, T.R. Elliott, M. Weaver, G.L. Glandon, J.L. Raper, J. N. Giger, Social support, social problem-solving abilities, and adjustment of family caregivers of stroke survivors, Arch. Phys. Med. Rehabil. 87 (2006) 343–350.

[20] I. Chi, K.L. Chou, Social support and depression among elderly Chinese people in Hong Kong, Int. J. Aging Hum. Dev. 52 (2001) 231–252.

[21] S. Chan, S. Jia, H. Chiu, W.T. Chien, D.R. Thomson, Y. Hu, et al., Subjective health-related quality of life of Chinese older persons with depression in Shanghai and Hong Kong: relationship to clinical factors, level of functioning and social support, Int. J. Geriatr. Psychiatry 24 (2009) 355–362.

[22] S.E. Hobfoll, A. Nadler, J. Leiberman, Satisfaction with social support during crisis: intimacy and self-esteem as critical determinants, J. Pers. Soc. Psychol. 51 (1986) 296–304.

[23] World Health Organization, The World Health Report 2004 – Changing History, World Health Organization, Geneva, 2004.

[24] K.L. Chou, I. Chi, Prevalence and correlates of depression in Chinese oldest-old, Int. J. Geriatr. Psychiatry 20 (2005) 41–50.

[25] W.J. Kop, D.S. Berman, H. Gransar, N.D. Wong, R. Miranda-Peats, M.D. White, et al., Social network and coronary artery calcification in asymptomatic individuals, Psychosom. Med. 67 (2005) 343–352.

[26] H.X. Wang, M.A. Mittleman, K. Orth-Gomer, Influence of social support on progression of coronary artery disease in women, Soc. Sci. Med. 60 (2005) 599–607.

[27] T. Miyazaki, T. Ishikawa, A. Nakata, T. Sakurai, A. Miki, O. Fujita, et al., Association between perceived social support and Th1 dominance, Biol. Psychol. 70 (2005) 30–37.

[28] M.J. Barrera, Distinctions between social support concept, measures and models, Am. J. Community Psychol. 14 (1986) 413–445.

[29] R.S. Lazarus, S. Folkman, Stress, Appraisal, and Coping, Springer, New York, 1984.

[30] B. Lakey, S. Cohen, Social support theory and measurement, in: S. Cohen, L.U. Gordon, L. Underwood, H. Gottlieb (Eds.), Social Support Measurement and Intervention: A Guide for Health and Social Scientists, Oxford University Press, New York, 2000, pp. 29–52.

[31] C. Dunkel-Schetter, T.L. Bennett, Differentiating the cognitive and behavioral aspects of social support, in: B.R. Sarason, I.G. Sarason, G.R. Pierce (Eds.), Social Support: An Interactional View, Wiley, New York, 1990, pp. 267–296.

[32] W.F. Lai, Nucleic acid therapy for lifespan prolongation: present and future, J. Biosci. 36 (2011) 725–729.

[33] M. Wadman, Funding crisis hits US ageing research, Nature 468 (2010) 148.

[34] A.D. de Grey, Hype and anti-hype in academic biogerontology research, Rejuvenat. Res. 13 (2010) 137–138.

[35] O.H. Franco, T.B. Kirkwood, J.R. Powell, M. Catt, J. Goodwin, J.M. Ordovas, et al., Ten commandments for the future of ageing research in the UK: a vision for action, BMC Geriatr. 7 (2007) 10.

Index

Note: Page numbers followed by "*f*" and "*t*" refer to figures and tables, respectively.

D

E

K

L

M

N

O

P

Printed in the United States
By Bookmasters